T0146054

"Horses, in one form or another, have been roaming our planet at will for at least 56 million years, but today many of the free-roaming populations are in danger of disappearing. Now for the first time ever we have, in Ransom and Kaczensky's much-needed *Wild Equids*, a comprehensive assessment of the world's wild equines, including feral horses, zebras, asses, onagers, and Przewalski's horses. Horses deserve a better deal—a deal that needs to be based on science—and this volume, years in the making, is an important step in making that happen."

—WENDY WILLIAMS, author of *The Horse: The Epic History of Our Noble Companion*

"The editors of this book have done a masterful job assembling a comprehensive review of scientific research on equid behavior. This is essential reading for wildlife biologists, conservation specialists, and policymakers in equid management. The chapters, including one on the meaning of 'wild,' display a tremendous range of knowledge."

—TEMPLE GRANDIN, author of *Animals in Translation* and *Genetics and the Behavior of Domestic Animals*

"People have interacted with equids for thousands of years, and domesticated horses and donkeys are very familiar parts of our lives. Yet most of the original wild equid species are highly threatened or even extinct. *Wild Equids* is the most comprehensive summary of all that we know about these remarkable animals: wild asses, wild horses, and zebras. Anyone who has a serious interest in these species needs to start with this book."

—SIMON N. STUART, PhD, Chair of the IUCN Species Survival Commission

"Finally, a book that synthesizes a vast array of information on horses, zebras, and asses. Ransom and Kaczensky have brought the experts together to pool their knowledge and summarize what we know about the phylogeny, social behavior, ecology, and conservation needs of these species. This book is a critical step in recovering populations, species, and the ecological function they provide of this important Family of ungulates."

—JOSHUA R. GINSBERG, PhD, President, Cary Institute of Ecosystem Studies and formerly Senior Vice President, Wildlife Conservation Society Global Program

"Equids and humans share thousands of years of common and very special history. Today, however, 5 out of 7 species are assessed as threatened with extinction and still basics of behaviour and ecology of wild species remain unknown. It was therefore high time for this compilation of cutting-edge science with a focus on ecology, management, and conservation."

—GERALD DICK, PhD, MAS, Executive Director, World Association of Zoos and Aquariums

"A world-class roster of authors provides a comprehensive treatment of diverse subject matter that is deep, thorough, and yet extremely accessible to multiple audiences."

—STEVE MONFORT, DVM, PhD, Director, John & Adrienne Mars at Smithsonian Conservation Biology Institute, National Zoological Park

WILD EQUIDS

JOHNS HOPKINS UNIVERSITY PRESS | BALTIMORE

WILD EQUIDS

Ecology, Management, and Conservation

EDITED BY
Jason I. Ransom and Petra Kaczensky

Printed in the United States of America on acid-free paper
9 8 7 6 5 4 3 2 1

Johns Hopkins University Press
2715 North Charles Street
Baltimore, Maryland 21218-4363
www.press.jhu.edu

Library of Congress Cataloging-in-Publication Data

Wild equids : ecology, management, and conservation / edited by
Jason I. Ransom and Petra Kaczensky.
pages cm
Includes bibliographical references and index.
ISBN 978-1-4214-1909-1 (hardcover : alk. paper) —
ISBN 978-1-4214-1910-7 (electronic) — ISBN 1-4214-1909-2
(hardcover : alk. paper) — ISBN 1-4214-1910-6 (electronic)
1. Equidae. 2. Wild horses. 3. Horses. 4. Zebras. I. Ransom, Jason,
editor. II. Kaczensky, Petra, editor.
QL737.U62W526 2016
599.665—dc23 2015022981

A catalog record for this book is available from the British Library.

Special discounts are available for bulk purchases of this book. For more information, please contact Special Sales at 410-516-6936 or specialsales@press.jhu.edu.

Johns Hopkins University Press uses environmentally friendly book materials, including recycled text paper that is composed of at least 30 percent post-consumer waste, whenever possible.

For Liam, Nicky, and Una:
Part of the next generation to share the earth with wild equids

Contents

Acknowledgments

THIS BOOK WOULD NOT HAVE BEEN POSSIBLE IF IT WEREN'T FOR THE participants and sponsors of the International Wild Equid Conference held in Vienna, Austria, in 2012. Many of the participants contributed to this book, but the work herein reflects the efforts of many talented scientists that continue to work for the conservation of wild equids. Not all could contribute to this effort, but their work is enormously appreciated. We must thank the sponsors of that conference for starting this work in progress and ultimately helping to fund publication: University of Veterinary Medicine–Vienna, Colorado State University's Center for Collaborative Conservation, Vienna Business Agency, Ms. Evelyn Haim-Swarovski, International Takhi Group, US Geological Survey, World Associations of Zoos and Aquariums, and Vectronic Aerospace. Thank you also to the Austrian Science Foundation (FWF) for funding wild equid research in the Mongolian Gobi since 2001. Many outcomes of that work contributed substantially to this book.

A special thanks is owed to the many peer reviewers who donated their time to provide critical comments for this book's chapters, as well as to the anonymous reviewers solicited by the publisher to critique the book as a whole. Your comments, criticisms, and insights have improved the content of this volume, and it is this kind of dispute that ultimately drives good science. Thank you also to Vincent Burke, Catherine Goldstead, and Kathryn Marguy at Johns Hopkins University Press for pursuing this book to publication and keeping the administration and technical aspects of the book in line. Thanks also to Ashleigh McKown for her careful and thoughtful editing contributions.

Lastly, a sincere thank-you to each of the friends, family, and colleagues who shared their work, time, opinions, insights, and often data, to form this book. We don't always agree on methodologies or conclusions, but move forward nonetheless. This book is a testament of your dedication to the academic pursuit of knowledge and your belief in conserving the world's equids. Thank you.

WILD EQUIDS

1

Equus: An Ancient Genus Surviving the Modern World

JASON I. RANSOM AND
PETRA KACZENSKY

The common history of equids and humans is a long and complicated one. Equids were important prey animals of early humans, but they also provided spiritual enlightenment, as evident from ancient rock carvings and cave paintings (Chauvet *et al.* 1996; Clottes 2002; Olsen 2003). The domestication of the horse and the donkey resulted in a new and even tighter relationship with humans. Horses and donkeys provided milk, meat, and labor, but it was their ability to carry a rider fast and far that allowed the horse to become so special to humankind (Walker 2008). Horses gave Dschingis Khan's army the speed and endurance needed to create the largest empire in history, helped humans to explore and lay claim on newly discovered continents like the Americas and Australia, and aided the swift exchange of information (e.g., the Pony Express) and goods (e.g., along the Silk Road) over long distances. Although modern technology has replaced the horse's services of old, our fascination with them remains. We admire their speed, grace, beauty, and stamina; even still measuring our mechanical engine strength in "horse power." Horses are among the most popular animals (Kellert 1989), and in our modern world they have become highly valued companion animals (Walsh 2009).

Wild equids, on the other hand, receive remarkably little attention from modern humans, and even the basic behavior and ecology of several species are unknown (Moehlman 2002). Wild equids are similar in appearance, often making it difficult for a layperson to tell them apart. Generic names add to the confusion, as they often do not allow clear differentiation between species (e.g., *Goor* in Farsi can mean Asiatic wild ass or zebra), or can even suggest a hybrid origin (e.g., *hemione*, meaning "half-ass"). Wild equids live in some of the most remote regions of Africa and Asia, many of which were largely inaccessible until recently (fig. 1.1). This is particularly true for Asia: it was not until 1878 that the Polish explorer Nikolai M. Przewalski "discovered" a new wild horse species for the Western world, which was to become known as Przewalski's horse (Mohr and Volf 1984; Boyd and Houpt 1994).

One of the first nature documentaries produced for cinemas was Bernhard and Michael Grzimek's groundbreaking 1959 film *Serengeti Shall Not Die* (Grzimek and Grzimek 1959). The movie raised broad public awareness of the Serengeti plains and the amazing migration of wildebeest and plains zebra. A suite of scientific studies were then initiated and have been steadily increasing our understanding of this unique

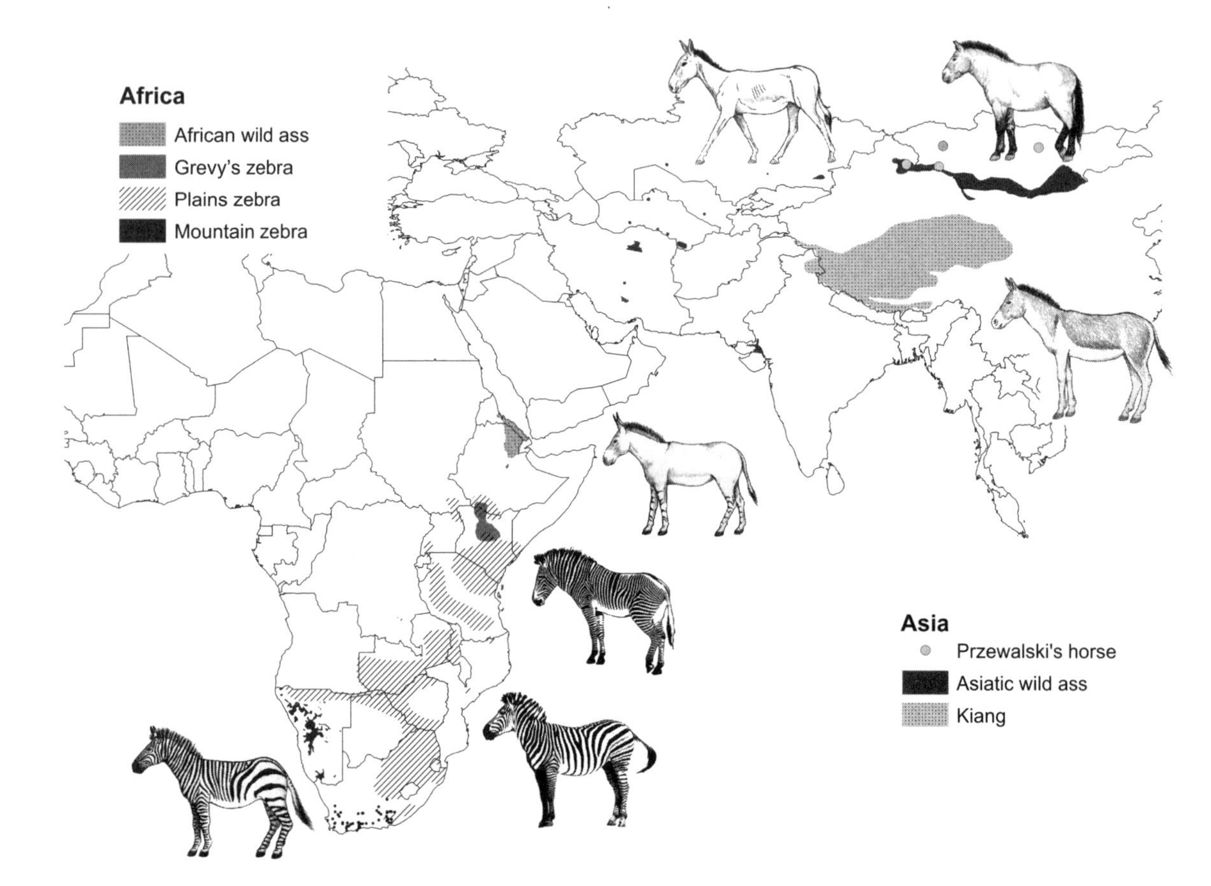

Fig. 1.1 Current distribution of wild equids.

ecosystem, the special role and social organization of the plains zebra, and the challenges of ecosystem conservation (Gwynne and Bell 1968; Klingel 1969; Sinclair and Norton-Griffiths 1979; Sinclair and Arcese 1995; Sinclair *et al.* 2008). The other zebra species remained less studied, and the African wild ass has nearly disappeared without leaving much of a trace in the early scientific literature.

The range of wild equids in Asia remained largely closed behind the Iron Curtain of the Soviet Union and Chairman Mao's Communist People's Republic of China until recently. Consequently, little had been heard about wildlife on the Tibetan Plateau, including the kiang or wild ass, before George Schaller's research in 1985 (Schaller 1998, 2012). Over the huge expanse of the Soviet Union and adjacent Mongolia, Russian scientists surveyed wild ass numbers, mapped their distribution, studied their basic ecology, and also initiated the first equid reintroductions. But little of their work reached the West or was translated into the English language (Bannikov 1981; Heptner *et al.* 1988).

Current research is picking up where past work stopped, introducing new scientific approaches such as GPS tracking technology, remote sensing, and stable isotope analysis, to name a few. Technical advances like whole genome sequencing (Jónsson *et al.* 2014) and ancient genome analysis (Orlando *et al.* 2013) are increasingly refining our knowledge of equid taxonomy. For this book, we align the text with the most conventionally accepted taxonomy for wild equids (see International Union for Conservation of Nature 2015), comprising three zebra species (plains zebra [*Equus quagga*], Grevy's zebra [*E. grevyi*], mountain zebra [*E. zebra*]), three ass species (African wild ass [*E. africanus*], Asiatic wild ass [*E. hemionus*], kiang [*E. kiang*]), and one horse species (Przewalski's horse [*E. ferus przewalskii*]). We also include the descendants of domestic horses (*E. caballus*) and domestic asses (*E. asinus*) (Integrated Taxonomic Information System 2014), which are free roaming and share many of the same ecological attributes and functions as their wild relatives. In this book, we refer to the free-roaming descendants of once-domesticated equids as "feral" simply for clarity in discerning between species.

This book brings together many of the world's experts on wild equids to help synthesize and understand the state of science regarding ecology, management, and conservation of these species. We aim to provide an overview of the most recent scientific advances to stimulate new and complementary research, as well as to raise awareness for a taxonomic group that is often overlooked. We have organized the volume into three larger themes: ecology, history and management, and conservation. The first section includes an overview of the complex social organization (chap. 2) and behavior (chap. 3). of equids This leads directly to a synthesis of equid habitat and diet selection (chap. 4), a conceptual evaluation of the fundamental and realized ecological niches that equid species occupy (chap. 5), and ultimately how equid populations change through time (chap. 6).

The next section of the book focuses on the often difficult and much-debated issues surrounding human dimensions, and how evolutionary history and naming conventions inform our stewardship actions. It begins with a comprehensive review of equine genetics (chap. 7), because genetic origins and taxonomy can provide a scientific foundation for balancing both our anthropogenic views and conservation priorities. The next contributions help us understand how equine domestication led to assisted global migration (chap. 8), how wild and feral equids are viewed in the modern cultural context, and how this influences our management actions (chap. 9). To end this section, we explore two special cases of equid management: managing feral horses (chap. 10) and caring for and breeding captive wild equids (chap. 11).

Last, we explore conservation—how we ensure that future generations will be able to experience life with horses, zebras, and asses still running wild in the world. This concluding section includes a current review of global equid status and conservation (chap. 12), the importance of migration and connectivity (chap. 13), and reintroductions as a last resort (chap. 14).

Despite the volume of knowledge we have brought together in this book, much remains to be learned about wild equids to fully understand and appreciate their role in the world's grazing ecosystems. Wild equids may look alike and follow similar life history patterns, but it is the nuances in behavior and physiology that define their ecological niches, making them fit for a life on high-altitude plains, low tidal salt marshes, hot and cold deserts, steppes, and savannahs. Although equid research has increased in recent years, only a better understanding of their ecology and regional threats will allow us to address the conservation concerns that most equid populations are facing in our quickly changing world. Communicating the uniqueness of each wild equid species and their habitats back to a wider public is also needed to raise awareness beyond the beloved domestic horse. In the words of Baba Dioum, speaking at the 1969 International Union for Conservation of Nature General Assembly in New Delhi, "We will conserve only what we love, we will love only what we understand, and we will understand only what we are taught."

REFERENCES

Bannikov, A.G. 1981. The Asian wild ass. [In Russian.] Lesnaya Promyshlennost, Moscow, Russia. [English translation by M. Proutkina, Zoological Society of San Diego, San Diego, CA, USA.]

Boyd, L., and K.A. Houpt. 1994. Przewalski's horse. State University of New York Press, Albany, USA.

Chauvet, J.-M., E.B. Deshamps, and C. Hillaire. 1996. Chauvet cave: The discovery of the world's oldest paintings. Thames and Hudson, London, UK.

Clottes, J. 2002. World rock art. Getty Publications, Los Angeles, CA, USA.

Grzimek, B., and M. Grzimek. 1959. Serengeti shall not die. [In German.] Asta Motion Pictures, West Germany.

Gwynne, M.D., and R.H.V. Bell. 1968. Selection of vegetation components by grazing ungulates in the Serengeti National Park. Nature 220:390–393.

Heptner, V.G., A.A. Nasimovich, and A.G. Bannikov. 1988. Kulan: *Equus* (Equus) *hemionus*. Pages 1011–1036 *in* Mammals of the Soviet Union, vol. 1: Artiodactyla and Perissodactyla. Smithsonian Institution Libraries and the National Science Foundation, Washington, DC, USA. [English translation of the original book published in 1961 by Vysshaya Shkola, Moscow, Russia.]

Integrated Taxonomic Information System. 2014. Integrated taxonomic information system. Available at www.itis.gov.

International Union for Conservation of Nature. 2015. Red list of threatened species. Version 2015.2. Gland, Switzerland. Available at www.iucnredlist.org.

Jónsson, H., M. Schubert, A. Seguin-Orlando, A. Ginolhac, L. Petersen, *et al.* 2014. Speciation with gene flow in equids despite extensive chromosomal plasticity. Proceedings of the National Academy of Sciences of the USA 111:18,655–18,660.

Kellert, S.R. 1989. Perceptions of animals in America. Pages 4–24 *in* R.J. Hoage (Ed.), Perceptions of animals in American culture. Smithsonian Press, Washington, DC, USA.

Klingel, H. 1969. The social organisation and population ecology of the plains zebra (*Equus quagga*). Zoologica Africana 4:249–263.

Moehlman, P. (Ed.). 2002. Equids: Zebras, asses and horses. Status survey and conservation action plan. International Union for Conservation of Nature, Gland, Switzerland.

Mohr, E., and J. Volf. 1984. Das Urwildpferd: Die Neue Brehm-Bücherei. A. Ziemsen Verlag, Wittenberg Lutherstadt, Germany.

Olsen, S.L. 2003. Horse hunters of the Ice Age. Pages 35–56 *in* S.L. Olsen (Ed.), Horses through time. Rowman and Littlefield, Lanham, MD, USA.

Orlando, L., A. Ginolhac, G. Zhang, D. Froese, A. Albrechtsen, *et al.* 2013. Recalibrating *Equus* evolution using the genome sequence of an early Middle Pleistocene horse. Nature 499:74–78.

Schaller, G.B. 1998. Wildlife of the Tibetan Steppe. University of Chicago Press, Chicago, USA.

Schaller, G.B. 2012. Tibet wild: A naturalist's journeys on the roof of the world. Island Press, Washington, DC, USA.

Sinclair, A.R.E., and P. Arcese (Eds.). 1995. Serengeti II: Dynamics management and conservation of an ecosystem. University of Chicago Press, Chicago, USA.

Sinclair, A.R.E., and M. Norton-Griffiths (Eds.). 1979. Serengeti: Dynamics of an ecosystem. University of Chicago Press, Chicago, USA.

Sinclair, A.R.E., C. Packer, S.A.R. Mduma, and J.M. Fryxell (Eds.). 2008. Serengeti III: Human impacts on ecosystem dynamics. University of Chicago Press, Chicago, USA.

Walker, E. 2008. Horse. Reaktion, London, UK.

Walsh, A. 2009. Human–animal bonds: I. The relational significance of companion animals. Family Process 48:462–480.

PART I ECOLOGY

2

Social Organization of Wild Equids

LEE BOYD, ALBERTO SCOROLLI, HANIYEH NOWZARI, AND AMOS BOUSKILA

An understanding of social organization is vital to the conservation and management of a species, and wild equids are no exception. All equids are highly social, and the extant species of equids exhibit two contrasting alternatives in their social structure and the stability of bonds between group members: harem-forming equids with female defense polygyny (Type I, as described by Klingel 1974*a*, 1975), and territorial equids with resource defense polygyny (Type II, Klingel 1974*a*, 1975; Rubenstein 1986).

Species that display female defense polygyny show long-term stability of social groups (Klingel 1975, 1982; Rubenstein 1986). These harem-forming species are adapted to mesic habitats and include plains zebra (*Equus quagga*), mountain zebra (*E. zebra*), feral horses (*E. caballus*), and Przewalski's horses (*E. ferus przewalskii*) (Klingel 1975, 1982; Rubenstein 1986; Linklater 2000). Feral asses (*E. asinus*) also form harems under mesic conditions (Moehlman 1998*b*). The main social unit, referred to as harem or band, consists of one or more adult males, one or more adult females, young animals of both sexes not yet dispersed, and the foals of the year (Linklater 2000). Nonbreeding males often form all-male bachelor groups (Klingel 1974*a*, 1975; Keiper 1986; Rubenstein 1986). All-female and mixed-sex groups have been described in feral horses on Assateague Island, USA (Keiper 1976), and in the Kaimanawa Mountains, New Zealand (Linklater *et al.* 2000), but these groups are uncommon and of short duration. Solitary animals are typically males, usually harem stallions that have lost their females, and are often old or sick (Klingel 1975; Rubenstein 1986; Boyd and Keiper 2005).

In contrast, the species that present resource defense polygyny, also called fission-fusion societies, have no permanent or long-lasting bonds between any two individuals other than between an adult female and her current foal (Klingel 1975, 1982, 1998; Rubenstein 1986, 1994; Kaczensky *et al.* 2008). Group membership is fluid: changes occur within days, or even hours. Nevertheless, recent research using network analysis reveals stronger bonding between some individuals (Sundaresan *et al.* 2007*a*). This system is exhibited by arid-adapted species such as Grevy's zebra (*E. grevyi*) (Klingel 1974*b*; Ginsberg 1987), the African wild ass (*E. africanus*) (Klingel 1977), and its feral counterpart in arid conditions (Woodward 1979; Moehlman 1998*b*). Asiatic wild asses (*E. hemionus*, including onagers, Indian khur, and khulan) and the kiang (*E. kiang*) show this same type of social organization (Klingel 1977; Bahloul *et al.* 2001; Feh *et al.* 2002; Shah 2002; Shah and Qureshi 2007; St-Louis and Côté 2009). A population of Asiatic wild ass in the Mongolian Gobi Desert exhibited a social structure that could be considered female defense polygyny, with harems lasting for two or more years (Feh *et al.* 1994, 2001). Some researchers (Neumann-Denzau and Denzau 2007; Kaczensky *et al.* 2008) have questioned the interpretation of this observation, however, and a thorough study of the social system of the Mongolian khulan is needed. More typically, Type II adult males are territorial, defending large portions of grasslands during the mating season. Sometimes they are alone, and sometimes they are with females of different reproductive states that form temporary associations with them when on their territories. Females are rarely alone, but the groups they form are often short-lived, and the proportion of females of the same reproductive status varies. Subadult males that are sexually mature, but not strong enough to defend territories, often join all-male bachelor groups. These groups, like female associations, are also temporary. Therefore Type II groups may contain adult males (bachelors or territorial breeders), adult females (nonlactating or lactating) with yearlings and foals, or some combination thereof (Klingel 1975, 1977; Rubenstein 1986; Nowzari *et al.* 2013).

An exceptional case has been documented in the Shackleford Banks feral horse population demonstrating that a single species, and even the same population, may present different social systems under different environmental conditions. Horses on the island were either organized in harems as expected, or in open-membership groups, or in harems in which the stallion was able to defend not only his mares but also a territory because the female-biased sex ratio reduced the number of potential rivals, the open habitat made detection of rivals easy, and the ocean reduced boundaries needing defense. When the vegetation later changed to a more uniform cover, all horses were organized in harems with overlapping home ranges (Rubenstein 1986, 1994).

Group Size

Some attributes of equid social groups, such as group size and number of adult females, have been widely studied and are considered descriptive of the social context group members experience (table 2.1). Total group size in wild equids is moderate compared to other ungulate species (Jarman 1974). Harems of harem-forming species have a similar mean size (table 2.1). Mean number of adult females per harem have been reported as 2.0–2.8 in plains and mountain zebra (Klingel 1967, 1968), 2.2–4.0 in unmanaged feral horses (Klingel 1982; Keiper 1986; Linklater 2000; Boyd and Keiper 2005; Scorolli 2007; Roelle *et al.* 2010), and 3.8–5.7 in female-biased populations of feral horses (Keiper 1986). Maximum group size is more variable; it is moderate (8–16) in plains and mountain zebra and higher (9–21) in feral horses. Exceptionally high values of 28 on Assateague Island (Keiper 1986) and 35 in Hato El Frío, Venezuela (Pacheco and Herrera 1997), occur in populations with strongly female-biased adult sex ratios.

Although territorial species do not form permanent groups, temporary mixed-sex assemblages are similar in size to those observed in harem-forming species (table 2.1). Maximum group size reaches values as high as 261 in kiang and >1,000 in Asiatic wild asses (Feh *et al.* 2001; Bhatnagar *et al.* 2006; Kaczensky *et al.* 2008; St-Louis and Côté 2009). The size of groups changes seasonally, with values much higher in winter than in summer (table 2.1). It can be difficult to determine whether these large assemblages are a single group or an aggregation of groups forming a herd.

Many extrinsic and intrinsic factors likely influence the size of social groups. Social systems are adaptive (Klingel 1975) and shaped by the environment. As ecological pressures change, group size also changes (Rubenstein 1986). The availability of resources such as food and water as well as predation risk are proposed as key factors (Rubenstein 1986). Obviously, reproduction influences group size, as offspring are born and later disperse (Klingel 1975), so that time of year when the count is made affects results. Some studies include juveniles in counts and others do not, adding to the difficulty of comparisons across populations. The population's adult sex ratio is important, as observed in feral horses at Toi-Cape, Japan, where harem size decreased as the adult male proportion increased (Kaseda and Khalil 1996). Endoparasite burden influences the group size of feral horses on Shackleford Island, USA (Rubenstein and Hohmann 1989). In Tornquist Park, Argentina, changes in population density, rainfall, and adult sex ratio did not influence feral horse group size. Stallion behaviors, mainly fighting and ability to defend females, are proposed as possible explanatory factors for future inquiry (Scorolli 2007, 2013).

Social Networks

The quantitative depiction of social structure as social networks has recently become common. In this approach, groups are described as networks of individuals (nodes) connected by the interactions among

Table 2.1. **Equid group size**

Harem-Forming Species	Band Size		Number of Adult Mares		References
	Mean	Maximum	Mean	Maximum	
E. zebra zebra	5.5	8	2.8	4	Klingel 1968
E. z. hartmannae	4.7	10	2.2	4	Klingel 1968
E. quagga	4.5-7.7	11-16	2.2-2.8	4	Klingel 1967
E. caballus (feral)	4-12.3	8-24	2-4	8	Berger 1986; Keiper 1986; Linklater *et al.* 2000; Scorolli 2007; Grange *et al.* 2009; Roelle *et al.* 2010
E. caballus (female-biased sex ratio)	14	18-35	5.6	8-22	Keiper 1986
E. f. przewalskii	8	10-22	2-4	8	King and Gurnell 2005; Hoesli *et al.* 2009; Zimmermann *et al.* 2009

Territorial Species	Group Size		Herd Size		References
	Mean	Maximum	Mean	Maximum	
E. grevyi	5.1	17			Sundaresan *et al.* 2007*a*
E. africanus	2-60	60			Klingel 1977; Moehlman 1998*b* (feral asses)
E. kiang	2.8 (2-74) 6.8 (summer) to 10 (winter)		160-500	1,000	Schaller 1998 (seasonal, cited in St-Louis and Côté 2009); Bhatnagar *et al.* 2006
E. h. hemionus	6.25-28.8 4-24 (summer) 8-39 (winter)	11-50	75-450	786-1,241	Feh *et al.* 2001; Reading *et al.* 2001; Kaczensky *et al.* 2008; Ransom *et al.* 2012
E. h. khur	14	129		~150 in Touran National Park	Shah 2002; Shah and Qureshi 2007
E. h. onager	5	47		~302 in Qatrouyeh National Park	Nowzari 2011; Nowzari *et al.* 2012; Hemami and Momeni 2013; Nowzari *et al.* 2013

them (edges), and various parameters are calculated to describe the position of individuals and the global properties of the network, providing additional information beyond group size (Wey *et al.* 2008; Croft *et al.* 2010). For equids, the interactions commonly used involve distance between individuals. In fission-fusion equids, social networks were utilized to compare the social structure of species or subspecies that inhabit different environments in order to understand how environment determines social structure. Sundaresan *et al.* (2007*a*) reported that Grevy's zebra form more stable cliques and associate more selectively with partners as compared to Asiatic wild asses from the desert in Gujarat, India, whose social structure is more fluid. In a separate comparison between Asiatic wild asses from the Negev Desert, Israel, with those from Gujarat, India, those from the Negev (where the environment is more variable) had fewer larger social components with tightly knit cliques (Rubenstein *et al.* 2007). Groups in fission-fusion species frequently change composition. Special tools for dynamic social networks were able to follow individuals changing affiliation among different groups in Grevy's zebra and separate occasional visits from a real change of affiliation (Tantipathananandh *et al.* 2007). In harem-forming equids, social matrices are formed separately within each band. In feral horses in Snowdonia National Park, North Wales, network parameters showed that colts became more central within the harem with time and that stronger affiliative bonds formed between mares and their juvenile sons than with their juvenile daughters (Stanley and Shultz 2012). In addition, social networks analysis highlights transmission processes that may be crucial in the life of equids, such as information sharing (Rubenstein *et al.* 2007) and disease transmission (Sundaresan *et al.* 2007*a*).

Herds

Within a region, groups have been observed to form a population known as a herd, with similar movement patterns, a common home range, and interband dominance hierarchies that affect access to scarce resources (Miller 1983). This higher-level social unit is

Fig. 2.1 Herd of Persian onager groups, Qatrouyeh National Park, Iran. Photo by Haniyeh Nowzari

documented in most species of equids (fig. 2.1). These herds are often in the vicinity of water sources or preferred grazing sites, creating a temporary aggregation of smaller groups drawn to the resource. In harem-forming species, herds have been reported for plains zebra (Klingel 1967; Rubenstein and Hack 2004; Fischhoff *et al.* 2007); feral horses in the western United States (Miller 1979); Konik horses in the Oostvaardersplassen Reserve, Netherlands (Wernicke and van Dierendonck 2003); and Przewalski's horses in the Gobi Desert (Ganbaatar and Enkhsaikhan 2012). In plains zebra, as many as 100 harems form cohesive herds of up to 400 animals (Rubenstein and Hack 2004), and feral horse herds may have more than 300 individuals (Wernicke and van Dierendonck 2003). Herd formation may reduce predation risk (Rubenstein 1986; Ganbaatar and Enkhsaikhan 2012), afford shelter against harsh climatic conditions (Ganbaatar and Enkhsaikhan 2012; Nowzari *et al.* 2013), and help avoid bachelor harassment of females (Rubenstein 1994; Rubenstein and Hack 2004).

Bachelor Groups

In most unmanaged equid populations, the sex ratio is 1:1 (see chap. 6) and all equids are polygynous, creating a surplus of males that frequently form groups and occasionally live solitarily. Bachelor groups, also called stallion groups or all-male groups, are social units observed in every equid species (Klingel 1975; Keiper 1986; Linklater 2000; Boyd and Keiper 2005). Males younger than 5 years old that have dispersed from their natal bands, but are not yet old enough to obtain mares, typically form such all-male groups. Often the males are of similar age, but older stallions that do not have mares may also join bachelor groups (Klingel 1975; Keiper 1986; Linklater *et al.* 2000; Cameron *et al.* 2001).

The bachelor group size of harem-forming species tends to be similar, with a mean of three individuals (range 2–17) (Klingel 1967, 1968; Penzhorn 1984; Berger 1986; Linklater 2000; Boyd and Keiper 2005). Unlike harems, bachelor groups are not stable (Keiper 1986; Linklater 2000). In populations where large numbers of bachelors were reliably identified and followed, however, dyads or core groups of up to three to four males have been observed to last for several months, or even years, among feral horses (Miller 1979; Linklater *et al.* 2000; Scorolli 2007), plains zebra (Klingel 1967), and mountain zebra (Penzhorn 1984). A clear linear dominance hierarchy was observed between members of a bachelor group of feral horses (Feist and McCullough 1976; Rubenstein 1981), but not in mountain zebra (Penzhorn 1984).

Territorial equid species have been comparatively little studied, but bachelor group size seems to be similar to that of harem-forming species (range 2–20 males). The bonds between males are even more ephemeral, and group composition therefore much more variable (Klingel 1974*b*, 1977; Woodward 1979; Moehlman 1998*a*; Shah 2002). An exception may be a population of Asiatic wild asses in the Gobi Desert of Mongolia, where bachelors groups lasting for months were observed (Feh *et al.* 1994, 2001); however, the ability to unambiguously identify individuals in this large population may have been overestimated.

Hypothesized functions of bachelor groups for young males include learning fighting skills and the opportunity to gain mares jointly while affording the protection of a group against predators (fig. 2.2). Older displaced harem stallions probably seek the security

BOX 2.1 DETERMINANTS OF GROUP COMPOSITION, DENSITY, AND SIZE IN PERSIAN ONAGERS

Persian onagers (*Equus hemionus onager*) are a fission-fusion species, and group size varies according to season, weather conditions, water location, vegetation quality, and reproductive status. Winter strongly affects juveniles, whose numbers decrease owing to high mortality, but the number of mature individuals increases in groups, probably because individuals can take advantage of others to block the wind or share their warmth. In spring, the numbers of lactating females and juveniles increase in groups because of breeding and recruitment; however, nonlactating females and subordinate males then leave groups to mate throughout the summer (Nowzari 2011; Nowzari *et al.* 2012).

The co-occurrence of wind and rain, especially during cold periods, intensifies the harshness of environmental conditions, and onagers—irrespective of age, sex, or reproductive state—mitigate the impacts of these conditions through their behavior. Access to valleys appears to reduce the impact of wind and rain; groups also tend to be relatively large under these conditions so that individuals can take advantage of others to block the wind or share their warmth (Nowzari *et al.* 2013).

Age class and reproductive state affect the distribution of onagers with respect to water. Juveniles and females with foals tend to reside close to water sources (Nowzari *et al.* 2013), which matches observations of lactating females in other fission-fusion equids, such as onagers in the Negev Desert (Rubenstein 1994) and Grevy's zebra in semiarid regions of Kenya (Rubenstein 1986).

Vegetation quality, as measured by Normalized Difference Vegetation Index (NDVI) values, affects the density of onagers found in particular locations, but in different ways depending on sex and female reproductive state. Densities and group sizes of females, both with and without juveniles, are highest in areas with highest NDVI values (Nowzari *et al.* 2013). Similar correlations between plant abundance and group size are found in some other Asiatic wild ass populations (Henley *et al.* 2006), but not all (Ransom *et al.* 2012). Males are never unaccompanied by females in high-quality areas, underscoring the fact that these territorial males are more successful in gaining mating opportunities than males in areas of lower vegetation quality. The presence of large groups of females with their juveniles near water and in areas with high-quality vegetation shows that these groups require both adequate water resources and high-quality food much of the year (Nowzari *et al.* 2013).

The size of onager groups in this population ranged from 2 to 47 individuals, with a median size of 5. Groups are large in high-quality areas because most females and juveniles are there and form large groups (Nowzari 2011; Nowzari *et al.* 2012, 2013).

Most territorial males are seen alone, without females or bachelor males, outside of the breeding season. During the mating season, however, Persian onager females with young foals come into postpartum estrus and are reproductively active (Rubenstein 1994). The most able and dominant males position themselves to increase their chances of mating with females proximal to water and those coming to water at irregular intervals (Rubenstein 2010). As females with young are found in areas with the highest NDVI values close to water, it is not surprising that sightings of lone males are rare in these locations. Territorial males near water are almost always with females, some with young and others that come to water from afar to drink. During the mating season, territorial males that are alone are the subordinate males found in areas where the vegetation is of intermediate quality, which is where most females without foals are also found. Studies of Grevy's zebra and other Asiatic wild asses (Rubenstein 1986, 1994; Ginsberg 1989; Sundaresan *et al.* 2007*b*) suggest that females without young occupy areas controlled by midranking males, some of which will be temporarily alone. Females move to these areas to seek large quantities of vegetation and to minimize competition with the high density of lactating females that must remain near water. Observations of the composition and location of Persian onager groups by relative size support this interpretation (Nowzari *et al.* 2013).

Fig. 2.2 Play-fighting among Przewalski's horse bachelors, Hustai National Park, Mongolia. Photo by Amos Bouskila

and familiarity that being in a group provides (Rubenstein 1986; Linklater 2000).

Stability of Harem Groups in Species with Female Defense Polygyny

In pioneering field research on plains zebra in Tanzania's Ngorongoro Crater, Klingel (1967) was the first to study the stability of social groups in a wild equid population. Individually identified plains zebra showed remarkable social stability, and after 2.5 years most harems (70%) had not changed their membership, and in the others, changes were mostly the addition of young mares to existing harems (Klingel 1969*a*). A more recent study in Zimbabwe reported much lower stability, probably caused by predation (Grange 2006). The two mountain zebra subspecies—Hartmann's zebra and Cape Mountain zebra—also showed a remarkable stability: mares that lost their stallion remained together as a unit (Klingel 1968, 1969*b*). A later long-term study of the same population of mountain zebra showed that most harems remained stable during 6–8 years, and stallion tenure reached 10.5 years (Penzhorn 1984). Some researchers hypothesize that social bonds in these zebra species could be lifelong and mares probably remain in the same harem until death (Klingel 1975; Penzhorn 1984), but there is no conclusive evidence of this as of yet.

Feral horse social organization has been studied worldwide under different environmental and demographic conditions, and year-round social stability is reported for all populations (Klingel 1982; Keiper 1986; Linklater 2000; Boyd and Keiper 2005). In studies of more than three years' duration, most harems were stable over many years, as observed in Atlantic barrier islands of the United States—such as Assateague Island (Keiper 1986), Shackleford Island (Rubenstein 1981), and Cumberland Island (Goodloe *et al.* 2000)—in the Great Basin Desert, USA (Berger 1986), in the Camargue region of France (Duncan 1992; Feh 1999), in the Kaimanawa Mountains of New Zealand (Linklater *et al.* 2000), and in the Pampas region of Argentina (Scorolli and Lopez Cazorla 2010).

Stallion tenure in their harem could be one indicator of stability. Mean feral horse stallion tenure was not lengthy on Cumberland Island (2.8 years) (Goodloe *et al.* 2000) or in the Great Basin (3.2 years) (Berger 1986), but was higher in Tornquist Park, Argentina (4.4 years) (Scorolli and Lopez Cazorla 2010). Only 30–37% of adult stallions reached maximum tenure, defined as holding a harem throughout the duration of the study (3 years, Cameron *et al.* 2001; 5 years, Berger 1986 and Goodloe *et al.* 2000; 8 years, Scorolli and Lopez Cazorla 2010). For 428 stallions from three feral horse populations in the western United States, the average harem tenure was 5 years, with 7% of the stallions holding a harem for more than 10 years (J.I. Ransom, unpublished data). Extremely long stallion tenure of 18 years was registered in the Camargue region of France (Feh 1999); however, this population was managed and the adult sex ratio was strongly female-biased, which probably reduced stallion competition for mares. But a stallion at Little Book Cliffs, USA, held a harem for 18 years in a population without a skewed sex ratio (J.I. Ransom, unpublished data).

The proportion of bands originally present that remain stable may also be used as a measure of social stability. Most long-term research on feral horses reported high stability, with less than a third of all harems dissolving during the study period (Berger 1986; Goodloe *et al.* 2000; Cameron *et al.* 2001). For example, the average annual percentage of bands dissolved was 18% in Argentina (Scorolli and Lopez Cazorla 2010). The percentage of adult mares that changed harems annually was low in some studies (2–12%) (Rubenstein 1981; Berger 1986; Duncan 1992; Feh 1999; Linklater *et al.* 2000) and higher (20–30%) in others (Franke-Stevens 1990; Goodloe *et al.* 2000; Scorolli and Lopez Cazorla 2010).

The social organization of Przewalski's horses was poorly known before the species became extinct in the wild in the 1960s. After their reintroduction into Mongolia, long-term studies of behavior and social organization showed that this species forms harems. Long-lasting bonds exist between band members, with stallion tenures of up to seven years observed (King and Gurnell 2005).

Many factors may influence the feature of social stability. The existence of two markedly different social organization strategies in wild equids has been correlated to the distribution and abundance of resources such as water and forage (Klingel 1975; Rubenstein 1986; Franke-Stevens 1990). Harem stability in one population of feral horses was related to winter food abundance (Franke-Stevens 1990). The formation of harems in an arid-adapted equid such as Mongolian khulan could be a response to abundant predators (Feh *et al.* 1994, 2001), or may be a temporary phenomenon and the manifestation of flexibility within the social organization along a gradient between Type I and Type II social organization. Predation and adult male harassment of mares have been suggested as important factors influencing social bonds in feral horses (Feh *et al.* 1994, 2001; Linklater *et al.* 1999), but conclusive evidence is still lacking.

Interband Dominance Hierarchies

Animals that interact with one another regularly form a dominance hierarchy that determines priority of access to resources. The presence of a hierarchy implies that participants recognize one another as individuals and remember the outcome of past interactions. Dominance hierarchies are a prominent aspect of equid social organization, existing both within and between groups. Intraband hierarchies are discussed in chapter 3, but interband hierarchies are a significant aspect of social organization. Groups of equids in a herd have similar movement patterns, increasing the probability of interactions, particularly at scarce resources such as water points. Larger groups generally have priority access (Berger 1977; Miller and Denniston 1979). Although groups already in possession of the resource have a defensive advantage (Franke-Stevens 1988), some low-ranking feral horse bands and bachelor groups in the Red Desert of Wyoming, USA, left the water hole at the approach of high-ranking bands, and waited as long as five hours for higher-ranking bands to depart before they were able to drink (Miller and Denniston 1979). Harem stallions typically move to the front of the band when approaching a water hole (Berger 1977; Miller 1979; Penzhorn 1984) and initiate contests with other groups for access, but the adult mares also participate through shoving, threats, bites, and kicks (Miller and Denniston 1979).

Reproductive Strategies

Harem Formation Strategies

Stallions of harem-forming species obtain mares in one of the following ways:

1. Acquiring unguarded females, such as females dispersing from their natal band, or adult mares that have separated from their harem or whose stallion has died (Keiper 1976; Salter 1978; Nelson 1979).
2. Defeating another harem stallion in combat, obtaining the entire harem (Klingel 1969*a*).
3. Raiding, resulting in abduction of part of a harem (Miller 1979; Nelson 1979).
4. Attaching themselves to a harem as a satellite male, perhaps eventually ousting the resident stallion or departing with a fraction of the mares (Welsh 1973; Salter 1978; Miller 1979; Nelson 1979; Rubenstein 1982; Khalil and Murakami 1999*a*).
5. Belonging to a bachelor group that obtains females in one of the ways above, typically by the first method listed. One bachelor, or an alliance of bachelors, eventually ousts the other males, or if several females were obtained, the group breaks up to form several harems (McCort 1979; Miller 1979).
6. Remaining in the natal band and inheriting it (Miller 1979; Rutberg and Keiper 1993; Roelle *et al.* 2010).

BOX 2.2 MARKING BEHAVIOR

Marking behavior is ubiquitous among male equids, assuming an important role in communication and social interaction. Males defecate on the dung of other males and mark the excretions of females with either urine or feces. By the time a juvenile is 3 years old, both forms of marking behavior have become part of his behavioral repertoire (Hoffmann 1985). Females and immature animals rarely exhibit marking behavior (Turner *et al.* 1981).

Harem stallions create dung piles, also known as stud piles, throughout their home range (Rubenstein 1981), especially in core areas (King and Gurnell 2007). These may be concentrated in areas devoid of vegetation, often along trails (Salter and Hudson 1982; King and Gurnell 2007) where they can be seen and encountered, especially if the trails are used by many bands, for example, leading to watering points (Mohr 1971; Feist and McCullough 1976). The dung piles of territorial males are also commonly on routes to water (Woodward 1979; Ginsberg 1987; Moehlman 1998*a*) and may also be observed along territorial boundaries (Klingel 1974*b*, 1977; Moehlman 1998*a*). Stud piles are utilized year-round (Woodward 1979; Hoffmann 1985). During marking, the stallion approaches a dung pile, sniffs it, steps over it, defecates, and then turns and sniffs his excretion (Klingel 1967, 1968; Moehlman 1998*a*; King and Gurnell 2007). No other stallion need be present, but the behavior is common during stallion encounters. More than 60% of all male feral horse interactions involve fecal deposition (Miller 1981; Salter and Hudson 1982). Feral horse bachelor stallions may defecate on dung piles in order of their rank, from least to most dominant (Feist and McCullough 1975), as was also observed in Camargue horse bachelors (von Goldschmidt-Rothschild and Tschanz 1978) and feral asses (McCort 1980). Feral horse stallions in multimale bands also typically defecate in order of rank (Feist and McCullough 1976; Miller 1981). Feces contain a more complex mixture of volatile substances than urine and may reflect an individual's gender, age, and reproductive status (Kimura 2001). Equids often dwell in open areas; these extensive piles create a visual and olfactory communication center that is not dependent upon vegetation to hold scent. Dung piles offer information about an individual's identity and dominance rank, advertise the male's presence (Ginsberg 1989; Rubenstein and Hack 1992), and may play a role in orienting territorial individuals whose habitat does not offer many other visual landmarks (Klingel 1974*a*, 1975; Moehlman 1998*a*).

Przewalski's horse stallion turns to sniff a dung pile after defecating, Hustai National Park, Mongolia. Photo by Lee Boyd

A second form of marking behavior is typical of male equids of all species. Stallions use their own urine, or sometimes feces, to mark the urine and feces of their mares. Marking the eliminations of females occurs more frequently during the breeding season than at other times of the year (Feist and McCullough 1976; Turner *et al.* 1981; Hoffmann 1985; Boyd and Kasman 1986; Kimura 2001) and is closely correlated with seasonal testosterone levels (Turner and Kirkpatrick 1986). When the stallion detects an elimination, he approaches, sniffs it, and may flehmen, defecate, or urinate on top of the elimination and perhaps smell it again prior to moving away (Klingel 1967; Joubert 1972; Welsh 1973; McCort 1984; Penzhorn 1984; Boyd and Kasman 1986). The response is more likely to contain all of these elements during the breeding season (Turner *et al.* 1981). Fatty acid concentrations are higher in the feces of estrous mares than nonestrous mares (Kimura 2001). The urine of stallions contains high levels of long-lasting pungent cresols. When males use urine to mark the feces of estrous mares, the fatty acid concentration decreases to levels found in feces of nonestrous mares. Thus stallions are masking the difference between the odor of estrous and nonestrous feces by urinating on them, reducing the likelihood that rival males will be attracted.

In harem-forming equids, stallions are ~5 years old when they obtain their first mares (Klingel 1969*a*; Penzhorn 1984; L. Boyd, unpublished data on Przewalski's horses). In the Misaki feral horse population, with its female-biased sex ratio, about 20% of the bachelors obtained females as early as their fourth year (Kaseda *et al.* 1984), and another 40% by their fifth year of life (Kaseda *et al.* 1997). In a later study of the same population, as the sex ratio approached 1:2, the age range at the time of first acquisition of mares was 3.8–7.6 years with a mean of 5.2 years (Khalil and Murakami 1999*a*); 70% of the stallions formed harems at the age of 5 years (Khalil and Murakami 1999*b*). In the Granite Range of Nevada, USA, only feral horse bachelors that were at least 6 years old were able to win mares through combat (Berger 1986).

Male Alliances or Coalitions

Among harem-forming equids, multiple-male bands have been reported in plains zebra (Klingel 1967), Przewalski's horses (Ganbaatar and Enkhsaikhan 2012), many feral horse populations, and also in feral asses that form harems (McCort 1980; Moehlman *et al.* 1998). In the Pryor Mountains of Montana, USA, 6.3% of the feral horse bands contained two, and in one case three, adult stallions (Roelle *et al.* 2010). Granite Range feral horse stallions that held single-male harems had longer tenure (Berger 1986). But if rivalry is intense and a stallion is not a skillful competitor, an alliance provides a way to obtain mares and sire offspring more quickly than by remaining a bachelor. In Camargue horses, single harem males had better reproductive success than dominant stallions in multimale bands, which in turn had a greater reproductive success than subordinate stallions. The latter had a higher reproductive success than bachelors that sneaked copulations (Feh 1999).

In the Camargue population, high-ranking stallions held single-male harems, but low-ranking sons of low-ranking mothers formed alliances with males of a similar age during their bachelor years, leading to the formation of multiple-male bands (Feh 1999). In the Shackleford Banks feral horse population, male alliances also formed prior to the attainment of a harem. Two paternal half-brothers dispersed together into the bachelor group and achieved much higher status than young males that arrived singly (Rubenstein 1982).

Sixty percent of the Camargue alliances lasted two to three years, after which the stallions separated to establish single-male bands. The remaining alliances lasted as long as 16 years, often until the death of one of the stallions. Alliances among some Pryor Mountain feral horses lasted five to six years, although most lasted less than two years (Roelle *et al.* 2010). In the Granite Range of Nevada, 17 alliances were observed between feral horse stallions, but most were short-lived (Berger 1986). Only two coalitions exceeded 7 months, but one of these lasted 2.5 years, and the other at least 4 years. In the 13 cases where kinship was known, the stallions were unrelated.

In the Rachel Carson Estuarine Sanctuary of North Carolina, USA, only the dominant stallion was observed mating in multiple-male bands (Franke-Stevens 1990). One stallion was dominant over the other in 15 of the 17 Granite Range alliances and performed all of the observed mating (Berger 1986). If there were more than two stallions in Kaimanawa harems, only the two highest-ranking males were observed breeding (Linklater and Cameron 2000). For multiple-male bands in the Red Desert, however, 49% of the mating was by the dominant stallion, 42% by one of the subordinate stallions, and 9% by stallions from other bands (Miller 1981). Three mating systems were apparent in multimale bands in the Red Desert: (1) the dominant male did the majority of the breeding; (2) the stallions mated the same mare serially during her estrous period with little aggression; and (3) in one harem, each of the three males consorted with one of three females that were in estrus at the same time.

The dominant male of an alliance may lose some matings to subordinates but may be compensated by a division of labor, resulting in additional females that the males together can hold. When a band flees from perceived danger, the dominant male is often observed leading the mares while the subordinate male is at the rear between the females and danger (Pacheco and Herrera 1997). The subordinate stallion confronts approaching rivals while the dominant stallion drives the mares away from the confrontation (Miller 1981; Rubenstein 1982; Berger 1986; Franke-Stevens 1990; Feh 1999; Linklater and Cameron 2000). This strategy appears effective, as Linklater and Cameron (2000) observed copulations by males from outside the harem to occur only in single-male bands.

Subordinate stallions may eventually inherit the band (Feh 1999), but Berger (1986) observed this only once in the Granite Range horses. Half of the methods of harem formation reported and described above involve some form of short- or long-term association among stallions, however, after which one male may ultimately "inherit" the harem. Subordinate males made feral horse bands in the Rachel Carson Estuarine Sanctuary more stable; these bands were less likely to lose females (Franke-Stevens 1990). There was a positive correlation between number of males and the number of females in Venezuelan feral horse bands (Pacheco and Herrera 1997). Both Perkins *et al.* (1979) and Miller (1981) reported that multiple-male bands were larger, and in the Red Desert population these bands were more stable (Miller 1981). Mares in stable bands produce more foals than mares from unstable bands (Berger 1986; Kaseda *et al.* 1995). Although Boyd (1980) recorded significantly higher natality in Red Desert single-male bands, the natality of multimale bands was less affected after a severe winter, perhaps because the larger band size provided warmth and combined effort to dig through snow. The Sable Island population also experiences harsh winters, and Welsh (1975) reported lower natality and survival in smaller bands, which he attributed in part to less huddling protection and also to the fact that small bands were often newer and less stable. Foal survival was higher in larger Pryor Mountain feral horse bands, but only 20% of the multimale bands were larger than average (Roelle *et al.* 2010).

The effects of alliances on reproductive success are deserving of further study, as there is little consistency between feral horse populations. Boyd (1980) found no difference in foal survival between single- and multiple-male bands in the Red Desert of Wyoming. But in the Camargue population, foal survival was higher in multimale bands (Feh 1999). In contrast to other studies, Linklater and Cameron (2000) and Roelle *et al.* (2010) observed no difference between single- and multiple-male harems of feral horses regarding the number of mares contained. Kaimanawa mares from single-male bands were in significantly better condition throughout the year than mares in multimale bands, and had higher foaling rates and lower rates of offspring mortality (Linklater *et al.* 1999; Linklater and Cameron 2000). An experiment in which the subordinate male was removed from two multimale bands for three weeks did not result in loss of the band by the dominant male (Linklater *et al.* 2013).

Evidence for Mate Choice by Females

Feral horse mares sometimes attempted to return to their original band for as long as two days after abduction by another stallion (Miller 1979). Whether this indicates a preference for the original stallion or for the other mares is not known. Plains zebra mares show similar fidelity (Klingel 1967, 1969*a*). Females may use the ability of the stallion to obtain and retain mares as a measure of the stallion's quality (Rutberg 1990). But female choice is not obviated by despotic behavior of stallions. Rutberg (1990) observed Assateague Island mares moving out of sight of the stallion in areas where dense cover could thwart the stallion's attempt to find them unless the mare cooperated. Assateague mares were also observed alone or accompanied only by immature offspring, implying that they were able to leave the stallion and remain unincorporated if they so chose. Gates (1979) observed that Exmoor pony mares sometimes temporarily broke away from their harem, and after a gathering brought harems into close proximity, some mares shifted their allegiance from one stallion to another. During the course of a year, some mares in the Kaimanawa feral horse population moved between several bands or away from their own band for several days at a time (Linklater *et al.* 1999). A significant portion of the Misaki harems broke up in winter and re-formed during the breeding season, giving mares the opportunity to change their allegiance. A few mountain zebra left their bands and were bred by other stallions and returned or joined new bands (Penzhorn 1984). Feral horse mares in Alberta, Canada, approached and initiated courtship with dominant stallions while rejecting approaches of subordinate stallions (Salter 1978). Older, experienced mares reject the attention of young, inexperienced males by biting and kicking them (Tyler 1972; Berger 1986). When Red Desert stallions did not achieve intromission, it was almost always because of the mare's aggression toward the stallion, suggesting that she was exerting some choice (Miller 1981).

Berger and Cunningham (1987) reported that half of the dispersing females that they observed copulating mated with the first male they met, but 33% mated with a subsequent male rather than the first male, and the remaining 17% mated with multiple males. Young plains zebra mares may change harems three to four times before settling (Klingel 1967).

Among territorial equids, female feral asses may copulate with several males (Moehlman 1998*a*). Ginsberg and Rubenstein (1990) reported 73% of Grevy's zebra mares to be polyandrous. In these situations the potential for sperm competition is high. Immediately after copulation, seminal fluid was discharged from the vagina of Grevy's zebra females (Ginsberg and Rubenstein 1990), but whether this behavior is related to female mate choice is unknown. Although Grevy's zebra stallions attempt to drive estrous females into their own territories, the mare generally decides where she is going and thus with which male she mates (Klingel

1975). See box 2.1 regarding mate choice by Persian onagers.

Inbreeding Avoidance

Klingel (1969*c*) and Penzhorn (1984) observed that young female plains and mountain zebra had an exaggerated urination stance during estrus that attracted abductors that might potentially remove them from their father's harem. Once they became part of a new harem, the stance of adult females was much more subtle, minimizing interference from other stallions.

Granite Range feral horse stallions did not breed daughters with which they were familiar, but familiarity waned if parent and offspring were separated for more than 18 months, so that 4% of copulations for which the genealogies of the participants were known were between fathers and daughters (Berger and Cunningham 1987). Polish feral horse stallions usually do not mate their daughters, nor do young males show interest in mounting their mothers (Jaworowska 1981). Among Konik horses in the Blauwe-Kamer Reserve, Netherlands, two stallions that were allies in the same band mated with their daughters, as determined by deoxyribonucleic acid (DNA) paternity analysis (Bouskila *et al.* 2012). Welsh (1975) reported only one case of father–daughter mating among the feral horses of Sable Island, Nova Scotia. There was no evidence of mother–son matings in Camargue horses, but father–daughter and sibling matings were observed (Duncan *et al.* 1984). The level of sexual activity was less than expected whenever these relatives were familiar with one another, however. Harem stallions were less closely related to mares in their own band than to mares in other bands, suggesting negative assortment with regard to kinship. Young females accepted the attention of males from other groups but rejected courtship by males that were kin from their natal group (Monard *et al.* 1996). Harem stallions do not always interfere when daughters that have not yet dispersed are courted and bred by other males (Feist and McCullough 1976; Boyd 1980; Berger and Cunningham 1987; Asa 1999). Additionally, dams interposed themselves when related males began to court their daughters but not when unrelated males approached. Monard and Duncan (1996) showed that dispersing feral horse females preferred to move to harems containing familiar mares similar to them in age but lacking familiar males. Only 30% of dispersing fillies joined harems headed by a stallion that was a close relative. Usually the stallion was unknown to them. In the few cases where the filly joined a familiar stallion, there were no other harems with unrelated males available.

On Assateague Island, 19% of young females remained in their natal harem after sexual maturity, and father–daughter matings resulted in live offspring (Rutberg and Keiper 1993). But the foaling percentage of only 22.7% compared with 36.8% for half-sibling matings and 61.8% for mating by unrelated individuals (Keiper and Houpt 1984). Inbreeding avoidance needs to be studied in territorial equids, as this social system lacks the inherent mechanisms that reduce the risk of inbreeding in harem-forming species.

Extrapair Copulations in Harem-Forming Species

Half of observed copulations by harem males in the feral horse population of Jicarilla, USA, were with females from other bands (Nelson 1980). In single-male Red Desert bands, 14% of the observed copulations were by outside stallions (Miller 1981). Two or more males bred the same mare during the same estrus in 2% of the copulations observed by Berger and Cunningham (1987). Of mares in feral horse bands headed by vasectomized stallions, 17–33% produced foals (Asa 1999). Eighty percent of these bands contained a subordinate stallion, with sneak breeding by bachelor males also a possibility. Paternity testing confirmed that 14–15% of Misaki foals were not sired by the harem stallion (Kaseda *et al.* 1982; Kaseda and Khalil 1996). A large amount of reproductive success could be attributed to subordinate and bachelor stallions in Great Basin feral horses (Ashley 2000); blood typing confirmed that approximately one-third of foals were not sired by the dominant harem stallion (Bowling and Touchberry 1990). In only about half of either the single- or multiple-male bands did one stallion sire all of the foals. Some of the populations had been disturbed by roundups, however, so it is possible that some pregnant mares had changed bands prior to this study. Extrapair copulation has also been reported in zebra; a mountain zebra left her band, was bred by another stallion, and then returned (Penzhorn 1984).

Conclusion

The social organization of equids is proving to be more complex and more plastic than previously thought. Mesic conditions can shift territorial species such as asses into a harem social system and, conversely, easily defensible geography permits harem-forming species to hold territories. A variety of factors influences the size of groups. Harem stallions adopt single- versus multiple-male strategies depending on circumstance. Complex social interactions such as alliances and hierarchies, both within and between groups in a population, create a rich social milieu. The sociality and adaptability of equids no doubt contribute to their continued survival, the facility with which they can be domesticated, the ability of domestic species to become feral, and the success of reintroductions in regions where wild equids have become extinct.

Many challenges still remain to researchers of wild equids. Little information is available about bachelor group size in Przewalski's horses, but reintroduced populations have now reached a size that would permit this parameter to be studied. Lifelong stability of harems remains little more than a hypothesis without generational studies of all harem-forming species. Whether predation and male harassment of females influence social bonds needs to be tested. We need to know what effect alliances have on reproductive success of the participants and whether the size of a group is important in coping with severe weather. How do territorial equids with fission-fusion societies avoid inbreeding with unfamiliar relatives? Most equid species deserve more thorough study. Long-term research on individually identified animals, involving extensive fieldwork, and novel analysis techniques are needed to bring about a deeper understanding of equid social organization.

ACKNOWLEDGMENTS

The authors thank Jason Ransom and Petra Kaczensky for organizing the International Wild Equid Conference, from whence this book sprung. Jason also provided some unpublished data that strengthened this chapter. The authors also thank the anonymous reviewers for improving the chapter, the pioneers of equid research for their inspirational dedication and tantalizing results, and current colleagues for their collaborative spirit. May these collective efforts help equids roam the planet for many years yet to come.

REFERENCES

Asa, C.S. 1999. Male reproductive success in free-ranging feral horses. Behavioral Ecology and Sociobiology 47:89–93.

Ashley, M.C. 2000. Feral horses in the desert: Population genetics, demography, mating, and management. Thesis, University of Nevada, Reno, USA.

Bahloul, K., O.B. Pereladova, N. Soldatova, G. Fisenko, E. Sidorenko, and A. Semperé. 2001. Social organization and dispersion of introduced kulans (*Equus hemionus kulan*) and Przewalski horses (*Equus przewalskii*) in the Bukhara Reserve, Uzbekistan. Journal of Arid Environments 47:309–323.

Berger, J. 1977. Organizational systems and dominance in feral horses in the Grand Canyon. Behavioral Ecology and Sociobiology 2:131–146.

Berger, J. 1986. Wild horses of the Great Basin: Social competition and population size. University of Chicago Press, Chicago, USA.

Berger, J., and C. Cunningham. 1987. Influence of familiarity on frequency of inbreeding in wild horses. Evolution 41:229–231.

Bhatnagar, Y.V., R. Wangchuk, H.H.T. Prins, S.E. Van Wieren, and C. Mishra. 2006. Perceived conflicts between pastoralism and conservation of the kiang *Equus kiang* in the Ladakh Trans-Himalaya, India. Environmental Management 38:934–941.

Bouskila, A., H. de Vries, Z.M. Hermans, and M. van Dierendonck. 2012. Insights from the pedigree on the social structure of free-roaming Konik horses (*Equus caballus*) in a Dutch reserve. Presented at the International Wild Equid Conference, 18–22 September 2012, Vienna, Austria.

Bowling, A.T., and R.W. Touchberry. 1990. Parentage of Great Basin feral horses. Journal of Wildlife Management 54:424–429.

Boyd, L. 1980. The natality, foal survivorship and mare–foal behavior of feral horses in Wyoming's Red Desert. Thesis, University of Wyoming, Laramie, USA.

Boyd, L., and L. Kasman. 1986. The marking behavior of male Przewalski's horses. Pages 623-626 *in* D. Duvall, D. Muller-Schwarze, and R. M. Silverstein (Eds.), Chemical signals in vertebrates, vol. 4: Ecology, evolution, and comparative biology. Plenum Press, New York, USA.

Boyd, L., and R.R. Keiper. 2005. Behavioral ecology of feral horses. Pages 55–82 *in* D.S. Mills and S. M. McDonnell (Eds.), The domestic horse: The origins, development, and management of its behaviour. Cambridge University Press, Cambridge, UK.

Cameron, E.Z., W.L. Linklater, E.O. Minot, and K.J. Stafford. 2001. Population dynamics, 1994–1998, and management of Kaimanawa wild horses. Science for Conservation Publication 171. Department of Conservation, Otago, New Zealand.

Croft, D.P., R. James, and J. Krause. 2010. Exploring animal social networks. Princeton University Press, Princeton, NJ, USA.

Duncan, P. 1992. Horses and grasses: The nutritional ecology of equids and their impact on the Camargue. Ecological Studies 87. Springer, New York, USA.

Duncan, P., C. Feh, J.C. Gleize, P. Malkas, and A.M. Scott. 1984. Reduction of inbreeding in a natural herd of horses. Animal Behaviour 32:520–527.

Feh, C. 1999. Alliances and reproductive success in Camargue stallions. Animal Behaviour 57:705–713.

Feh, C., T. Boldsukh, and C. Tourenq. 1994. Are family groups in equids a response to cooperative hunting by predators? The case of Mongolian khulans (*Equus hemionus* Luteus Matschie). Revue d'Ecologie (Terre Vie) 49:11–20.

Feh, C., S. Munkhtuya, S. Enkhbold, and T. Sukhbataar. 2001. Ecology and social structure of the Gobi Khulan *Equus hemionus* sbsp. in the Gobi B National Park, Mongolia. Biological Conservation 101:51–61.

Feh, C., N. Shah, M. Rowen, R. Reading, and S.P. Goyal. 2002. Status and action plan for the Asiatic wild ass (*Equus hemionus*). Pages 62–71 *in* P.D. Moehlman (Ed.), Equids: Zebras, asses and horses. Status survey and conservation action plan. International Union for Conservation of Nature, Gland, Switzerland.

Feist, J.D., and D.R. McCullough. 1975. Reproduction in feral horses. Journal of Reproduction and Fertility 23:13–18.

Feist, J.D., and D.R. McCullough. 1976. Behaviour patterns and communication in feral horses. Zeitschrift für Tierpsychologie 41:337–371.

Fischhoff, I.R., S.R. Sundaresan, J. Cordingley, H.M. Larkin, M. Sellier, and D.I. Rubenstein. 2007. Social relationships and reproductive state influence leadership roles in movements of plains zebra, *Equus burchelli*. Animal Behaviour 73:825–831.

Franke-Stevens, E. 1988. Contests between bands of feral horses for access to fresh water: The resident wins. Animal Behaviour 36:1851–1853.

Franke-Stevens, E. 1990. Instability of harems of feral horses in relation to season and presence of subordinate stallions. Behaviour 112:149–161.

Ganbaatar, O., and N. Enkhsaikhan. 2012. Herd structure, behaviour, and social organization—Takhin Tal. Pages 97–100 *in* N. Bandi and O. Dorjaraa (Eds.), Takhi: Back to the wild. International Takhi Group, Ulaanbaatar, Mongolia.

Gates, S. 1979. A study of the home ranges of free-ranging Exmoor ponies. Mammalian Review 9:3–18.

Ginsberg, J.R. 1987. Social organization and mating strategies of an arid adapted equid: The Grevy's zebra. Thesis, Princeton University, Princeton, NJ, USA.

Ginsberg, J.R. 1989. The ecology of female behaviour and male mating success in Grevy's zebra. Symposium of the Zoological Society London 61:89–110.

Ginsberg, J.R., and D.I. Rubsenstein. 1990. Sperm competition and variation in zebra mating behavior. Behavioral Ecology and Sociobiology 26:427–434.

Goodloe, R.B., R.J. Warren, D.A. Osborn, and C. Hall. 2000. Population characteristics of feral horses on Cumberland Island, Georgia, and their management implications. Journal of Wildlife Management 64:114–121.

Grange, S. 2006. The great dilemma of equids: Living with grazing bovids and avoiding large predators. Thesis, Université de Poitiers, Poitiers, France.

Grange, S., P. Duncan, and J.-M. Gaillard. 2009. Poor horse traders: Large mammals trade survival for reproduction during the process of feralization. Proceedings of the Royal Society of London B: Biological Sciences 1663:1911–1919.

Hemami, M.R., and M. Momeni. 2013. Estimating abundance of the endangered onager *Equus hemionus onager* in Qatruiyeh National Park, Iran. Oryx 47:266–272.

Henley, S.R., D. Ward, and I. Schmidt. 2006. Habitat selection by two desert-adapted ungulates. Journal of Arid Environments 70:39–48.

Hoesli, T., T. Nikowitz, C. Walzer, and P. Kaczensky. 2009. Monitoring of agonistic behaviour and foal mortality in free-ranging Przewalski's horse harems in the Mongolian Gobi. Equus 2009 (Praha Zoo):113–138.

Hoffmann, R. 1985. On the development of social behavior in immature males of a feral horse population (*Equus przewalskii* f. caballus). Zeitschrift für Säugetierkunde 50:302–314.

Jarman, P.J. 1974. The social organization of antelope in relation to their ecology. Behaviour 48:215–267.

Jaworowska, M. 1981. Die Fortpflanzung primitiver polnischer Pferde, die frei in Waldschutzgebiet leben. Säugetierkundliche Mitteilungen 29:46–71.

Joubert, E. 1972. The social organization and associated behavior in the Hartmann zebra *Equus zebra hartmannae*. Madoqua 1:17–56.

Kaczensky, P., O. Ganbaatar, H.V. Wehrden, and C. Walzer. 2008. Resource selection by sympatric wild equids in the Mongolian Gobi. Journal of Applied Ecology 45: 1762–1769.

Kaseda, Y., and A.M. Khalil. 1996. Harem size and reproductive success of stallions in Misaki feral horses. Applied Animal Behaviour Science 47:163–173.

Kaseda, Y., A.M. Khalil, and H. Ogawa. 1995. Harem stability and reproductive success of Misaki feral mares. Equine Veterinary Journal 27:368–372.

Kaseda, Y., K. Nozawa, and K. Mogi. 1982. Sire-foal relationships between harem stallions and foals in Misaki horses. Japanese Journal of Zootechnical Science 53:822–830.

Kaseda, Y., K. Nozawa, and K. Mogi. 1984. Separation and independence of offsprings from the harem groups in Misaki horses. Japanese Journal of Zootechnical Science 55:852–857.

Kaseda, Y., H. Ogawa, and A.M. Khalil. 1997. Causes of natal dispersal and emigration and their effects on harem formation in Misaki feral horses. Equine Veterinary Journal 29:262–266.

Keiper, R.R. 1976. Social organization of feral ponies. Proceedings of the Pennsylvania Academy of Science 50:69–70.

Keiper, R.R. 1986. Social structure. Pages 465–484 *in* K.A. Houpt and S. Crowell-Davis (Eds.), Veterinary clinics of North America. Equine Practice 2. W.B. Saunders, Philadelphia, USA.

Keiper, R., and K. Houpt. 1984. Reproduction in feral horses: An eight-year study. American Journal of Veterinary Research 45:991–995.

Khalil, A.M., and N. Murakami. 1999*a*. Factors affecting the harem formation process by young Misaki feral stallions. Journal of Veterinary Medical Science 61:667–671.

Khalil, A.M., and N. Murakami. 1999*b*. Effect of natal dispersal on the reproductive strategies of the young Misaki feral stallions. Applied Animal Behaviour Science 62:281–291.

Kimura, R. 2001. Volatile substances in feces, urine and urine-marked feces of feral horses. Canadian Journal of Animal Science 81:411–420.

King, S.R.B., and J. Gurnell. 2005. Habitat use and spatial dynamics of takhi introduced to Hustai National Park, Mongolia. Biological Conservation 124:277–290.

King, S.R.B., and J. Gurnell. 2007. Scent-marking behavior by stallions: An assessment of function in a reintroduced population of Przewalski horses (*Equus ferus przewalskii*). Journal of Zoology 272:30–36.

Klingel, H. 1967. Soziale Organisation und Verhalten freilebender Steppenzebras. Zeitschrift für Tierpsychologie 24:580–624.

Klingel, H. 1968. Soziale Organisation und Verhaltenweisen von Hartmann-und Bergzebras (*Equus zebra hartmannae und E. z. zebra*). Zeitschrift für Tierpsychologie 25:76–88.

Klingel, H. 1969*a*. The social organization and population ecology of the plains zebra (*Equus quagga*). Zoologica Africana 4:249–263.

Klingel, H. 1969*b*. Dauerhafte Sozialverbände beim Bergzebra. Zeitschrift für Tierpsychologie 26:965–966.

Klingel, H. 1969*c*. Reproduction in the plains zebra, *Equus burchelli boehmi*: Behaviour and ecological factors. Journal of Reproductive Fertility Supplement 6:339–345.

Klingel, H. 1974*a*. A comparison of the social behavior of the Equidae. Pages 124–132 *in* The behavior of ungulates and its relation to management: Papers of the IUCN Symposium. University of Calgary, Calgary, AB, Canada.

Klingel, H. 1974*b*. Soziale Organisation und Verhalten des Grevy-Zebras (*Equus grevyi*). Zeitschrift für Tierpsychologie 36:37–70.

Klingel, H. 1975. Social organization and reproduction in equids. Journal of Reproduction and Fertility Supplement 23:7–11.

Klingel, H. 1977. Observations on social organization and behaviour of African and Asiatic wild asses (*Equus africanus* and *Equus hemionus*). Applied Animal Behaviour Science 60:103–113.

Klingel, H. 1982. Social organization of feral horses. Journal of Reproduction and Fertility Supplement 32:89–95.

Klingel, H. 1998. Observations on social organization and behaviour of African and Asiatic wild asses (*Equus africanus* and *Equus hemionus*). Applied Animal Behaviour Science 60:103–113.

Linklater, W.L. 2000. Adaptive explanation in socio-ecology: Lessons from the Equidae. Biological Reviews 75:1–20.

Linklater, W.L., and E.Z. Cameron. 2000. Tests for cooperative behaviour between stallions. Animal Behaviour 60:731–743.

Linklater, W.L., E.Z. Cameron, E.O. Minot, and K.J. Stafford. 1999. Stallion harassment and the mating system of horses. Animal Behaviour 58:295–306.

Linklater, W.L., E.Z. Cameron, K.J. Stafford, and C.J. Veltman. 2000. Social and spatial structure and range use by Kaimanawa wild horse (*Equus caballus*: Equidae). New Zealand Journal of Ecology 24:139–152.

Linklater, W.L., E.Z. Cameron, K.J. Stafford, and E.O. Minot. 2013. Removal experiments indicate that subordinate stallions are not helpers. Behavioural Processes 94:1–4.

McCort, W.D. 1979. The feral asses (*Equus asinus*) of Ossabaw Island, Georgia: Mating system and the effects of vasectomies as a population control procedure. Pages 71–83 *in* R.H. Denniston (Ed.), Symposium on the ecology and behaviour of wild and feral equids. University of Wyoming, Laramie, USA.

McCort, W.D. 1980. The behavior and social organization of feral asses (*Equus asinus*) on Ossabaw Island, Georgia. Thesis, Pennsylvania State University, State College, USA.

McCort, W.D. 1984. Behavior of feral horses and ponies. Journal of Animal Science 58:493–499.

Miller, R. 1979. Band organization and stability in Red Desert feral horses. Pages 113–128 *in* R.H. Denniston (Ed.), Symposium on the ecology and behaviour of wild and feral equids. University of Wyoming, Laramie, USA.

Miller, R. 1981. Male aggression, dominance, and breeding behavior in Red Desert feral horses. Zeitschrift für Tierpsychologie 57:340–351.

Miller, R. 1983. Seasonal movement patterns and home ranges of feral horses in Wyoming's Red Desert. Journal of Range Management 36:199–201.

Miller, R., and R. H. Denniston. 1979. Interband dominance in feral horses. Zeitschrift für Tierpsychologie 51:41–47.

Moehlman, P.D. 1998*a*. Behavioral patterns and communication in feral asses (*Equus africanus*). Applied Animal Behaviour Science 60:125–169.

Moehlman, P.D. 1998*b*. Feral asses (*Equus africanus*): Intraspecific variation in social organization in arid and mesic habitats. Applied Animal Behaviour Science 60:171–195.

Moehlman, P.D., L.E. Fowler, and J.H. Roe. 1998. Feral asses (*Equus africanus*) of Volcano Alcédo, Galapagos: Behavioral ecology, spatial distribution, and social organization. Applied Animal Behaviour Science 60:197–210.

Mohr, E. 1971. The Asiatic wild horse. J.A. Allen, London, UK.

Monard, A.-M., and P. Duncan. 1996. Consequences of natal dispersal in female horses. Animal Behaviour 52:565–579.

Monard, A.-M., P. Duncan, and V. Boy. 1996. The proximate mechanisms of natal dispersal in female horses. Behaviour 133:1095–1124.

Nelson, K.J. 1979. On the question of male-limited population growth in feral horses (*Equus caballus*). Thesis, New Mexico State University, Las Cruces, USA.

Nelson, K.J. 1980. Sterilization of dominant males will not limit feral horse populations. Forest Service Research Paper RM-226. US Department of Agriculture, Washington, DC, USA.

Neumann-Denzau, G., and H. Denzau. 2007. Remarks on the social system of the Mongolian wild ass (*Equus hemionus hemionus*). Exploration into the Biological Resources of Mongolia (Halle/Saale) 10:177–187.

Nowzari, H. 2011. Population size, population structure and habitat use of the Persian onager (*Equus hemionus onager*) in Qatrouyeh National Park, Fars, Iran. Thesis, Science and Research Branch, Islamic Azad University, Tehran, Iran.

Nowzari, H., M.R. Hemami, M. Karami, M.M. Kheirkhah Zarkesh, B. Riazi, and D.I. Rubenstein. 2012. Population parameters of Persian wild ass (*Equus hemionus onager*) in Qatrouyeh National Park, Iran. Presented at the International Wild Equid Conference, 18–22 September 2012, Vienna, Austria.

Nowzari, H., M.R. Hemami, M. Karami, M.M. Kheirkhah Zarkesh, B. Riazi, and D.I. Rubenstein. 2013. Habitat use by the Persian onager, *Equus hemionus onager* (Perissodactyla, Equidae), in Qatrouyeh National Park, Fars, Iran. Journal of Natural History 47:2795–2814, doi:10.1080/00222933.2013.802040.

Pacheco, M.A., and E.A. Herrera. 1997. Social structure of feral horses in the llanos of Venezuela. Journal of Mammalogy 78:15–22.

Penzhorn, B.L. 1984. A long-term study of social organization and behavior of Cape Mountain zebras *Equus zebra zebra*. Zeitschrift für Tierpsychologie 64:97–146.

Perkins, A., E. Gevers, J.W. Turner Jr., and J.F. Kirkpatrick. 1979. Age characteristics of feral horses in Montana. Pages 51–55 *in* R. H. Denniston (Ed.), Symposium on the ecology and behavior of wild and feral equids. University of Wyoming, Laramie, USA.

Ransom, J.I., P. Kaczensky, B.C. Lubow, O. Ganbaatar, and N. Altansukh. 2012. A collaborative approach for estimating terrestrial wildlife abundance. Biological Conservation 153:219–226.

Reading, R.P., H.M. Mix, B. Lhagvasuren, C. Feh, D.P. Kane, S. Dulamtseren, and S. Enkhbold. 2001. Status and distribution of khulan (*Equus hemionus*) in Mongolia. Journal of Zoology London 254:381–389.

Roelle, J.E., F.J. Singer, L.C. Zeigenfuss, J.I. Ransom, L. Coates-Markle, and K.A. Schoenecker. 2010. Demography of the Pryor Mountain wild horses 1993–2007. Scientific Investigations Report 2010-5125. US Geological Survey, Reston, VA, USA.

Rubenstein, D.I. 1981. Behavioural ecology of island feral horses. Equine Veterinary Journal 13:27–34.

Rubenstein, D.I. 1982. Reproductive value and behavioral strategies: Coming of age in monkeys and horses. Pages 469–487 *in* P.P.G. Bateson and P.H. Klopfer (Eds.), Perspectives in ethology, vol. 5. Plenum Press, New York, USA.

Rubenstein, D.I. 1986. Ecology and sociality in horses and zebras. Pages 282–302 *in* D. I. Rubenstein and R.W. Wrangham (Eds.), Ecological aspects of social evolution. Princeton University Press, Princeton, NJ, USA.

Rubenstein, D.I. 1994. The ecology of female social behaviour in horses, zebras and asses. Pages 13–28 *in* P. Jarman and A. Rossiter (Eds.), Animal societies: Individuals, interactions and organizations. Kyoto University Press, Kyoto, Japan.

Rubenstein, D.I. 2010. Ecology, social behaviour, and conservation in zebras. Pages 231–258 *in* R. Maedo (Ed.), Advances in the study of behavior, vol. 42: Behavioral ecology of tropical animals. Elsevier, Oxford, UK.

Rubenstein, D.I., and M.A. Hack. 1992. Horse signals: The sounds and scents of fury. Evolutionary Ecology 6:254–260.

Rubenstein, D.I., and M. Hack. 2004. Natural and sexual selection and the evolution of multi-level societies: Insights from zebras with comparison to primates. Pages 266–279 *in* P.M. Kappeler and C.P. van Schaik (Eds.), Sexual selection in primates: New and comparative perspectives. Cambridge University Press, Cambridge, UK.

Rubenstein, D.I., and M.E. Hohmann. 1989. Parasites and social behavior of island feral horses. Oikos 55:312–320.

Rubenstein, D.I., S. Sundaresan, I. Fischhoff, and D. Saltz. 2007. Social networks in wild asses: Comparing patterns and processes among populations. Exploration into the Biological Resources of Mongolia (Halle/Saale) 10: 159–176.

Rutberg, A.T. 1990. Inter-group transfer in Assateague pony mares. Animal Behaviour 40:945–952.

Rutberg, A.T., and R.R. Keiper. 1993. Proximate causes of natal dispersal in feral ponies: Some sex differences. Animal Behaviour 46:969–975.

Salter, R.E. 1978. Ecology of feral horses in western Alberta. Thesis, University of Alberta, Edmonton, AB, Canada.

Salter, R.E., and R.J. Hudson. 1982. Social organization of feral horses in western Canada. Applied Animal Ethology 8:207–223.

Schaller, G.B. 1998. Wildlife of the Tibetan Steppe. University of Chicago Press, Chicago, USA.

Scorolli, A.L. 2007. Dinámica poblacional y organización social de caballos cimarrones en el Parque Provincial Ernesto Tornquist. Thesis, Universidad Nacional del Sur, Bahía Blanca, Argentina.

Scorolli, A.L. 2013. Organización social de caballos cimarrones, un estudio a largo plazo en el Parque Tornquist. Presented at the XXVI Meeting of the Argentine Society of Study of Mammals, November 2013, Mar del Plata, Argentina.

Scorolli, A.L., and A.C. Lopez Cazorla. 2010. Feral horse social stability in Tornquist Park, Argentina. Mastozoología Neotropical 17:391–396.

Shah, N. 2002. Status and action plan for the kiang (*Equus kiang*). Pages 72–81 *in* P.D. Moehlman (Ed.), Equids: Zebras, asses and horses. Status survey and conservation action plan. International Union for Conservation of Nature, Gland, Switzerland.

Shah, N., and Q. Qureshi. 2007. Social organization and determinants of spatial distribution of Khur (*Equus hemionus khur*). Exploration into the Biological Resources of Mongolia (Haale/Saale) 10:189–200.

Stanley, C.R., and S. Shultz. 2012. Mummy's boys: Sex differential maternal-offspring bonds in semi-feral horses. Behaviour 149:251–274.

St-Louis, A., and S. Côté. 2009. *Equus kiang* (Perissodactyla: Equidae). Mammalian Species 835:1–11.

Sundaresan, S.R., I.R. Fischhoff, J. Dushoff, and D.I. Rubenstein. 2007*a*. Network metrics reveal differences in social organization between two fission-fusion species, Grevy's zebra and onager. Oecologia 151:140–149.

Sundaresan, S.R., I.R. Fischhoff, and D.I. Rubenstein. 2007*b*. Male harassment influences female movements and associations in Grevy's zebra (*Equus grevyi*). Behavioural Ecology 18:860–865.

Tantipathananandh, C., T. Berger-Wolf, and D. Kempe. 2007. A framework for community identification in dynamic social networks. Pages 717–726 *in* Proceedings of the 13th ACM SIGKDD international conference on knowledge discovery and data mining. San Jose, CA, USA.

Turner, J.W., Jr., and J.F. Kirkpatrick. 1986. Hormones and reproduction in feral horses. Journal of Equine Veterinary Science 6:250–258.

Turner, J.W., Jr., A. Perkins, and J.F. Kirkpatrick. 1981. Elimination marking behavior in feral horses. Canadian Journal of Zoology 59:1561–1566.

Tyler, S.J. 1972. The behaviour and social organization of New Forest ponies. Animal Behaviour Monographs 5:85–196.

von Goldschmidt-Rothschild, B., and B. Tschanz. 1978. Soziale Organisation und Verhalten einer Jungtierherde beim Camargue-Pferd. Zeitschrift für Tierpsychologie 46:372–400.

Welsh, D.A. 1973. The life of Sable Island's wild horses. Nature Canada (Ottawa) 2:7–14.

Welsh, D.A. 1975. Population, behavioural and grazing ecology of the horses of Sable Island, Nova Scotia. Thesis, Dalhousie University, Halifax, NS, Canada.

Wernicke, R., and M.C. van Dierendonck. 2003. Soziale Organisation und Ernährungszustand der Konik-Pferdeherde des Naturreservates Oostvaardersplassen (NL) im Winter: Eine Lehrstunde durch wild lebende Pferde. KTBL-Schrift 418:78–85.

Wey, T., D.T. Blumstein, W. Shen, and F. Jordán. 2008. Social network analysis of animal behaviour: A promising tool for the study of sociality. Animal Behaviour 75:333–344.

Woodward, S.L. 1979. The social system of feral asses (*Equus asinus*). Zeitschrift für Tierpsychologie 49:304–316.

Zimmermann, W., K. Brabender, and L. Kolter. 2009. A Przewalski's horse population in a unique European steppe reserve—The Hortobágy National Park in Hungary. Equus 2009 (Praha Zoo): 257–288.

Behavior of Horses, Zebras, and Asses

SARAH R.B. KING, CHERYL ASA,
JAN PLUHÁČEK,
KATHERINE HOUPT, AND
JASON I. RANSOM

Equid behavior is fascinating to observe because of its often subtle nature and complex relationship to fitness and ecology. Living in groups has advantages in terms of protection from predation, but it also carries costs, such as competition for forage (Silk 2007). Equids have evolved to balance these costs and benefits. Species living in more mesic environments, where there is less competition for food, tend to live in relatively stable family groups, whereas those living in more arid environments are found in small, unstable groups (Rubenstein 2010). What makes equids different from many other gregarious mammals is that, regardless of social system (territorial or harem groups), both sexes disperse at maturity: females either join a breeding group or move to find the best forage, and males join bachelor groups until they acquire females. These movements demand that individuals navigate a complex social landscape among both familiar and unknown individuals and groups. To maximize fitness while maintaining this flexibility of social systems, equids have developed an array of multifaceted social behaviors that have both similarities and differences across species.

Group living requires careful communication between members, and it has been shown that equids can recognize familiar individuals by sight, sound, and smell (Proops *et al.* 2009*b*; Krueger and Flauger 2011; Murray *et al.* 2013). Recognition of individuals enables expression of social behavior to conspecifics, which encourages stability and cohesion within groups while also mediating competition and harassment of reproductive females by males (Linklater 2000). Affiliative behaviors strengthen bonds between individuals, both kin and nonkin, and thus encourage them to move together. While agonistic behaviors act to maintain distance between individuals, they also encourage group stability, as individuals joining a new group receive more aggression (Rutberg and Greenberg 1990; Linklater 2000). Stable relationships between mares and stallions can reduce agonistic behavior and increase reproductive success (Linklater *et al.* 1999; Cameron *et al.* 2003; Sundaresan *et al.* 2007). Examining the behavioral ecology of equids can aid their conservation through our understanding of how groups function and individuals disperse (Rubenstein 2010), and also informs management and welfare of captive and domestic animals. It is therefore critical that behavioral research of equids continues to deepen our knowledge of the intricate social networks and behaviors of these species.

Perception

In order to properly investigate the communication methods and cognitive competence of equids, we should understand their perceptual abilities. Perception of the world shapes behavior, as it affects how individuals respond to each other and how they respond to external threats and opportunities. Understanding how an equid perceives the world therefore sheds light on how they interact with it. Most of the research on equid perception has been conducted on domestic horses (*Equus caballus*), but because of their relatively close phylogeny, it is likely that many attributes are similar across equid species.

Researchers have measured visual acuity, stereoscopic, and color vision in domestic horses. Their visual acuity is 20–23 cycles/degree, similar to cattle (Timney and Macuda 2001). The horse has a 60°–70° anterior binocular field of view, with up to a 215° field of view out of each eye (Waring 2003), using stereoscopsis to determine depth (Timney and Macuda 2001). Equids are dichromates, like most nonprimate mammals (Timney and Macuda 2001). They can see color, especially blue and red, but some horses cannot distinguish green and yellow from gray. Equids can perform cross-modal identification: asses can identify one another visually (Murray *et al.* 2013), and horses can tell which neigh is associated with which individual (Proops *et al.* 2009*b*).

There is some evidence that horses process sounds from familiar individuals in one hemisphere of the brain, but they do so much differently for unfamiliar sounds (Basile *et al.* 2009). Horses can hear sounds ranging from 55 Hz to 34 KHz, a higher frequency than humans can hear. Horses are poor at sound localization; they can detect differences no smaller than 20°, in comparison to the cat, which can detect 5° of separation, or the human, who can perceive fractions of a degree difference in location (Timney and Macuda 2001).

As in most other mammals, smell plays a crucial part in how equids gain information about the physical world and the state of conspecifics. Owing to their physiology, equids can move large volumes of air through their nose in one breath, trapping large numbers of molecules containing scent information (Saslow 2002). Equid stallions respond to female urine by sniffing, which typically includes contact of the nares with the fluid. In addition to volatile odors, contact facilitates drawing nonvolatile compounds (pheromones) into the vomeronasal organ, which communicates with the accessory olfactory bulb and subsequently to brain centers that coordinate sexual response (Asa 1986; Houpt and Boyd 1994). Stallions urine-mark and fecal-mark on feces, both on dung piles of other males year-round and on mare excretions, particularly during the breeding season (King and Gurnell 2006). This behavior could be either to notify other stallions of their consortship (King and Gurnell 2006) or possibly to mask their scent (Kimura 2001). Volatile substances in feces and urine differed in horses dependent on age, sex, and stage in the reproductive process (Kimura 2001). It has been shown that horses can differentiate between conspecifics based on urine odor (Hothersall *et al.* 2010), and they can differentiate their own feces from that of others and may be able to recognize the relative dominance of the source of the feces (Krueger and Flauger 2011).

Cognition

There are few comparative studies of equid cognition. Horses appear better able to remember which of a pair of symbols will result in a food reward than asses or zebras (Giebel 1958). In that study, the horse was able to learn all 20 pairs, the ass learned 13, and the zebra learned 10. Horses can remember which symbols were rewarded for at least one year, doing better than elephants. Mules are superior to horses and asses in visual discrimination: mules learned 16 pairs of patterns, the ponies 11, and the asses 6 (Proops *et al.* 2009*a*). Horses can learn to avoid noxious stimuli and to perform a response to gain a positive reinforcement such as food, freedom from a stall, and access to another horse or to a cribbing bar (Elia *et al.* 2010; Lee *et al.* 2011; Houpt 2012). They do not respond as well as dogs to human gestures and gaze, although they can respond to simple pointing (Maros *et al.* 2008). Horses, at least Arabian horses, can learn a concept such as triangularity (Sappington and Goldman 1994), but they are not good at solving barrier problems, tending to attempt to go through barriers rather than around them. More recently, it has been found that horses are able to learn a variety of conceptual tasks such as discriminating objects by relative size or filled versus open symbols; two horses were able to remember the tasks six to ten years later (Hanggi and Ingersoll 2009). Horses can select the greater of two quantities (two from one or three from two), but fail when the numbers are larger (four vs. six) (Uller and Lewis 2009).

Communication

Equids communicate using their ears, tails, facial expressions, general postures, and voices. Only the auditory signals differ markedly among species. Horses neigh, asses and Grevy's zebra (*E. grevyi*) bray, and plains zebra (*E. quagga*) bark. Equids that live in social groups have shorter calls than the nonsocial species, but in all species these vocalizations are a contact call made when the animal hears another equid (in the solitary species) or when it is separated from its group (in the social species) (Policht *et al.* 2011). Horses nicker: a low-decibel, low-frequency call to encourage close contact. Nickering is usually used by dams to attract the foal, but also by domestic horses to attract a caretaker. In addition to the bray, asses can produce a grunt, growl, snort, and whuffle vocalization given in the same context as a nicker by a horse (Moehlman 1998*a*). Equids also snort or blow out of frustration, fear, or to clear their nostrils, and may squeal during agonistic encounters (Houpt and Boyd 1994).

Ears forward indicates interest; flattened ears indicate aggression. The facial expressions vary from the most obvious, flehmen, to the subtle tightening of the facial muscles of the anxious horse or of the jaw muscles of the horse in pain. Fear is often reflected in the eye when the sclera becomes visible (showing the whites of the eyes), when sympathetic stimulation widens the palpebral fissure. Immature horses and mares in estrus exhibit snapping behavior (also called chewing, clapping, yawing, or champing), in which the mouth opens and closes when they approach or are approached by a more dominant individual. Fearful equids clamp the tail close to the rump, lash it when exhibiting aggression, and hold it straight up when aroused. General body posture changes from the relaxed posture, with head lowered and hind limb supported by the stay apparatus while the other limb is flexed, to the tense, head-up and ears-erect posture of the alert horse.

Time Budgets

Understanding how equids pass their time opens a window into many facets of their behavior and ecology, and aids in the management of domestic horses. Time budgets have therefore been well researched, although most attention has focused on domestic, feral, or Przewalski's horses (*E. ferus przewalskii*), with only a few studies on zebras and one recent study on Asiatic wild asses (*E. hemionus*) (Xia *et al.* 2013). A time budget consists of a day in the life of an animal broken down into different behaviors based on an ethogram (i.e., a list of behaviors) (e.g., McDonnell 2003; Ransom and Cade 2009).

The daily rhythm of horse behaviors is linked to sunrise and sunset (Berger *et al.* 1999), affected by ambient temperature (Souris *et al.* 2007), changes over the seasons (Duncan 1985; Mayes and Duncan 1986) arising from differences in availability and quality of forage (Arnold 1984; Xia *et al.* 2013), and presence of insect pests (King and Gurnell 2010). Although few data are available on how predators affect the time budgets of equids, they are likely to affect the time of day at which equids conduct their activities (Blom 2009) (e.g., plains zebra are more vigilant at dusk; Simpson *et al.* 2012). Horses tend to be most active during the crepuscular period, which is when most grazing and social behavior occurs (King 2002). During the night, horses continue to be active, with one study finding no difference compared to their behavior during day (Berman 1991), and others reporting more resting behavior and less feeding (e.g., Mayes and Duncan 1986).

As large nonruminant herbivores, equids evolved as grazers of relatively low-quality forage (see chap. 4); thus they spend a large part of their life feeding, often for at least half the time they were observed (table 3.1). There is a remarkably similar pattern of grazing behavior across species. Grazing typically takes place in two pulses: in the early morning and late afternoon, and also during the night. Generally, when equids are not feeding they are resting, which is often the second most frequent activity after grazing, tending to occur during the middle of the day. Although recumbent rest has been observed at any time of day (King 2002), horses often have a sleeping bout between 2:00 a.m. and dawn (Keiper and Keenan 1980; Boyd *et al.* 1998).

Like most mammals, equids have a seasonal change in their metabolism, with heart rate in Przewalski's horses being slower during the cold of winter than during the breeding season and summer (Arnold *et al.* 2006). This metabolic change corresponds to intake of dry matter (Kuntz *et al.* 2006) and is reflected in a change in the time budget of equids over the seasons. Equids tend to spend more time feeding during the day in winter than in summer, owing to the paucity or low quality of the forage in that season (Kaseda 1983; Berger *et al.* 1999; Xia *et al.* 2013) or increased metabolic rate for heat generation (McBride *et al.* 1985). It is also possible that they are able to spend less time feeding during the summer because of the high quality of the forage during that season (Berman 1991; King 2002). Conversely, equids spend more of the day resting in summer than in winter (Keiper *et al.* 1980), normally in exposed or dusty locations where they can escape from flies (Hughes *et al.* 1981; Rutberg 1987; Powell *et al.* 2006; King and Gurnell 2010).

Movement makes up a relatively small amount of the time budget of wild equids. In many populations,

Table 3.1. **Percentage of time spent in various behavioral activities by wild equids**

Species	Grazing	Stand Resting	Standing	Recumbent	Moving	Location	Source
E. f. przewalskii	52.3 ± 3.01	27.1 ± 5.37	6.1 ± 1.12	4.2 ± 1.17	15.0 ± 2.18	Germany, Mongolia, United Kingdom	Boyd 1988*b*; King 1996; Berger *et al.* 1999; Boyd and Bandi 2002; King 2002
E. caballus	62.9 ± 4.79	19.6 ± 2.76	16.4 ± 1.77	5.0 ± 1.95	9.5 ± 2.45	Australia, France, Japan, United States	Duncan 1979; Keiper and Keenan 1980; Kaseda 1983; Berman 1991; Ransom *et al.* 2010; Madosky 2011
E. quagga	54.3 ± 6.86		26.6 ± 3.78	6.3 ± 2.04	12.7 ± 1.97	Namibia	Neuhaus and Ruckstuhl 2002
E. hemionus	53.2	32.4			<9.0	China	Xia *et al.* 2013

Note: Values shown as mean ± standard error. Means across studies are weighted by sample size but include fine-scale differences in season, location, and time of year. See text for descriptions.

the amount of time spent moving is related to the distance between prime feeding areas and water sources. The need for water is greatest during dry conditions (Scheibe *et al.* 1998) and varies among individuals, with lactating females needing to drink most frequently (Fischhoff *et al.* 2007). Most populations are reported to drink at least once per day (Pellegrini 1971; Feist and McCullough 1976; Penzhorn and Novellie 1991; Scheibe *et al.* 1998); however, feral horses in Australia and zebras in Africa may drink as little as once every 48 hours owing to the distance between good forage and water sources (Berman 1991; Brooks 2005; Hampson *et al.* 2010). In more mesic conditions, or where water sources are more abundant, equids spend less time moving. When there is snow or temporary water pools formed from rain, equids do not need established water sources (Feist and McCullough 1976; Salter 1979). Horses also travel between feeding and resting spots, whether it is between salt marsh and dry ground to sleep lying down on barrier islands (Zervanos and Keiper 1979; Keiper *et al.* 1980), or to places where they can stand and rest in refuge from insects (Rutberg 1987; Powell *et al.* 2006; King and Gurnell 2010). In places where resources change with the seasons, equids may cover vast distances (see chap. 13).

The time budget of equids changes as they become adults. During their first month, foals stay close to their mother because they are entirely dependent on her for nutrition (Tyler 1972; Boy and Duncan 1979; Waring 1983). By the time they are 2 months old, foals are beginning to forage and interact more with other herd members; at 5 months old the time budget of foals is more similar to that of adult horses (Boyd 1988*a*). Foals continue to spend more time in recumbent rest than adults (Boy and Duncan 1979; Penzhorn 1984; King 2002), with juvenile time budgets differing from adults also in the amount of time spent lying down (Duncan 1979; Boyd 1988*a*).

Both Przewalski's horse stallions (Boyd 1988*b*; King 2002) and feral horse stallions (Ransom and Cade 2009; Ransom *et al.* 2014*a*) spend more time moving and less time feeding than the mares in their harem. In Camargue horses (Duncan 1992) and plains zebra (Neuhaus and Ruckstuhl 2002), however, there was little difference in activity times between the sexes. There are conflicting reports on whether fertility control affects time budget (Ransom *et al.* 2010; Madosky 2011), but differences are most likely reflections of impacts on body condition: mares in poor condition spent more time grazing than mares in good condition (Ransom *et al.* 2010). Lactation status also affects the time budget of mares, as they tend to spend more time feeding than nonlactating females (Duncan 1985; Boyd 1988*b*; Neuhaus and Ruckstuhl 2002; Madosky 2011).

While time budgets are similar across equid species, they are also similar among horses within the same group (Feist and McCullough 1976; Linklater *et al.* 1999), with synchrony among the sexes being related to lack of sexual dimorphism in equids (Neuhaus and Ruckstuhl 2002). Studies on Przewalski's horses that examined time budgets of members in a group found that they are synchronous over half of the time (van Dierendonck *et al.* 1996, 50–89%; King 2002, 75%; Souris *et al.* 2007, 87–91%), similar to that observed in plains zebra (73–78% synchronous depending on age and sex; Neuhaus and Ruckstuhl 2002). Reintroduced Przewalski's horses had a different time budget before and after release, spending more time moving and

Fig. 3.1 Przewalski's horses mutual grooming (also known as allogrooming), Hustai National Park, Mongolia. Photo by Sarah King

grazing when wild than when in acclimatization enclosures (Boyd 1998; Boyd and Bandi 2002). The greater amount of time spent moving is likely a result of exploration, or movement between food patches (Boyd and Bandi 2002). Decreasing time spent resting over the years following reintroduction could be a result of the animals recovering from the initial stress of release (King 2002). Studies of synchronization of behavior are important for an introduced species, as it is likely to be closely related to group cohesiveness, thus reducing the chance of individuals dispersing from the reintroduction site. Remaining in groups also reduces predation risk (Scheel 1993; Valeix *et al.* 2009).

Affiliative Behavior

Affiliative behaviors are nonagonistic interactions between two individuals that appear to promote stable relationships in group-living species. Observations of overt social interactions among wild equids are relatively uncommon, with some populations showing more agonistic than affiliative interactions (Lloyd and Rasa 1989; King 2002; Asa *et al.* 2011), and others exhibiting more affiliative interactions (Keiper and Receveur 1992; Linklater *et al.* 1999; Sundaresan *et al.* 2007; Ransom *et al.* 2010). Rates of affiliative behaviors are normally low, although there is some variation across studies: free-living domestic horse, 0.004–0.24 acts/hour (Sigurjónsdóttir *et al.* 2003; Heitor *et al.* 2006*a*); Przewalski's horse, 0.04–0.22 acts/hour (Mooring *et al.* 2000; King 2002); Grevy's zebra, 0.47 acts/hour (Mooring *et al.* 2000); feral ass (*E. asinus*), 0.3 acts/hour (Moehlman 1998*b*). Most affiliative behavior is expressed as mutual grooming (also referred to as allogrooming; fig. 3.1), but it can also be expressed by dyads standing or grazing close to each other (fig. 3.2), or even putting their heads on another's body while resting (King 2002; Ransom and Cade 2009).

Allogrooming has been well studied in a variety of taxa from carnivores to primates and is generally considered to promote bonding or reconciliation between individuals (e.g., Newton-Fisher and Lee 2011). In equids, this behavior consists of two individuals standing in reverse parallel position and using their teeth to scratch or pluck at the other's coat, normally around the withers (Feh and de Mazières 1993). Unlike in primates, mutual grooming is rarely not reciprocated on initiation (Houpt and Boyd 1994; Moehlman 1998*b*; King 2002), and it is only observed between individuals that are familiar with each other, normally from the same social group (Wells and von Goldschmidt-Rothschild 1979; Sigurjónsdóttir *et al.* 2003; Fischhoff *et al.* 2007). So far, no one has identified the signal that conveys an intention to mutually groom, but presumably there is one.

Mutual grooming is important for coat care in equids, as it acts to remove loose hair and parasites from places that an individual cannot groom itself except by rolling. The behavior appears to be stimulus driven (Mooring *et al.* 2000) but may also arise from a

Fig. 3.2 Close relationships can develop between individuals, whether kin or not, allowing physical proximity without aggression. Photo by Jan Pluháček

social need (van Dierendonck and Spruijt 2012). Some studies have shown most mutual grooming in the summer (Tyler 1972; Crowell-Davis *et al.* 1986; Kimura 1998), while another observed it more in the spring and autumn, when the animals were molting (King 2002). Mutual grooming tends to occur during periods of the day when the animals are more active (Kimura 1998; King 2002), while stand-resting close together occurs during the middle of the day in summer because of the effect of flies (King and Gurnell 2010).

In equids, several studies have shown the social function of mutual grooming through maintenance and enhancement of bonds between members of a group. Stronger bonds between individuals aids social cohesion of a group, with both stable mare–stallion and mare–mare relationships helping enhance reproductive success (Linklater *et al.* 1999; Cameron *et al.* 2009). Stable family groups can also be more effective at protecting foals from predators (Simpson *et al.* 2012). Socially cohesive groups move together, and although movement initiation was not related to affiliative relationships within a group of horses (Krueger *et al.* 2014), the combination of social relationships and reproductive state in plains zebra shaped movements of their groups (Fischhoff *et al.* 2007). Allogrooming has been reported to appease individuals, either to ameliorate aggression (Feist and McCullough 1976; Feh and de Mazières 1993), to reduce social tension by reducing the heart rate (Andersen 1992; Feh and de Mazières 1993), or to reduce weaning conflict (Penzhorn 1984; Keiper 1988). Captive conditions affect the amount of affiliative behavior expressed in domestic horses (Christensen *et al.* 2002*a,b*), Przewalski's horses (Hogan *et al.* 1988; Keiper 1988), and plains zebra (Andersen 1992), with the greater amount of affiliative interactions seen in more confined spaces being likely as a result of a need to relieve stress between individuals.

Within most equid groups, individuals have preferred partners for affiliative interactions, and these are also the animals that tend to remain closest during other activities (Clutton-Brock *et al.* 1976; Carson and Wood-Gush 1983; Feh 1999). The presence of a stallion affects the occurrence of mutual grooming, changing the behavior of mares from primarily grooming with their offspring (Feist and McCullough 1976; Wells and von Goldschmidt-Rothschild 1979; King 2002) to grooming with contemporary mares (Tyler 1972; Clutton-Brock *et al.* 1976), and at a greater rate (Sigurjónsdóttir *et al.* 2003). Mutual grooming among foals or juvenile siblings is most common, although it is seen between most dyads in a social group (Wells and von Goldschmidt-Rothschild 1979; Crowell-Davis *et al.* 1986; Boyd 1988*a*; Keiper and Receveur 1992). In other mammals, most affiliative interactions occur between related individuals, but besides mothers and their young offspring or juvenile siblings, this pattern does not seem as defined in equids. More interactions between adult kin were observed in some groups—Przewalski's horses (Keiper 1988); domestic horses (van Dierendonck *et al.* 2004)—but not in others—Przewalski's horses (King 2002); domestic horses (Clutton-Brock *et al.* 1976 and van Dierendonck *et al.* 2004). Generally, animals of the same rank tend to associate more (Ellard and Crowell-Davis 1989; Kimura 1998; Sigurjónsdóttir *et al.* 2003).

Most observers report that individuals in a group tend to have preferred associates, quantified by the proportion of time that they are "nearest neighbors" (Clutton-Brock *et al.* 1976; Kimura 1998). These associates are normally the same individuals with which mutual grooming occurs (Sigurjónsdóttir *et al.* 2003; van Dierendonck *et al.* 2004), although not always (Kimura 1998). A foal has its mother as nearest neighbor until about 2 years old (Crowell-Davis *et al.* 1986; Boyd 1988*a*; van Dierendonck *et al.* 2004), but as with mutual grooming, preferred associates outside a foal–dam relationship tend to be those that are similar in age and rank (van Dierendonck *et al.* 1995; Kimura 1998; Sigurjónsdóttir *et al.* 2003), although similarly also have an unclear relationship with kinship—most often with related individuals (van Dierendonck *et al.* 2004 and Heitor *et al.* 2006*a*) or no clear relationship (King 2002). Linkages between individuals create social networks, which animals are motivated to defend (van Dierendonck *et al.* 2009). These social networks can inform patterns in the wider social structure of different species: equids living in more variable environments tend to have larger, more cohesive social groups (Sundaresan *et al.* 2006, Rubenstein *et al.* 2007).

Play

Play has been observed in most populations that include young equids and has been well described in ethograms (McDonnell and Poulin 2002; Ransom and Cade 2009). Play consists of many of the behaviors shown by adult horses (such as running and fighting behavior) while also including exploratory behavior in the form of object play (Crowell-Davis *et al.* 1987; McDonnell and Haviland 1995; Simpson *et al.* 2012). It has been observed among most age and sex classes (Wells and von Goldschmidt-Rothschild 1979; Hogan *et al.* 1988; Moehlman 1998*a*; Sigurjónsdóttir *et al.* 2003). Play is quite rare among adult mares (Sigurjónsdóttir *et al.* 2003) but is common among bachelor stallions (Bourjade *et al.* 2009*c*), and while harem stallions play at a lower frequency than bachelors (Boyd *et al.* 1988), they have been observed to play with their own foals and to leave their group to go and play with bachelors (Berger 1986; Kolter and Zimmermann 1988; McDonnell and Poulin 2002). Play serves to maintain social relationships (Feh 1988); horses tend to have preferred partners for playing, which are often the same as those with which they mutual groom (Sigurjónsdóttir *et al.* 2003; van Dierendonck *et al.* 2009).

Self-play in Przewalski's horses began on the second day of life, with mutual play observed in the third week (Boyd 1988*a*). The highest frequency of play behavior was when foals were 3–10 months old, with foals beginning to play with adults at 8–12 months (Tyler 1972; Carson and Wood-Gush 1983; Crowell-Davis *et al.* 1987; Boyd 1988*a*). In feral asses, self-play was observed up to 6 months of age with mutual play only seen after 12 months, probably owing to the low rate of association between foals (Moehlman 1998*a*,*b*). Most studies reported that male foals played more than females (Boyd *et al.* 1988; Sigurjónsdóttir *et al.* 2003; Kurvers *et al.* 2006; Simpson *et al.* 2012). Other studies reported no difference in rate of play between the sexes, but that males either initiated play more often (Cameron *et al.* 2008) or exhibited a different type of play (Crowell-Davis *et al.* 1987), tending to play fight more than females (Tyler 1972; Carson and Wood-Gush 1983; Cameron *et al.* 2008).

Play is energetically costly, which is probably why it is more rarely seen among equids in arid environments (Moehlman 1998*a*), and why it decreased in a Przewalski's horse stallion after release (Boyd 1998). Play is a multifunctional behavior that helps hone motor skills and repertoire of social behavior (Waring 1983; Khalil and Kaseda 1998; Cameron *et al.* 2008), although it is surprisingly not related to fighting skill (Bourjade *et al.* 2009*c*). Individuals that played more had better survival and condition as a yearling despite weaning earlier (Cameron *et al.* 2008), either because of being in better condition or because of experiences learned through play, or both.

Agonistic Behavior

Agonistic behavior in equids has been defined as interactions resulting in an increased distance between two individuals, either through defensive or aggressive behaviors (Feh 1988). Ethograms have been developed to describe the range of agonistic behaviors observed (McDonnell and Haviland 1995; Ransom and Cade 2009), which range from mild threats to bite or kick, to chases, to biting or kicking with front or hind feet (fig. 3.3). Although these latter behaviors are obvious, it is likely that many signals between equids are too subtle or quick for human observers to reliably identify. Most agonistic behaviors are also observed when equids play, but a primary difference indicating aggression is that the ears are pinned back (Ransom and Cade 2009). Some authors suggest that head threats and rear threats should be viewed as two separate categories of behavior, with the first being offensive and the latter defensive (Wells and von Goldschmidt-Rothschild 1979; Feh 1988; van Dierendonck *et al.* 1995); however, other authors have suggested that rear threats or kicks may also be offensive and so all behaviors may be viewed as aggression (Tilson *et al.* 1988; Ellard and Crowell-Davis 1989; King 2002).

Agonistic behavior has been observed between and among all age and sex classes, but there do appear to be some species differences. Outside of territorial behavior (Klingel 1998), there are few reports of agonistic behavior from non-harem-forming equids—with the exception of Moehlman (1998*a*) and Asa *et al.* (2011)—and it appears to be very rare in wild plains zebra (Fischhoff *et al.* 2009; Simpson *et al.* 2012). Where reported,

Fig. 3.3 Hind leg kick by an African wild ass. Photo by Cheryl Asa

rates of aggression tend to be low: feral horses, 0.01–3.5 acts/hour (Wells and von Goldschmidt-Rothschild 1979; Ransom *et al.* 2010); free-ranging Przewalski's horses, 0.35 acts/hour (King 2002). Rates of agonistic behavior are affected by available space, owing to its function of regulating distance between individuals (Heitor *et al.* 2006*b*); populations with larger areas available (e.g., King 2002; Ransom *et al.* 2010) tend to have lower rates. Levels of all social interaction were inversely correlated with enclosure size for captive animals (Hogan *et al.* 1988; Andersen 1992), which may explain why studies of captive Przewalski's horses reported higher rates of aggression than feral domestic horses (Feh 1988; Christensen *et al.* 2002*b*); a study of wild Przewalski's horses reported the opposite (King 2002). Band size has also been reported to affect rates of aggression, with less aggression in larger groups (Penzhorn 1984; Ransom *et al.* 2010). Among mares, aggression may indicate competition for resources, especially when directed toward pregnant group members or those with new foals (Rutberg and Greenberg 1990; Rho *et al.* 2004).

Dominance hierarchies have been observed in most populations of harem equid species (Linklater 2000), even though the methods used to calculate them vary across studies and results are not always consistent, but see Feist and McCullough (1976), Berger (1977), and Arnold and Grassia (1982) for examples of populations where no hierarchy was discerned. Little behavioral work has been done on nonharem species, which appear to show less aggression outside of territoriality and therefore have not been observed to have a dominance hierarchy. Where a linear hierarchy has been detected among individuals in an equid group, there is generally a correlation between rank and aggression, with more aggressive horses being higher ranked (Houpt *et al.* 1978; King 2002; Heitor *et al.* 2006*b*). But other studies have reported more aggression among individuals of a similar rank as they vie to rise in position (Klimov 1988; Tilson *et al.* 1988; Ellard and Crowell-Davis 1989), or no correlation between rank and aggression at all (van Dierendonck *et al.* 1995). Older equids tend to be more dominant within a group (Andersen 1992; Keiper and Receveur 1992; van Dierendonck *et al.* 1995; King 2002; Rho *et al.* 2004; Pluháček *et al.* 2006; Heitor *et al.* 2006*b*; Powell 2008). As older animals have generally been in the group longest, there is also a correlation between residency and dominance rank (Wells and von Goldschmidt-Rothschild 1979; van Dierendonck *et al.* 1995; King 2002). The tallest (Ellard and Crowell-Davis 1989; Rutberg and Greenberg 1990) and heaviest (Houpt *et al.* 1978; Houpt and Wolski 1980) individuals in a group tend to be most dominant. Thus, within a group in general, the oldest, largest, or simply most aggressive individual is likely to be most dominant. Once dominance hierarchies have been established, they tend to be relatively stable over time (Joubert 1971; Houpt and Wolski 1980; Rutberg and Greenberg 1990; Rasa and Lloyd 1994). Some studies have reported that offspring rank is correlated with maternal rank (Houpt *et al.* 1978; Houpt and Wolski 1980; Feh 1990), although it is not known whether this is simply through inheritance of aggressiveness or size or other factors (van Dierendonck *et al.* 1995).

In many harem groups the stallion is dominant over mares, which are dominant over juveniles (Houpt *et al.* 1978; Penzhorn 1984; Andersen 1992; King 2002; Pluháček *et al.* 2006). The expression of adult stallion dominance over mares is usually through herding behavior (fig. 3.4; Joubert 1971; Feist and McCullough 1976; Miller 1981; Schilder 1992; Ransom *et al.* 2010), whereas dominance among mares and juveniles is expressed through threats and agonistic acts (e.g., Sigurjónsdóttir *et al.* 2003). In some studies the stallion was not the most dominant member of the group (Houpt and Keiper 1982; Keiper and Receveur 1992; Schilder 1992; Heitor *et al.* 2006*b*). In the case of captive herds this is likely a result of the artificial nature of their formation: an immature stallion placed with mature mares. For feral populations it was thought to be because stallions were simply less aggressive than their mares. Serious fights are rare among mares but are occasionally seen among stallions (Berger 1981; Penzhorn 1984; Tilson *et al.* 1988), although ritualization of greetings (Joubert 1971; Moehlman 1998*b*; Bourjade

Fig. 3.4 Feral horse stallion displaying herding behavior: the ears are pinned back and the head is low to the ground. Photo by Jason I. Ransom

et al. 2009*c*) and knowledge of relative fighting ability (Rubenstein and Hack 1992) means that these fights rarely lead to severe injury. Injuries and fatalities do occur, however. Bandi and Usukhjargal (2013) attributed 18.8% of mortality in male Przewalski's horses to injuries sustained in fights.

Historically, there has been little distinction in the literature between dominance in the classic behavioral context versus leadership in the social network context, which is important for understanding complex social structure and behavioral inputs (Chase *et al.* 2002). Despite the male-centric behavioral dominance widely reported in groups of harem-forming equids, some studies have also indicated that leadership of group actions may not be as despotic. Stallions are frequently at the tail end of group movements even when not herding (Joubert 1972*a*; Berger 1977; Miller 1979; Wells and von Goldschmidt-Rothschild 1979), and also mate choice by females is more of a determinant of individual membership in a harem than being "stolen" by stallions (e.g., Nuñez *et al.* 2009). While older studies reported a "lead mare" (Feist and McCullough 1976; Miller 1979; Klimov 1988) as being responsible for direction of movements, recent studies focused on leadership have not found any evidence that a single individual is always in the lead (Bourjade 2007; Bourjade *et al.* 2009*b*; Krueger *et al.* 2014), or any evidence of consistent leadership within a group (Joubert 1972*a*; Penzhorn 1982; Klingel 1998; Grinder *et al.* 2006). Any member of a group can initiate movement (Penzhorn 1982; Bourjade *et al.* 2009*b*; Krueger *et al.* 2014), with the individual leading tending to be of higher social rank (Krueger *et al.* 2014) or having elevated resource needs arising from pregnancy and lactation (Fischhoff *et al.* 2007). Stallion dominance has been correlated with male endocrine changes that arise with access to sexually receptive females in both Grevy's and plains zebra, despite their different organizational strategies (Chaudhuri and Ginsberg 1990).

Many studies that looked at dominance hierarchies also examined the linearity of the hierarchy, which is the assumption that one horse will always be dominant over the next, which is dominant over another, and that there are no triangular relationships. Many studies have found linearity (e.g., van Dierendonck *et al.* 1995), but many have also found variability in its presence year to year or within a population, and the presence of triangular relationships (Houpt *et al.* 1978; Penzhorn 1984; King 2002). Combined with the rarity of agonistic interactions in many populations, this finding suggests that linearity of dominance hierarchies within a group is more of a human construct than of importance to equids (as in canids; Mech 1999), and that relative rank between any two individuals is more biologically meaningful than where they reside in a continuum. When individuals have a long relationship with each other, they are undoubtedly aware of their relative personality and fighting ability (Rubenstein and Hack 1992). Creating affiliative interactions may be at least as important for a stable group as dominance rank order: while more dominant individuals have been reported to have greater reproductive success (Feh 1990; Rasa and Lloyd 1994; Powell 2008), social integration and stability of a group may have a bigger effect on overall fitness (Cameron *et al.* 2009).

The function of agonistic behaviors, as with affiliative behaviors, is related to maintaining group stability (Linklater 2000). Although there was no correlation between band turnover and aggression in feral horses (Rutberg 1990), when mares joined a new group, they generally received more aggression (Rutberg and Greenberg 1990). Heightened aggression would therefore be expected to select against frequent dispersal, and for the formation of stable breeding groups (Linklater 2000; Bourjade *et al.* 2009*a*). Equid society appears to have evolved to mediate stallion harassment of mares, as not only do stallions provide signals about relative fighting ability, which should prevent incursion by other males (Rubenstein and Hack 1992; King and Gurnell 2006), but also stable relationships between stallions and mares can lead to increased reproductive success and lower levels of agonistic behavior (Linklater *et al.* 1999; Cameron *et al.* 2003; Sundaresan *et al.* 2007).

Reproductive Behavior

Sexual Behavior

There are several basic features of courtship and mating behavior that all wild equid species share. Most basic perhaps is the estrous posture or stance of the female, which comprises holding the head low, lips pulled back, ears back, hind legs abducted, and tail deflected to the side or raised away from the perineum (Klingel 1969, 1974; Penzhorn 1984; Boyd and Houpt 1994; Pagan *et al.* 2009; Ransom and Cade 2009). The posture serves as a visual signal to males and may be especially important in young mares of the nonharem species to attract unrelated males (Klingel 1969). It is essentially an exaggerated urination posture, which is in keeping with the increased frequency of urination during estrus reported for mares of most equid species.

Urination during estrus is accompanied by repeated, rapid eversion of the genital labia that exposes the clitoris, called "clitoral winking," which is not restricted to passage of urine. Urination with winking could be a visual signal, since the tail is raised and the genital area exposed, but also likely incorporates labial secretions into the urine as part of olfactory signaling, as described for domestic horses (Asa 1986) and Asiatic wild asses (Rashek 1964*a*). Another feature often seen is the facial expression during estrus, called the "estrous face": features include ears back (Rashek 1964*a*; Klingel 1977), lips pulled back, mouth either open (i.e., teeth bared) or closed, with head and neck stretched low. Based on context, the common motivation seems to be appeasement or submission. Most reports of the estrous face indicate movement of the mouth (snapping), as seen in juveniles when approaching an adult (Rashek 1964*a*; Klingel 1969, 1974; Pagan *et al.* 2009). Notably, mouth movements are not part of the estrous face in Przewalski's horse mares (Houpt and Boyd 1994).

Either the mare or the stallion may initiate courtship interactions, perhaps dependent on the stage of estrus and the stallion's experience. The mare may turn and present her genitals for inspection, even backing up to the stallion, or the stallion may approach and sniff her genital area, often followed by flehmen. Before mounting, most stallions rest their heads on the female's rump, as perhaps a final test of her readiness to stand for copulation. In some species, stallions mount without erection (Klingel 1974, 1998; Penzhorn 1984), and erection develops before or during subsequent mounts, something that is also typical of domestic asses (Henry *et al.* 1991).

During copulation, stallions may bite the mare's neck (fig. 3.5); biting other body areas during court-

Fig. 3.5 An African wild ass stallion preparing to mount a mare, which is "snapping." Photo by Cheryl Asa

ship has been reported for African wild ass (*E. africanus*) stallions (Pagan *et al.* 2009). Instead of biting, plains zebra stallions may groom the mare's neck, flanks, and rump before mounting (Klingel 1969). This is also reported for domestic horses (Asa 1986) but not for other wild equids. Sexually unreceptive mares will not stand for mounting and may kick the stallion, sometimes forcefully (Klingel 1969, 1974; Penzhorn 1984). But kicking—especially small jumps with the hind legs, as if threatening to kick—may be part of proceptive courtship, at least in Cape mountain zebra (*E. zebra zebra*) and African wild asses (Penzhorn 1984; Pagan *et al.* 2009). Other aggressive behaviors include herding during courtship by territorial stallions, something that Klingel (1998) did not observe in nonterritorial equid species.

Reports of the frequency of copulation (intercopulatory interval) vary considerably among species and among studies, but a range of between one and three hours was common (Rashek 1964*a*; Klingel 1969, 1974; Joubert 1972*b*). Ginsberg and Rubenstein (1990) reported a correlation between social organization and copulatory frequency: monandrous plains zebra mares (harem breeders / female defense strategy) copulated fewer times per hour and were in estrus fewer days than polygynandrous Grevy's zebra mares (territorial breeders / resource defense strategy). The role of mating system could be important in other aspects of mating rituals. For example, there were more mentions of aggression in studies of territorial breeders than harem breeders.

Nursing Behavior

The four main behavioral parameters of suckling behavior are initiation of suckling, suckling bout duration,

Fig. 3.6 Grevy's zebra mare nursing her foal. Photo by Jan Pluháček

suckling frequency, and suckling bout termination. Suckling bout duration and frequency are important characteristics, as they reflect the amount of maternal care in current offspring (Mendl and Paul 1989; Therrien *et al.* 2008; Pluháček *et al.* 2010*b*; Bartošová *et al.* 2011), while suckling rejection and termination indicate any conflict over milk between mother and offspring (Cameron *et al.* 2003; Pluháček *et al.* 2010*a*). In all equid species, the foal almost always initiates the suckling bout (Tyler 1972). When suckling, the foal most often adopts a reverse-parallel position (fig. 3.6), with some individuals preferring to stand either to the right or left side of the mother (Komárková and Bartošová 2013; Pluháček *et al.* 2013).

Suckling bout duration differs among equid species and populations (Becker and Ginsberg 1990; Pluháček *et al.* 2014). The average suckling bout lasts from 40 seconds (wild plains zebra; Becker and Ginsberg 1990) to 120 seconds (mountain zebra; Penzhorn 1984). It decreases with increasing age of the foal in some populations (Joubert 1972*b*; Rashek 1976; Becker and Ginsberg 1990; Pluháček *et al.* 2014), while remaining unchanged in others (Barber and Crowell-Davis 1994; Cameron *et al.* 1999*a*). In all captive zebra species, suckling bouts terminated by the foal lasted longer than those terminated by the mare or by a herd mate; in plains zebra, bouts terminated by a herd mate were shorter than those terminated by the mare (Pluháček *et al.* 2012).

In all equid species, suckling bout frequency decreases sharply with the age of the foal. In zebras it decreases from about 6 bouts in 180 minutes after birth to <2 bouts at 1 year old (Pluháček *et al.* 2014). In plains zebra and Przewalski's horses, the suckling reflex starts within 60 minutes after birth, and the first successful suckling occurs within the first 2 hours (Klingel 1969; Boyd 1988*a*). Suckling frequency varies among species; the highest suckling bout frequency was observed in feral horses and asses, followed by plains and mountain zebras, then Grevy's zebra (Becker and Ginsberg 1990; Pluháček *et al.* 2014).

The foal ends most suckling bouts (at least 50%) (Tyler 1972; Duncan *et al.* 1984; Pluháček *et al.* 2010*a*, 2012), but the mother may terminate suckling by moving away, flexing a hind leg, blocking the nipple with a hind leg, or displaying direct aggression (Penzhorn 1984; Crowell-Davis 1985). Moving by the mare is the most common termination behavior in both zebras and horses (Tyler 1972; Duncan *et al.* 1984; Pluháček *et al.* 2012). The proportion of suckling bouts terminated by the mother decreases with increasing age of the foal (Tyler 1972; Duncan *et al.* 1984; Pluháček *et al.* 2012). In captive conditions, mothers of mountain and Grevy's zebras reject and terminate a lower rate of suckling bouts/attempts than those of plains zebra, which originally came from a more mesic environment (Pluháček *et al.* 2012). Thus zebra mothers from the arid-adapted species seem to be more tolerant of their foals' nursing attempts, at least in captive conditions.

One of the main factors affecting lactation length in mammalian species, especially those in which the lactating female conceives shortly after birth, is subsequent pregnancy (Bateson 1994). In most equid species, pregnant mares nurse their foals for a shorter period than nonpregnant dams (Tyler 1972; Rashek 1976; Penzhorn 1979; Duncan *et al.* 1984; Pluháček *et al.* 2007). In feral horses, the duration of lactation lasted longer for primiparous than for multiparous mares (Duncan *et al.* 1984; Cameron and Linklater 2000). In captive plains zebra, the sex of the fetus affected natural weaning: if the fetus was a male, the mare tended to wean her current suckling offspring earlier than if the fetus was female (Pluháček *et al.* 2007). But the age at which weaning occurred did not differ between male and female foals (Duncan *et al.* 1984; Cameron and Linklater 2000; Pluháček *et al.* 2007), and weaning dates did not vary in relation to the condition of the mare (Cameron and Linklater 2000), or with the number of stallions in the group (Cameron *et al.* 2003). The earliest weaning age of a feral horse foal in the literature was 140 days old (Duncan *et al.* 1984), whereas Asiatic wild ass weaned their foals at 236 days (Rashek 1976), plains zebra at 243 days (Pluháček *et al.* 2007), and mountain and Grevy's zebra at 300 and 330 days,

BOX 3.1 ADOPTION

Although adoption (care of offspring from other parents) occurs in most mammalian species, it is a rare event (Riedman 1982). The biggest cost of adoption is allonursing, defined as nursing offspring that are not the female's own (Roulin 2002). This has been recorded in all equid species in both the wild (Rashek 1976; Lloyd and Harper 1980; Penzhorn 1984; Cameron *et al.* 1999*b*; Nuñez *et al.* 2013) and in captivity (Lang 1983; Pluháček *et al.* 2011; Olléová *et al.* 2012); all of these studies are case reports including no more than two adoptions in each. Adoption in equids can be divided into three different scenarios: (1) two closely related mares care for one foal (Cameron *et al.* 1999*b*); (2) a mare loses her own offspring and adopts another foal, either an orphan or a foal acquired by attacking another mare (Rashek 1964*b*; Lloyd and Harper 1980; Nuñez *et al.* 2013); or (3) a mare adopts an orphaned foal while also nursing her own foal (Lang 1983; Pluháček *et al.* 2011; Olléová *et al.* 2012). If the foal is already several months old at the time of adoption, the mare shows preference to her own foal (Olléová *et al.* 2012); however, if the foal is adopted shortly postpartum, no difference in care toward a mare's own and adopted offspring is apparent in terms of nursing behavior (Pluháček *et al.* 2011; Nuñez *et al.* 2013).

The reason for adoption in equids remains unclear. Kin selection as well as other relevant hypotheses (misdirected care, reciprocity, milk evacuation, parenting; Roulin 2002) do not seem applicable (Pluháček *et al.* 2011; Olléová *et al.* 2012; Nuñez *et al.* 2013). The most plausible explanation could be a mare's postpartum physiological state (Nuñez *et al.* 2013), but this does not explain all reported cases (Lang 1983; Olléová *et al.* 2012). Adoption remains a rare situation in equids, with allosuckling therefore also being rare, associated exclusively with adoption in all species except Grevy's zebra (Olléová *et al.* 2012). Social organization of females is apparently an important factor affecting the occurrence of allosuckling, with Grevy's zebra mares apparently being more tolerant of nonfilial foals than other equid species (Olléová *et al.* 2012).

respectively (Joubert 1972*b*; Penzhorn 1984; Becker and Ginsberg 1990). Time between weaning and birth of a new foal in feral horse and captive plains zebra varied from 21 to 185 days, averaging 105 to 138 days (Duncan *et al.* 1984; Pluháček *et al.* 2007). Because the time to the next delivery among various equid species does not differ as much as the weaning age of the foal, the interspecific differences in weaning age could be based on differences in their gestation periods (Jones 1976).

Conclusion

This chapter has provided some insight into the social complexities of life as an equid. Despite the wide range of environments in which they are found, many aspects of equid social behavior are similar across species; for instance, although some populations live in lush grasslands and others travel large distances between water and forage resources, the actual amount of time spent grazing remains surprisingly consistent. Being more active during crepuscular periods is a similar feature across equid populations, despite differences in predation threat. Affiliative behavior and the expression of agonistic behavior were similar in all populations examined, with the same patterns of reproductive behavior also observed in terms of both sexual and nursing behavior. Recent studies have dispelled prevalent myths such as the importance of a linear dominance hierarchy, despotism of stallions, and presence of a lead mare for the function of equid society, while highlighting how important affiliative behaviors are for the cohesion of groups and improving reproductive success.

The research presented here has also revealed gaps in our knowledge about species that are generally considered well understood. There is a lack of information on the perception and cognition of wild species, probably owing to the difficulty of carrying out experiments with nondomesticated animals. Although we can assume that the species are similar, the phylogenetic distance between horses and zebras is sufficient enough that this cannot be taken for granted. Although horse and zebra species are well represented in literature on time budgets, and a recent study has shown how Asiatic wild asses follow the same pattern, there is no published information on time budgets of African wild asses or kiang (*E. kiang*). Similarly, there is a marked lack of data on behavior of territorial species, which may reveal important differences among the equids that are independent of environmental conditions. Although there have been many studies on equid social behavior, the mechanism of bond development between individuals still needs more research to understand why individuals choose to move between groups or to stay with familiar individuals.

Our management tools and actions can have unforeseen and far-reaching implications for wildlife populations (e.g., Ransom *et al.* 2014*b*). Understanding the behavior of equids can therefore aid their conservation in the wild and improve management of domestic and captive populations. Equid research has been dominated by studies on horses, however—domestic, feral, and Przewalski's—with less work on zebras and little on African and Asiatic wild ass species. There are sufficient data on zebra behavior that we understand their basic social structure and behavior enough to begin aiding their conservation in the wild and management in zoos, but the paucity of information on the behavior of African and Asiatic wild asses may hamper conservation efforts. Across all equid species, understanding the intricate physiology of perception and cognition, how fidelity among groups develops and is maintained, and how animals choose to use the landscape are important advances needed for conservation on our rapidly changing planet. As climate changes and human populations spread, integrating behavioral science into mitigation efforts will be one of our strongest tools toward effectively ameliorating the effects of habitat fragmentation and minimizing resource competition, and ultimately conserving wild equids.

ACKNOWLEDGMENTS

The authors would like to thank those who have spent countless hours watching equids in all weather conditions to gather the data represented in this chapter. Thanks also to the reviewers who provided insightful comments toward improving this synthesis.

REFERENCES

Andersen, K.F. 1992. Size, design and interspecific interactions as restrictors of natural behaviour in multi-species exhibits. 1. Activity and intraspecific interactions of plains zebra (*Equus burchelli*). Applied Animal Behaviour Science 34:157–174.

Arnold, G.W. 1984. Comparison of the time budgets and circadian patterns of maintenance activities in sheep, cattle and horses grouped together. Applied Animal Behaviour Science 13:19–30.

Arnold, G.W., and A. Grassia. 1982. Ethogram of agonistic behaviour for thoroughbred horses. Applied Animal Ethology 8:5–25.

Arnold, W., T. Ruf, and R. Kuntz. 2006. Seasonal adjustment of energy budget in a large wild mammal, the Przewalski horse (*Equus ferus przewalskii*): II. Energy expenditure. Journal of Experimental Biology 209:4566–4573.

Asa, C.S. 1986. Sexual behavior of mares. Veterinary Clinics of North America: Equine Practice 2:519–534.

Asa, C.S., F. Marshall, and M. Fischer. 2011. Affiliative and aggressive behavior in a group of female Somali wild ass (*Equus africanus somalicus*). Zoo Biology 31:87–97.

Bandi, N., and D. Usukhjargal. 2013. Reproduction and mortality of re-introduced Przewalski's horse *Equus przewalskii* in Hustai National Park, Mongolia. Journal of Life Sciences 7:624–631.

Barber, J.A., and S.L. Crowell-Davis. 1994. Maternal behavior of Belgian (*Equus caballus*) mares. Applied Animal Behaviour Science 41:161–189.

Bartošová, J., M. Komárková, J. Dubcová, L. Bartoš, and J. Pluháček. 2011. Concurrent lactation and pregnancy: Pregnant domestic horse mares do not increase mother-offspring conflict during intensive lactation. PLoS ONE 6:e22068.

Basile, M., S. Boivin, A. Boutin, C. Blois-Heulin, M. Hausberger, and A. Lemasson. 2009. Socially dependent auditory laterality in domestic horses (*Equus caballus*). Animal Cognition 12:611–619.

Bateson, P. 1994. The dynamics of parent-offspring relationships in mammals. Trends in Ecology and Evolution 9:399–403.

Becker, C., and J.R. Ginsberg. 1990. Mother-infant behaviour of wild Grevy's zebra: Adaptations for survival in semidesert East Africa. Animal Behaviour 40:1111–1118.

Berger, A., K.M. Scheibe, K. Eichorn, A. Scheibe, and J. Streich. 1999. Diurnal and ultradian rhythms of behaviour in a mare group of Przewalski horse (*Equus ferus przewalskii*), measured through one year under semi-reserve conditions. Animal Behaviour 64:1–17.

Berger, J. 1977. Organizational systems and dominance in feral horses in the Grand Canyon. Behavioural Ecology and Sociobiology 2:131–146.

Berger, J. 1981. The role of risks in mammalian combat: Zebra and onager fights. Zietschrift für Tierpsychologie 56:297–304.

Berger, J. 1986. Wild horses of the Great Basin. University of Chicago Press, Chicago, USA.

Berman, D.M. 1991. The ecology of feral horses in central Australia. Thesis, University of New England, Armidale, NSW, Australia.

Blom, S. 2009. Diurnal behaviour of mother-young pairs of Plains zebras (*Equus burchelli*) in Maasai Mara National Reserve, Kenya. Swedish University of Agricultural Sciences.

Bourjade, M. 2007. Sociogenèse et expression des comportements individuels et collectifs chez le cheval. Thesis, Université Louis Pasteur, Strasbourg, France.

Bourjade, M., A. de Boyer des Roches, and M. Hausberger. 2009*a*. Adult-young ratio, a major factor regulating social behaviour of young: A horse study. PLoS ONE 4:e4888.

Bourjade, M., B. Thierry, M. Maumy, and O. Petit. 2009*b*. Decision-making in Przewalski horses (*Equus ferus przewalskii*) is driven by the ecological contexts of collective movements. Ethology 115:321–330.

Bourjade, M., L. Tatin, S.R.B. King, and C. Feh. 2009*c*. Early reproductive success, preceding bachelor ranks and their behavioural correlates in young Przewalski's stallions. Ethology Ecology and Evolution 21:1–14.

Boy, V., and P. Duncan. 1979. Time-budgets of Camargue horses: I. Developmental changes in the time-budgets of foals. Behaviour 71:187–202.

Boyd, L. 1988*a*. Ontogeny of behaviour in Przewalski horses. Applied Animal Behaviour Science 21:41–69.

Boyd, L. 1988*b*. Time budgets of adult Przewalski horses: Effects of sex, reproductive status and enclosure. Applied Animal Behaviour Science 21:19–39.

Boyd, L. 1998. The 24-h time budget of a takh harem stallion (*Equus ferus przewalskii*) pre- and post-reintroduction. Applied Animal Behaviour Science 60:291–299.

Boyd, L., and N. Bandi. 2002. Reintroduction of takhi, *Equus ferus przewalskii*, to Hustai National Park, Mongolia: Time budget and synchrony of activity pre- and post-release. Applied Animal Behaviour Science 78:87–102.

Boyd, L., and K.A. Houpt. 1994. Activity patterns. Pages 195–228 *in* L. Boyd and K.A. Houpt (Eds.), Przewalski's horse: The history and biology of an endangered species. State University of New York Press, Albany, USA.

Boyd, L., D. Carbonaro, and K.A. Houpt. 1988. The 24-hour time budget of Przewalski horses. Applied Animal Behaviour Science 21:5–17.

Brooks, C.J. 2005. The foraging behaviour of Burchell's zebra (*Equus burchelli antiquorum*). Thesis, University of Bristol, Bristol, UK.

Cameron, E.Z., and W.L. Linklater. 2000. Individual mares bias investment in sons and daughters in relation to their condition. Animal Behaviour 60:359–367.

Cameron, E.Z., K.J. Stafford, W.L. Linklater, and C.J. Veltman. 1999*a*. Suckling behaviour does not measure milk intake in horses, *Equus caballus*. Animal Behaviour 57:673–678.

Cameron, E.Z., W.L. Linklater, K.J. Stafford, and E.O. Minot. 1999*b*. A case of co-operative nursing and offspring care by mother and daughter feral horses. Journal of Zoology 249:486–489.

Cameron, E.Z., W.L. Linklater, K.J. Stafford, and E.O. Minot. 2003. Social grouping and maternal behaviour in feral horses (*Equus caballus*): The influence of males on maternal protectiveness. Behavioural Ecology and Sociobiology 53:92–101.

Cameron, E.Z., W.L. Linklater, K.J. Stafford, and E.O. Minot. 2008. Maternal investment results in better foal condition through increased play behaviour in horses. Animal Behaviour 76:1511–1518.

Cameron, E.Z., T.H. Setsaas, and W.L. Linklater. 2009. Social bonds between unrelated females increase reproductive success in feral horses. Proceedings of the National Academy of Sciences of the USA 106:13,850–13,853.

Carson, K., and D.G.M. Wood-Gush. 1983. Equine behavior: I. A review of the literature on social and dam-foal behaviour. Applied Animal Ethology 10:165–178.

Chase, I.D., C. Tovey, D. Spangler-Martin, and M. Manfredonia. 2002. Individual differences versus social dynamics in the formation of animal dominance hierarchies. Proceedings of the National Academy of Sciences of the USA 99:5744–5749.

Chaudhuri, M., and J.R. Ginsberg. 1990. Urinary androgen concentrations and social status in two species of free ranging zebra (*Equus burchelli* and *E. grevyi*). Journal of Reproduction and Fertility 88:127–133.

Christensen, J.W., J. Ladewig, E. Søndergaard, and J. Malmkvist. 2002*a*. Effects of individual versus group stabling on social behaviour in domestic stallions. Applied Animal Behaviour Science 75:233–248.

Christensen, J.W., T. Zharkikh, J. Ladewig, and N. Yasinetskaya. 2002*b*. Social behaviour in stallion groups (*Equus przewalskii* and *Equus caballus*) kept under natural and domestic conditions. Applied Animal Behaviour Science 76:11–20.

Clutton-Brock, T.H., P.J. Greenwood, and R.P. Powell. 1976. Ranks and relationships in Highland ponies and Highland cows. Zietschrift für Tierpsychologie 41:202–216.

Crowell-Davis, S.L. 1985. Nursing behaviour and maternal aggression among Welsh ponies (*Equus caballus*). Applied Animal Behaviour Science 14:11–25.

Crowell-Davis, S.L., K.A. Houpt, and C.M. Carini. 1986. Mutual grooming and nearest-neighbor relationships among foals of *Equus caballus*. Applied Animal Behaviour Science 15:113–123.

Crowell-Davis, S.L., K.A. Houpt, and L. Kane. 1987. Play development in Welsh pony (*Equus caballus*) foals. Applied Animal Behaviour Science 18:119–131.

Duncan, P. 1979. Time-budgets of Camargue horses: II. Time-budgets of adult horses and weaned sub-adults. Behaviour 72:26–49.

Duncan, P. 1985. Time-budgets of Camargue horses: III. Environmental influences. Behaviour 92:188–208.

Duncan, P. 1992. Horses and grasses: The nutritional ecology of equids and their impact on the Camargue. Springer-Verlag, New York, USA.

Duncan, P., P.H. Harvey, and S.M. Wells. 1984. On lactation and associated behaviour in a natural herd of horses. Animal Behaviour 32:255–263.

Elia, J.B., H.N. Erb, and K.A. Houpt. 2010. Motivation for hay: Effects of a pelleted diet on behavior and physiology of horses. Physiology and Behavior 101:623–627.

Ellard, M., and S.L. Crowell-Davis. 1989. Evaluating equine dominance in draft mares. Applied Animal Behaviour Science 24:55–75.

Feh, C. 1988. Social behaviour and relationships of Przewalski horses in Dutch semi-reserves. Applied Animal Behaviour Science 21:171–187.

Feh, C. 1990. Long-term paternity data in relation to different aspects of rank for Camargue stallions, *Equus caballus*. Animal Behaviour 40:995–996.

Feh, C. 1999. Alliances and reproductive success in Camargue stallions. Animal Behaviour 57:705–713.

Feh, C., and J. de Mazières. 1993. Grooming at a preferred site reduces heart rate in horses. Animal Behaviour 46: 1191–1194.

Feist, J.D., and D.R. McCullough. 1976. Behavior patterns and communication in feral horses. Zietschrift für Tierpsychologie 41:337–371.

Fischhoff, I.R., S.R. Sundaresan, J.E. Cordingley, H.M. Larkin, M.-J. Sellier, and D.I. Rubenstein. 2007. Social relationships and reproductive state influence leadership roles in movements of plains zebra, *Equus burchelli*. Animal Behaviour 73:825–831.

Fischhoff, I.R., S.R. Sundaresan, H.M. Larkin, M.-J. Sellier, J.E. Cordingley, and D.I. Rubenstein. 2009. A rare fight in female plains zebra. Journal of Ethology 28:201–205.

Giebel, H. 1958. Visuelles lernvermogen bei einhufern. Zoologische Jahrbücher 67:487–520.

Ginsberg, J.R., and D.I. Rubenstein. 1990. Sperm competition and variation in zebra mating behavior. Behavioural Ecology and Sociobiology 26:427–434.

Grinder, M.I., P.R. Krausman, and R.S. Hoffmann. 2006. *Equus asinus*. Mammalian Species 794:1–9.

Hampson, B.A., M.A. de Laat, P.C. Mills, and C.C. Pollitt. 2010. Distances travelled by feral horses in "outback" Australia. Equine Veterinary Journal 42:582–586.

Hanggi, E.B., and J.F. Ingersoll. 2009. Long-term memory for categories and concepts in horses (*Equus caballus*). Animal Cognition 12:451–462.

Heitor, F., M. do Mar Oom, and L. Vicente. 2006*a*. Social relationships in a herd of Sorraia horses: Part II. Factors affecting affiliative relationships and sexual behaviours. Behavioural Processes 73:231–239.

Heitor, F., M. do Mar Oom, and L. Vicente. 2006*b*. Social relationships in a herd of Sorraia horses; Part I. Correlates of social dominance and contexts of aggression. Behavioural Processes 73:170–177.

Henry, M., S.M. McDonnell, L.D. Lodi, and E.L. Gastal. 1991. Pasture mating behaviour of donkeys (*Equus asinus*) at natural and induced oestrus. Journal of Reproduction and Fertility Supplement 44:77–86.

Hogan, E.S., K.A. Houpt, and K. Sweeney. 1988. The effect of enclosure size on social interactions and daily activity patterns of the captive Asiatic wild horse (*Equus przewalskii*). Applied Animal Behaviour Science 21:147–168.

Hothersall, B., P. Harris, L. Sörtoft, and C.J. Nicol. 2010. Discrimination between conspecific odour samples in the horse (*Equus caballus*). Applied Animal Behaviour Science 126:37–44.

Houpt, K.A. 2012. Motivation for cribbing by horses. Animal Welfare 21:1–7.

Houpt, K.A., and L. Boyd. 1994. Social behaviour. Pages 229–254 *in* L. Boyd and K. A. Houpt (Eds.), Przewalski's horse: The history and biology of an endangered species. State University of New York, Albany, USA.

Houpt, K.A., and R.R. Keiper. 1982. The position of the stallion in the equine dominance hierarchy of feral and domestic ponies. Journal of Animal Science 54:945–950.

Houpt, K.A., and I. Wolski. 1980. Stability of equine hierarchies and the prevention of dominance related aggression. Equine Veterinary Journal 12:15–18.

Houpt, K.A., K. Law, and V. Martinisi. 1978. Dominance hierarchies in domestic horses. Applied Animal Ethology 4:273–283.

Hughes, R.D., P. Duncan, and J. Dawson. 1981. Interactions between Camargue horses and horseflies (Diptera: Tabanidae). Bulletin of Entomological Research 71:227–242.

Jones, D.M. 1976. The husbandry and veterinary care of wild horses in captivity. Equine Veterinary Journal 8:140–146.

Joubert, E. 1971. Ecology, behaviour and population dynamics of the Hartmann Zebra *Equus zebra hartmannae* Matschie, 1898 in southwest Africa. Thesis, University of Pretoria, Pretoria, South Africa.

Joubert, E. 1972*a*. The social organization and associated behaviour in the Hartmann zebra *Equus zebra hartmannae*. Madoqua Series 1:17–56.

Joubert, E. 1972*b*. Activity patterns shown by Hartmann zebra *Equus zebra hartmannae* in southwest Africa with reference to climatic factors. Zoologica Africana 7:309–331.

Kaseda, Y. 1983. Seasonal changes in time spent grazing and resting of Misaki horses. Japanese Journal of Zootechnical Sciences 54:464–469.

Keiper, R.R. 1988. Social interactions of the Przewalski horse (*Equus przewalskii* Poliakov, 1881) herd at the Munich Zoo. Applied Animal Behaviour Science 21:89–97.

Keiper, R.R., and M.A. Keenan. 1980. Nocturnal activity patterns of feral ponies. Journal of Mammalogy 61:116–118.

Keiper, R.R., and H. Receveur. 1992. Social interactions of free-ranging Przewalski horses in semi-reserves in the Netherlands. Applied Animal Behaviour Science 33:303–318.

Keiper, R.R., M. Moss, and S. Zervanos. 1980. Daily and seasonal patterns of feral ponies on Assateague Island. Proceedings of the Second Conference on Scientific Research in the National Parks 8:369–381.

Khalil, A.M., and Y. Kaseda. 1998. Early experience affects developmental behaviour and timing of harem formation in Misaki horses. Applied Animal Behaviour Science 59:253–263.

Kimura, R. 1998. Mutual grooming and preferred associate relationships in a band of free-ranging horses. Applied Animal Behaviour Science 59:265–276.

Kimura, R. 2001. Volatile substances in feces, urine and urine-marked feces of feral horses. Canadian Journal of Animal Science 81:411–420.

King, S.R.B. 1996. The social behaviour of a bachelor group of Przewalski horses under free-ranging conditions. B.S. Honors Thesis, Queen Mary and Westfield College, University of London, London, UK.

King, S.R.B. 2002. Behavioural ecology of Przewalski horses (*Equus przewalskii*) reintroduced to Hustai National Park, Mongolia. Thesis, Queen Mary, University of London, London, UK.

King, S.R.B., and J. Gurnell. 2006. Scent-marking behaviour by stallions: An assessment of function in a reintroduced population of Przewalski horses (*Equus ferus przewalskii*). Journal of Zoology 272:30–36.

King, S.R.B., and J. Gurnell. 2010. Effects of fly disturbance on the behaviour of a population of reintroduced Przewalski horses (*Equus ferus przewalskii*) in Mongolia. Applied Animal Behaviour Science 125:22–29.

Klimov, V.V. 1988. Spatial-ethological organisation of the herd of Przewalski horses (*Equus przewalskii*) in Askania Nova. Applied Animal Behaviour Science 21:99–115.

Klingel, H. 1969. Reproduction in the plains zebra, *Equus burchelli boehmi*: Behaviour and ecological factors. Journal of Reproduction and Fertility Supplement 6:339–345.

Klingel, H. 1974. Soziale organisation und verhalten des Grevy-zebras (*Equus grevyi*). Zeitschrift für Tierpsychologie 36:37–70.

Klingel, H. 1977. Observations on social organization and behaviour of African and Asiatic wild asses (*Equus africanus* and *Equus hemionus*). Zietschrift für Tierpsychologie 44:323–331.

Klingel, H. 1998. Observations on social organization and behaviour of African and Asiatic wild asses (*Equus africanus* and *Equus hemionus*). Applied Animal Behaviour Science 60:103–113.

Kolter, L., and W. Zimmermann. 1988. Social behaviour of Przewalski horses (*Equus p. przewalskii*) in the Cologne Zoo and its consequences for management and housing. Applied Animal Behaviour Science 21:117–145.

Komárková, M., and J. Bartošová. 2013. Lateralized suckling in domestic horses (*Equus caballus*). Animal Cognition 16:343–349.

Krueger, K., and B. Flauger. 2011. Olfactory recognition of individual competitors by means of faeces in horse (*Equus caballus*). Animal Cognition 14:245–257.

Krueger, K., B. Flauger, K. Farmer, and C.K. Hemelrijk. 2014. Movement initiation in groups of feral horses. Behavioural Processes 103:91–101.

Kuntz, R., C. Kubalek, T. Ruf, F. Tataruch, and W. Arnold. 2006. Seasonal adjustment of energy budget in a large wild mammal, the Przewalski horse (*Equus ferus przewalskii*): I. Energy intake. Journal of Experimental Biology 209:4557–4565.

Kurvers, C.M.H.C., P.R. van Weeren, C.W. Rogers, and M.C. van Dierendonck. 2006. Quantification of spontaneous locomotion activity in foals kept in pastures under various management conditions. American Journal of Veterinary Research 67:1212–1217.

Lang, E.M. 1983. Die Somaliwildesel, *Equus asinus somalicus*, im Basler Zoo. Der Zoologische Garten 53:73–80.

Lee, J., T. Floyd, H. Erb, and K.A. Houpt. 2011. Preference and demand for exercise in stabled horses. Applied Animal Behaviour Science 130:91–100.

Linklater, W.L. 2000. Adaptive explanation in socio-ecology: Lessons from the Equidae. Biological Reviews 75:1–20.

Linklater, W.L., E.Z. Cameron, E.O. Minot, and K.J. Stafford. 1999. Stallion harassment and the mating system of horses. Animal Behaviour 58:295–306.

Lloyd, P.H., and D.A. Harper. 1980. A case of adoption and rejection of foals in Cape mountain zebra, *Equus zebra zebra*. South African Journal of Wildlife Research 10:61–62.

Lloyd, P.H., and O.A.E. Rasa. 1989. Status, reproductive success and fitness in Cape mountain zebra (*Equus zebra zebra*). Behavioural Ecology and Sociobiology 25:411–420.

Madosky, J.M. 2011. Factors that affect harem stability in a feral horse (*Equus caballus*) population on Shackleford Banks Island, NC. Thesis, University of New Orleans, New Orleans, LA, USA.

Maros, K., M. Gácsi, and A. Miklosi. 2008. Comprehension of human pointing gestures in horses (*Equus caballus*). Animal Cognition 11:457–466.

Mayes, E., and P. Duncan. 1986. Temporal patterns of feeding behaviour in free-ranging horses. Behaviour 96:105–129.

McBride, G.E., R.J. Christopherson, and W. Sauer. 1985. Metabolic rate and plasma thyroid hormone concentrations of mature horses in response to changes in ambient temperature. Canadian Journal of Animal Science 65:375–382.

McDonnell, S.M. 2003. A practical field guide to horse behavior: The equid ethogram. CAB International, Cambridge, MA, USA.

McDonnell, S.M., and J.C.S. Haviland. 1995. Agonistic ethogram of the equid bachelor band. Applied Animal Behaviour Science 43:147–188.

McDonnell, S.M., and A. Poulin. 2002. Equid play ethogram. Applied Animal Behaviour Science 78:263–290.

Mech, L.D. 1999. Alpha status, dominance, and division of labor in wolf packs. Canadian Journal of Zoology 77: 1196–1203.

Mendl, M., and E.S. Paul. 1989. Observation of nursing and sucking behaviour as an indicator of milk transfer and parental investment. Animal Behaviour 37:513–515.

Miller, R. 1979. Band organization and stability in Red Desert feral horses. Pages 113–128 *in* R.H. Denniston (Ed.), Symposium on the ecology and behaviour of wild and feral equids. University of Wyoming, Laramie, USA.

Miller, R. 1981. Male aggression, dominance and breeding behaviour in Red Desert feral horses. Zietschrift für Tierpsychologie 57:340–351.

Moehlman, P.D. 1998*a*. Behavioural patterns and communication in feral asses (*Equus africanus*). Applied Animal Behaviour Science 60:125–169.

Moehlman, P.D. 1998*b*. Feral asses (*Equus africanus*): Intraspecific variation in social organization in arid and mesic habitats. Applied Animal Behaviour Science 60:171–195.

Mooring, M.S., J.E. Benjamin, C.R. Harte, and N.B. Herzog. 2000. Testing the interspecific body size principle in ungulates: The smaller they come the harder they groom. Animal Behaviour 60:35–45.

Murray, L.M.A., K. Byrne, and R.B. D'Eath. 2013. Pair-bonding and companion recognition in domestic donkeys, *Equus asinus*. Applied Animal Behaviour Science 143: 67–74.

Neuhaus, P., and K.E. Ruckstuhl. 2002. The link between sexual dimorphism, activity budgets, and group cohesion: The case of the plains zebra (*Equus burchelli*). Canadian Journal of Zoology 80:1437–1441.

Newton-Fisher, N.E., and P.C. Lee. 2011. Grooming reciprocity in wild male chimpanzees. Animal Behaviour 81:439–446.

Nuñez, C.M.V., J.S. Adelman, and C. Mason. 2009. Immunocontraception decreases group fidelity in a feral horse population during the non-breeding season. Applied Animal Behaviour Science 117:74–83.

Nuñez, C.M.V., J.S. Adelman, and D.I. Rubenstein. 2013. A free-ranging, feral mare *Equus caballus* affords similar maternal care to her genetic and adopted offspring. American Naturalist 182:674–681.

Olléová, M., J. Pluháček, and S.R.B. King. 2012. Effect of social system on allosuckling and adoption in zebras. Journal of Zoology 288:127–134.

Pagan, O., F. von Houwald, C. Wenker, and B.L. Steck. 2009. Husbandry and breeding of Somali wild ass *Equus africanus somalicus* at Basel Zoo, Switzerland. International Zoo Yearbook 43:198–211.

Pellegrini, S. 1971. Home range, territoriality and movement patterns of wild horses in the Wassuk Range of western Nevada. Thesis, University of Nevada, Reno, USA.

Penzhorn, B.L. 1979. Social organisation of the Cape mountain zebra *Equus z. zebra* in the Mountain Zebra National Park. Koedoe 22:115–156.

Penzhorn, B.L. 1982. Habitat selection by Cape mountain zebras in the Mountain Zebra National Park. South African Journal of Wildlife Research 12:48–54.

Penzhorn, B.L. 1984. A long term study of social organisation and behaviour of Cape mountain zebra *Equus zebra zebra*. Zietschrift für Tierpsychologie 64:97–146.

Penzhorn, B.L., and P.A. Novellie. 1991. Some behavioural traits of Cape mountain zebras (*Equus zebra zebra*) and their implications for the management of a small conservation area. Applied Animal Behaviour Science 29:293–299.

Pluháček, J., L. Bartoš, and L. čulík. 2006. High-ranking mares of captive plains zebra *Equus burchelli* have greater reproductive success than low-ranking mares. Applied Animal Behaviour Science 99:315–329.

Pluháček, J., L. Bartoš, M. Doležalová, and J. Bartošová-Víchová. 2007. Sex of the foetus determines the time of weaning of the previous offspring of captive plains zebra (*Equus burchelli*). Applied Animal Behaviour Science 105:192–204.

Pluháček, J., J. Bartošová, and L. Bartoš. 2010*a*. Suckling behavior in captive plains zebra (*Equus burchellii*): Sex differences in foal behavior. Journal of Animal Science 88:131–136.

Pluháček, J., L. Bartoš, and J. Bartošová. 2010*b*. Mother–offspring conflict in captive plains zebra (*Equus burchellii*): Suckling bout duration. Applied Animal Behaviour Science 122:127–132.

Pluháček, J., J. Bartošová, and L. Bartoš. 2011. A case of adoption and allonursing in captive plains zebra (*Equus burchellii*). Behavioural Processes 86:174–177.

Pluháček, J., M. Olléová, J. Bartošová, and L. Bartoš. 2012. Effect of ecological adaptation on suckling behaviour in three zebra species. Behaviour 149:1395–1411.

Pluháček, J., M. Olléová, J. Bartošová, J. Pluháčková, and L. Bartoš. 2013. Laterality of suckling behaviour in three zebra species. Laterality: Asymmetries of Body, Brain and Cognition 18:349–364.

Pluháček, J., M. Olléová, L. Bartoš, and J. Bartošová. 2014. Time spent suckling is affected by different social organization in three zebra species. Journal of Zoology 292:10–17.

Policht, R., A. Karadžos, and D. Frynta. 2011. Comparative analysis of long-range calls in equid stallions (Equidae): Are acoustic parameters related to social organization? African Zoology 46:18–26.

Powell, D.M. 2008. Female-female competition or male mate choice? Patterns of courtship and breeding behaviour among feral horses (*Equus caballus*) on Assateague Island. Journal of Ethology 26:137–144.

Powell, D.M., D.E. Danze, and M.A. Gwinn. 2006. Predictors of biting fly harassment and its impact on habitat use by feral horses (*Equus caballus*) on a barrier island. Journal of Ethology 24:147–154.

Proops, L., F. Burden, and B. Osthaus. 2009*a*. Mule cognition: A case of hybrid vigour? Animal Cognition 12:75–84.

Proops, L., K. McComb, and D. Reby. 2009*b*. Cross-modal individual recognition in domestic horses (*Equus caballus*). Proceedings of the National Academy of Sciences of the USA 106:947–951.

Ransom, J.I., and B.S. Cade. 2009. Quantifying equid behavior—A research ethogram for free-roaming feral horses. Techniques and Methods 2-A9. US Geological Survey, Reston, VA, USA.

Ransom, J.I., B.S. Cade, and N.T. Hobbs. 2010. Influences of immunocontraception on time budgets, social behavior, and body condition in feral horses. Applied Animal Behaviour Science 124:51–60.

Ransom, J.I., J.G. Powers, H.M. Garbe, M.W. Oehler Sr., T.M. Nett, and D.L. Baker. 2014*a*. Behavior of feral horses in response to culling and GnRH immunocontraception. Applied Animal Behaviour Science 157:81–92.

Ransom, J.I., J.G. Powers, N.T. Hobbs, and D.L. Baker. 2014*b*. Ecological feedbacks can reduce population-level efficacy of wildlife fertility control. Journal of Applied Ecology 51:259–269.

Rasa, O.A.E., and P.H. Lloyd. 1994. Incest avoidance and attainment of dominance by females in a Cape mountain zebra (*Equus zebra zebra*) population. Behaviour 128:169–188.

Rashek, V.A. 1964*a*. Reproduction of *Equus hemionus onager* Boddaert, in Barsa-Kelmes Island (Aral Sea). Acta Societatis Zoologicae Bohemoslovenicae 1:89–95.

Rashek, V.A. 1964*b*. Twenty-four hours of the onager, *Equus hemionus onager* Boddaert, and its behaviour in Barsa-Kelmes Island (Aral Sea). Acta Societatis Zoologicae Bohemoslovenicae 1:96–104.

Rashek, V.A. 1976. Details of feeding and suckling behaviour in young wild asses on Barsa-Kelmes island (Aral Sea). Zoologicheshskii Zhurnal 55:784–787.

Rho, J.R., R.B. Srygley, and J.C. Choe. 2004. Behavioral ecology of the Jeju pony (*Equus caballus*): Effects of maternal age, maternal dominance hierarchy and foal age on mare aggression. Ecological Research 19:55–63.

Riedman, M.L. 1982. The evolution of alloparental care and adoption in mammals and birds. Quarterly Review of Biology 57:405–435.

Roulin, A. 2002. Why do lactating females nurse alien offspring? A review of hypotheses and empirical evidence. Animal Behaviour 63:201–208.

Rubenstein, D.I. 2010. Ecology, social behavior, and conservation in zebras. Advances in the Study of Behavior 42:231–258.

Rubenstein, D.I., and M.A. Hack. 1992. Horse signals: The sounds and scents of fury. Evolutionary Ecology 6:254–260.

Rubenstein, D.I., S.R. Sundaresan, I.R. Fischhoff, and D. Saltz. 2007. Social networks in wild asses: Comparing patterns and processes among populations. Exploration into the Biological Resources of Mongolia (Halle/Saale) 10:159–176.

Rutberg, A.T. 1987. Horse fly harassment and the social behaviour of feral ponies. Ethology 75:145–154.

Rutberg, A.T. 1990. Inter-group transfer in Assateague pony mares. Animal Behaviour 40:945–952.

Rutberg, A.T., and S. Greenberg. 1990. Dominance, aggression frequencies and modes of aggressive competition in feral pony mares. Animal Behaviour 40:322–331.

Salter, R.E. 1979. Biogeography and habitat-use behavior of feral horses in western and northern Canada. Pages 129–141 *in* R.H. Denniston (Ed.), Symposium on the ecology and behaviour of wild and feral equids. University of Wyoming, Laramie, USA.

Sappington, B.F., and L. Goldman. 1994. Discrimination learning and concept formation in the Arabian horse. Journal of Animal Science 72:3080–3087.

Saslow, C.A. 2002. Understanding the perceptual world of horses. Applied Animal Behaviour Science 78:209–224.

Scheel, D. 1993. Watching for lions in the grass: The usefulness of scanning and its effects during hunts. Animal Behaviour 46:695–704.

Scheibe, K.M., K. Eichorn, B. Kalz, W.J. Streich, and A. Scheibe. 1998. Water consumption and watering behaviour of Przewalski mares (*Equus ferus przewalskii*) in a semireserve. Zoo Biology 17:181–192.

Schilder, M.B.H. 1992. Stability and dynamics of group composition in a herd of captive plains zebras. Ethology 90:154–168.

Sigurjónsdóttir, H., M.C. van Dierendonck, S. Snorrason, and A.G. Thorhallsdóttir. 2003. Social relationships in a group of horses without a mature stallion. Behaviour 140:783–804.

Silk, J.B. 2007. The adaptive value of sociality in mammalian groups. Philosophical Transactions of the Royal Society of London B: Biological Sciences 362:539–559.

Simpson, H.I., S.A. Rands, and C.J. Nicol. 2012. Social structure, vigilance and behaviour of plains zebra (*Equus burchellii*): A 5-year case study of individuals living on a managed wildlife reserve. Acta Theriologica 57:111–120.

Souris, A., P. Kaczensky, R. Julliard, and C. Walzer. 2007. Time budget-, behavioral synchrony- and body score development of a newly released Przewalski's horse group *Equus ferus przewalskii*, in the Great Gobi B Strictly Protected Area in SW Mongolia. Applied Animal Behaviour Science 107:307–321.

Sundaresan, S.R., I.R. Fischhoff, J. Dushoff, and D.I. Rubenstein. 2006. Network metrics reveal differences in social organization between two fission–fusion species, Grevy's zebra and onager. Oecologia 151:140–149.

Sundaresan, S.R., I.R. Fischhoff, and D.I. Rubenstein. 2007. Male harassment influences female movements and associations in Grevy's zebra (*Equus grevyi*). Behavioral Ecology 18:860–865.

Therrien, J.F., S.D. Côté, M. Festa-Bianchet, and J.P. Ouellet. 2008. Maternal care in white-tailed deer: Trade-off between maintenance and reproduction under food restriction. Animal Behaviour 75:235–243.

Tilson, R., K. Sweeny, G. Binclik, and N. Reindl. 1988. Buddies and bullies: Social structure of a bachelor group of Przewalski horses. Applied Animal Behaviour Science 21:169–185.

Timney, B., and T. Macuda. 2001. Vision and hearing in horses. Journal of the American Veterinary Medical Association 218:1567–1574.

Tyler, S. 1972. The behaviour and social organisation of the New Forest ponies. Animal Behaviour Monographs 5:85–196.

Uller, C., and J. Lewis. 2009. Horses (*Equus caballus*) select the greater of two quantities in small numerical contrasts. Animal Cognition 12:733–738.

Valeix, M., H. Fritz, A.J. Loveridge, Z. Davidson, J.E. Hunt, F. Murindagomo, and D.W. Macdonald. 2009. Does the risk of encountering lions influence African herbivore behaviour at waterholes? Behavioural Ecology and Sociobiology 63:1483–1494.

van Dierendonck, M.C., and B.M. Spruijt. 2012. Coping in groups of domestic horses—Review from a social and neurobiological perspective. Applied Animal Behaviour Science 138:194–202.

van Dierendonck, M.C., H. de Vries, and M.B.H. Schilder. 1995. An analysis of dominance, its behavioural parameters and possible determinants in a herd of Icelandic horses in captivity. Journal of Zoology 45:362–385.

van Dierendonck, M.C., N. Bandi, D. Batdorj, S. Dugerlham, and B. Munkhtsog. 1996. Behavioural observations of reintroduced takhi or Przewalski horses (*Equus ferus przewalskii*) in Mongolia. Applied Animal Behaviour Science 50:95–114.

van Dierendonck, M.C., H. Sigurjónsdóttir, B. Colenbrander, and A.G. Thorhallsdóttir. 2004. Differences in social behaviour between late pregnant, post-partum and barren mares in a herd of Icelandic horses. Applied Animal Behaviour Science 89:283–297.

van Dierendonck, M.C., H. de Vries, M.B.H. Schilder, B. Colenbrander, A.G. Thorhallsdóttir, and H. Sigurjónsdóttir. 2009. Interventions in social behaviour in a herd of mares and geldings. Applied Animal Behaviour Science 116:67–73.

Waring, G.H. 1983. Horse behaviour: The behavioural traits and adaptations of domestic and wild horses, including ponies. Noyes, Saddle River, NJ, USA.

Waring, G.H. 2003. Horse behavior, 2nd ed. Noyes, Norwich, NY, USA.

Wells, S.M., and B. von Goldschmidt-Rothschild. 1979. Social behaviour and relationships in a herd of Camargue horses. Zietschrift für Tierpsychologie 49:363–380.

Xia, C., W. Liu, W. Xu, W. Yang, F. Xu, and D.A. Blank. 2013. Diurnal time budgets and activity rhythm of the Asiatic wild ass *Equus hemionus* in Xinjiang, western China. Pakistan Journal of Zoology 45:1241–1248.

Zervanos, S., and R.R. Keiper. 1979. Seasonal home ranges and activity patterns of feral Assateague Island ponies. Pages 3–14 *in* R.H. Denniston (Ed.), Symposium on the ecology and behaviour of wild and feral equids. University of Wyoming, Laramie, USA.

4

Habitat and Diet of Equids

KATHRYN A. SCHOENECKER,
SARAH R.B. KING,
MEGAN K. NORDQUIST,
DEJID NANDINTSETSEG,
AND QING CAO

The subgenus *Equus* (order Perissodactyla; Groves 1974) appeared ~4.5 million years ago in North America (Orlando *et al.* 2013) and successfully colonized North America, Asia, Europe, and Africa (see chap. 8). Equids were once the most abundant medium-sized grazing animals of the world's grasslands and steppes (Duncan 1992). During the Late Pleistocene and Holocene, all North American equids went extinct for reasons that are not well understood (Berger 1986); however, the evolution of ruminant digestion is thought to have led to competition and reduction in diversity of perissodactyls coincident with the radiation of artiodactyls (Van Soest 1994). Modern equid species inhabit, and have adapted to, a wide range of habitats. From arid hot and cold deserts to temperate forests, and from the hottest, driest landscapes to subzero cold temperatures, they survive and thrive in some of the harshest terrestrial habitats on earth.

Today's wild and feral equids are predominantly grazers. Their digestive system is specialized to consume large quantities of low-quality foliage as compared to ruminant mammals. They can extract nutrients out of most types of vegetation, even consuming it twice in the practice of coprophagy, or the consumption of feces (fig. 4.1). Equids have an enlarged colon and cecum to accommodate their cellulose-rich diet, allowing the indigestible material to remain in contact with a commensal microbial population for digestion (called "cecal digesters," or "hind-gut fermenters"; Van Soest 1994). The jaw, high-crowned (hypsodont) teeth, and flexible lips of equids are well adapted for processing tough (cellulose and silica-rich) plant material. Being fairly nonselective (roughage) feeders on grasses that are low in protein:fiber ratios, these nonruminants sustain faster rates of food passage and higher intake rates than ruminants. Ruminants of equivalent sizes that feed on the same low-quality diet as equids are at a disadvantage compared with cecal digesters because lignified food requires energy for ruminating and time for processing (Van Soest 1994).

In this chapter, we present information from studies of equids and their habitat use across various habitat types. We provide a synthesis of the scientific literature on equid habitat selection, home range and movements, water needs, and diet.

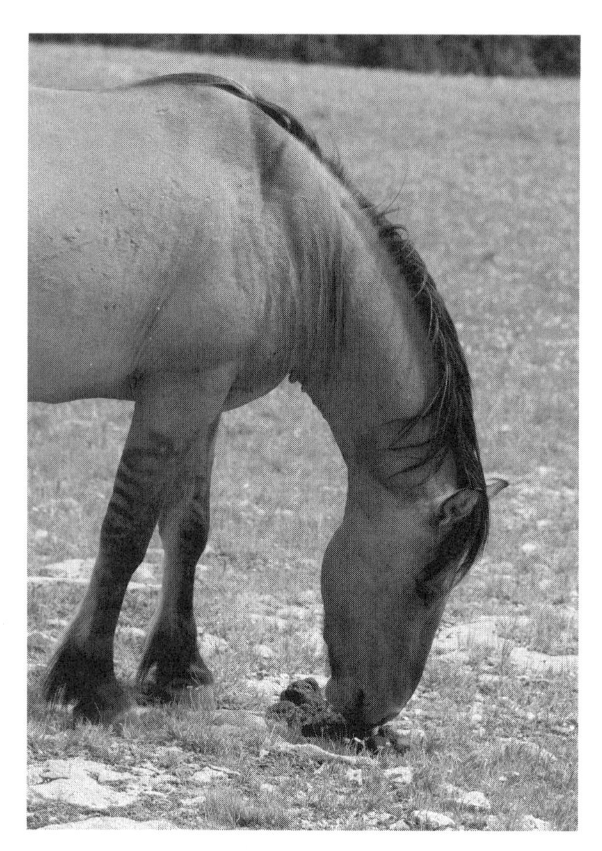

Fig. 4.1 A feral horse in the Pryor Mountains, Montana, USA, consumes feces as a food supplement during winter. Coprophagy can help equids gain more nutrients per effort when forage is minimal. Dung of feral horses in the Namib Desert, Namibia, for example, contains almost three times more fat (1.99%) than the area's dry grass (*Stipagrostis obtuse*, 0.7%) and almost twice as much protein (6.1% vs. 3.1%) (Greyling *et al.* 2007). Photo by Jason I. Ransom

Habitat Selection

Resource selection in arid-adapted equids is primarily influenced by forage availability and also quality (Henley *et al.* 2007; Sundaresan *et al.* 2007*b*). The best predictor of equid habitat use is abundance of forage (King and Gurnell 2005; St-Louis and Côté 2014). Przewalski's horses (*Equus ferus przewalskii*) show a preference for the most productive communities in their habitats in Mongolia, which are riparian communities and areas of high biomass production in the Great Gobi B Strictly Protected Area (SPA) and lowland vegetation in Hustai National Park (King and Gurnell 2005; Kaczensky *et al.* 2008). Plant community is the strongest predictor for resource selection for Przewalski's horses in the Gobi, with slope, biomass, distance to water, and elevation having only limited predictive ability (Kaczensky *et al.* 2008). In free-ranging Przewalski's horses in Kalamaili, China, plant community, forage abundance, slope, and soil types determine species distributions in spring, summer, and fall (in winter they are in an enclosure; Chen 2008). Liu (2012) reported that Przewalski's horses in Kalamaili prefer lower, gentle slopes, facing east or west, with 10–30% vegetation cover and short distance to water. Daily use of habitat for Przewalski's horses consists of movements between preferred grazing areas, watering areas, and refuges from weather or flies (Bahloul *et al.* 2001; King and Gurnell 2005; Kaczensky *et al.* 2008).

Kiangs (*E. kiang*) select for forage abundance, but more so in summer when plants are greener (St-Louis and Côté 2014). Kiangs inhabit high-altitude rangelands, specifically the steppes of the Tibetan Plateau, where harsh climatic conditions prevail and overall plant production is low (Schaller 1998). They occur from 3,500 to 6,000 m, but at slopes $<15°$ (Shah 1994, 1996; Sharma *et al.* 2004; St-Louis and Côté 2009). Kiangs in Pakistan are found along river and stream basins in patches of *Myricaria*, willow, and *Hippophe* (Rasool 1992), and occur in largest numbers in xeric south-facing basins (Harris and Miller 1995). In India, kiangs are commonly found around high-altitude lakes and along the Indus and Hanley Rivers (Shah 1996). They are nonmigratory but have large local movements (Schaller 1998).

Grevy's zebra (*E. grevyi*) select areas with higher Normalized Difference Vegetation Index values (NDVI, otherwise known as "greenness") and highest biomass of food, regardless of quality, as they need to consume more bulk compared to ruminants (Williams 1998, 2002; Hostens 2009). Grevy's zebra are found in Kenya and Ethiopia in stony semideserts (Mungall 1979; Churcher 1993; Williams 2002), and like plains zebra (*E. quagga*) tend to avoid forest, instead preferring habitat with open shrub cover and herbaceous vegetation (Williams 2002; Hostens 2009; Mwazo 2012).

Plains zebra are found in open savanna with an abundance of grass and the presence of some trees or open woodland (Hirst 1975; Mungall 1979; Mwangi and Western 1998; Traill and Bigalke 2006; Averbeck *et al.* 2009; Thaker *et al.* 2010; Arsenault and Owen-Smith 2011; Kleynhans *et al.* 2011; Macandza *et al.* 2012), and they avoid habitat where they have an increased risk of predation, especially at night (Fischhoff *et al.* 2007; Valeix *et al.* 2009; Burkepile *et al.* 2013). Plains zebra are found in midproductive grasslands in areas where there is the highest green standing crop (McNaughton 1985), selecting areas of the highest grass biomass regardless of quality (Brooks 2005; Groom and Harris 2010).

Research on plains zebra habitat use has largely focused on their role within the herbivore community (Arsenault and Owen-Smith 2011), examining competition and facilitation among grazers resulting from their high overlap in habitat use (Dekker 1997). Zebra mitigate competition with ruminants through their bulk diet and tend to consume a greater proportion of low-palatability species in the dry season (McNaughton 1985). Zebra have the highest overlap in diet and habitat use with species of a similar body weight early in the wet season, and least during the height of the wet season (Shem *et al.* 2013). Plains zebra are most often associated with wildebeest (*Connochaetes taurinus*), a species of similar size (Hirst 1975; McNaughton 1985; de Boer and Prins 1990; Kgathi and Kalikawe 1993; Dekker *et al.* 1996; Groom and Harris 2010; Hopcraft *et al.* 2010; Kleynhans *et al.* 2011). They occur together more often than would be expected by chance (de Boer and Prins 1990), particularly during the wet season (McNaughton 1985; Voeten and Prins 1999; Kleynhans *et al.* 2011). But they avoid competition by using different movement cues for migration (Hopcraft *et al.* 2010), spatial and temporal separation within a habitat (Hirst 1975; McNaughton 1985), and use of a different sward height (de Boer and Prins 1990).

Plains zebra track resource abundance across seasons (Young *et al.* 2005). At a large scale, zebra follow the long grass that grows after the rains (Bell 1971). At a finer scale, they move to maximize intake of food of sufficient quality while minimizing time spent in habitats where they may encounter predators (Hopcraft *et al.* 2010). Not all zebra herds migrate (Stelfox *et al.* 1986; Georgiadis *et al.* 2003), however, and different herds respond differently to changing conditions (Owen-Smith 2013). In the Okavango Delta of Botswana, only 55% of zebra make the 588-km round trip migration (Bartlam-Brooks *et al.* 2011). Migration allows zebra to optimize their nutrition by moving to prime grasslands during the wet season (Bartlam-Brooks *et al.* 2013), selecting higher-quality resources rather than absolute abundance of grass (Bartlam-Brooks *et al.* 2011).

Mountain zebra (*E. zebra*) are adapted to mountainous habitat in southern Africa (Joubert 1973), with the two subspecies (Cape mountain zebra and Hartmann's mountain zebra) historically being separated by a low plain with less suitable or marginal habitat (Novellie *et al.* 2002). Mountain zebra are found in grassland vegetation communities, or in shrub communities that contain a high biomass of grass (Winkler and Owen-Smith 1995). Mountain zebra select habitat types with >50% grass cover, with the best habitat found in low-lying areas that have nutrient-rich soils and a high abundance of palatable grasses, and they prefer burned over unburned veld (Watson *et al.* 2005, 2011; Watson and Chadwick 2007; Kraaij and Novellie 2010).

Asiatic wild ass (*E. hemionus*) in India (referred to as khur) inhabit the Little Rann of Kutch, Gujarat, India. The habitat is a saline desert with extremely sparse vegetation cover and very specific flora. Monsoon rains from July to September drop an average of 517.8 mm, transforming the habitat into grassy meadows with saline pools (Smielowski and Raval 1988). Asiatic wild ass in the southwest part of the Mongolian Gobi (referred to as khulan) do not appear to have a strong preference for any particular plant species but use plant communities almost relative to their availability (Kaczensky *et al.* 2008). Regression analyses of khulan abundance as a function of slope, elevation, distance to water, vegetation type, and NDVI did not reveal evidence that khulan select habitat on the basis of these factors (Ransom *et al.* 2012). Habitat use analysis suggests that khulan are well adapted to cope with unpredictable resource distribution, covering long distances in search of water and pasture (Kaczensky *et al.* 2008), and they track patches of green-up, similar to zebras and other large ungulates in semiarid and arid ecosystems of Africa and Asia. When snow cover is higher (20–550 cm), khulan in Mongolia use lower-elevation mountain slopes, and when snow partly melts, they move to the plains (Feh *et al.* 2001).

African wild ass (*E. africanus*) are found in arid and semiarid bushland and grassland (Mungall 1979; Moehlman *et al.* 2008) and are thought to be predominantly grazers. They often cover large distances to fulfill resource requirements (Mungall 1979; Moehlman 2002) and are arid adapted, moderating their temperature through efficient water use (Moehlman 2002; Grinder *et al.* 2006) and use of terrain (Mungall 1979).

By far the most numerous and widespread present-day equids are feral horses (*E. caballus*) and feral asses (*E. asinus*), also called "feral burros." Their distribution is tied to, and has been facilitated by, the spread of human culture (see chap. 8). Today, feral equids are found in a diverse range of habitats and environmental conditions in North America, South America, Europe, Asia, Africa, New Zealand, and Australia. The term "feral" describes equids that escaped or were released by humans and reverted to a wild condition. Feral equids are well adapted to harsh climatic and habitat conditions and thus are coincidently often found where human density is low.

The availability of preferred forage for feral horses is the primary determinant of habitat use in all seasons

(Salter and Hudson 1979). In Wyoming, USA, feral horses select for streamsides, bogs, or meadows that have high standing crop, and mountain sagebrush, whereas lowland sagebrush is avoided and no selection is shown for grassland or coniferous forest (Crane *et al.* 1997). Feral horses in northern sagebrush steppe exhibit no preference for a particular plant community, instead making use of the most prevalent habitat type (Ganskopp and Vavra 1986). Other studies also indicate that feral horses utilize certain meadow types on a yearlong basis in proportion to their availability (Salter 1978; Berger 1986; Crane *et al.* 1997). Likewise, feral horses in Alberta, Canada, select for habitats covering 14% of a study area while avoiding 42% of habitats, and the largest percentage of areas (44%) are used in proportion to their availability (Girard *et al.* 2013).

Like equids elsewhere, feral horses adapt seasonally to utilize what forage is available, accessible, and of suitable quality within the limits of their range (Girard *et al.* 2013). In a coniferous forest ecosystem, feral horses select for lowland grasslands across all seasons, with shrublands increasingly selected in spring and summer (Girard *et al.* 2013). During fall, they select grasslands, but at a lower level than summer, and avoid conifer forests. In Alberta winters, feral horses select for conifer patches because of warmer ambient temperatures and available forage in early winter. Shrublands are selected in spring because feral horses can access these areas as snow melts, with taller shrubs being some of the only forage available after winter and prior to spring green-up (Girard *et al.* 2013). In Wyoming, feral horses select high elevations, areas close to water, flatter slopes, south-facing aspects, low forest canopies, and open nonforested vegetation types in summer, whereas in winter they select flatter slopes, lower forest canopies, and riparian vegetation (Wockner *et al.* 2004). Conversely, in New Zealand, feral horses are more likely to occupy north-facing aspects, steeper slopes in exotic and red tussock grasslands, and flush zones in winter, but in spring and summer they occupy lower altitudes on gentler slopes (Linklater *et al.* 2000). Feral horses shift to river basin and stream valley floors in spring and higher altitudes in autumn and winter, coinciding with the beginning of foaling and mating in spring and formation of frost inversion layers in winter (Linklater *et al.* 2000).

Snow limits feral horse habitat use in some temperate zones, despite their aptitude for obtaining forage beneath snow (Salter and Hudson 1979). In winter, feral horses in the Australian Alps are limited in distribution by deep snow at higher elevations (Walter 2002), and in Wyoming, feral horses use drier habitats of low elevation with less snow cover (Wockner *et al.* 2004).

In a desert ecosystem in California, feral burros spend 60–78.8% of their time in interfluves (a region between valleys of adjacent watercourses, typically in a dissected upland) in winter (Woodward and Ohmart 1976). In spring, burro habitat use is predominantly in washes, and in summer, when temperatures approach 48°C, much of their time is spent in densely shaded pockets of vegetation (Woodward and Ohmart 1976). Feral burros spend little time foraging in arroyos and use washes almost strictly as travel paths to the Colorado River for water. After drinking, feral burros often remain briefly near water to consume the salt-encrusted soil beneath salt cedars.

Insect pests influence habitat selection by equids. In Mongolia, bloodsucking insects such as tabanids and mosquitoes influence habitat use by khulan (Feh *et al.* 2001). In the Xihu semireserve, where habitat is predominantly sand dunes and oases, Wang *et al.* (2012) found that from late spring to early fall, Przewalski's horses stay away from wetlands to avoid mosquitoes and instead graze on higher ground.

Other factors that influence habitat selection include human settlement and pastoralism. Areas closer than 5 km to human settlement are assumed unsuitable for kiang because of their intolerance of human disturbance (Sharma *et al.* 2004), and onager (*E. hemionus onager*) reside far from roads in all seasons to avoid humans (Nowzari *et al.* 2012). As human activities develop across their range, Grevy's zebra are increasingly constrained by the presence of livestock, especially as their movements often take them outside protected areas (Hostens 2009; see chap. 13). During the wet season, Grevy's zebra have less available habitat owing to the presence of livestock, which can spatially exclude them and cause zebra to eat a less nutritious diet (Kleine 2010; Kebede *et al.* 2012). Grevy's zebra avoid areas with high livestock density (Sundaresan *et al.* 2007*b*; Hostens 2009; Kebede *et al.* 2012), yet they overlap spatially and temporally: 40–50% of daylight observations were in association with livestock (Low *et al.* 2009). Lactating females are particularly observed in proximity of livestock owing to their need for access to water (Low *et al.* 2009).

Habitat use is influenced by forage quality and water availability but also differs by sex and reproductive class. For example, female onager with juveniles in Iran are found closer to water points and use plains with high-quality vegetation, whereas females without young and solitary males use intermediate-quality vegetation farther from water (Nowzari *et al.* 2013). Lactating Grevy's zebra females require access to the most nutritious vegetation, even if it is accompanied by a higher predation risk, whereas nonreproductive individuals can moderate their risk with less nutritious forage. All reproductive classes of Grevy's zebra prefer shorter and greener grass, but lactating females and bachelors use areas with greener, shorter grass and medium-dense bush more than nonlactating females or territorial males, which are found in areas of deeper grass and more open habitats (Sundaresan *et al.* 2007*b*).

Home Range and Movements

Home range size of equids is highly dependent on resource availability. Przewalski's horses in Mongolia have smaller home ranges in the mesic steppe grassland of Hustai National Park, with a kernel home range of 0.75–12 km^2 (King 2002) and 2.84–78 km^2 (Nandintsetseg and Boldgiv 2007), compared to the arid Great Gobi B SPA, with a minimum convex polygon home range of 152–826 km^2 (Kaczensky *et al.* 2008) and 290–1,357 km^2 (Lugauer 2010), and arid Kalamaili in China, with a home range of 120–660 km^2 (Chen 2008). In China, home range sizes increased over time since release (Chen 2008); this was not seen in Hustai National Park, although home ranges gradually moved farther from release sites (King 2002). In Hustai, there was no difference in home range size of harems across seasons, although they tended to be slightly larger in winter (King and Gurnell 2005; Nandintsetseg and Boldgiv 2007). While there is little spatial overlap between bands, there is no evidence of exclusive range use in Hustai (King and Gurnell 2005) or the Great Gobi B SPA (Kaczensky *et al.* 2008). Home ranges of Przewalski's horses are centered on the most productive habitat types but are only about a tenth of the size of sympatric Asiatic wild ass (Kaczensky *et al.* 2008). Home ranges of Asiatic wild asses are from 18,186 to 69,988 km^2 in southeast Gobi and 4,889 to 7,368 km^2 in southwest Gobi (Great Gobi B SPA; Kaczensky *et al.* 2011). They show an almost linear relationship between travel distance and time interval between successive locations, suggesting they are highly mobile, with mean daily traveled distance of 8,451 m (Lugauer 2010).

Resources determine space use and movements of zebras (Williams 1998): when conditions are severe, Grevy's zebra can move >80 km and have home ranges of 10,000 km^2 (Moehlman *et al.* 2013). Home range of plains zebra varies across the continent, determined by seasonal vegetation (Smuts 1975), and varies by group composition (Klingel 1969; Rubenstein 2010). In East Africa, plains zebra home ranges in Ngorongoro range from 80 km^2 to 250 km^2 in different parts of the crater (Klingel 1969). In the Serengeti, range size varies by season; 300–400 km^2 in the wet season and 400–600 km^2 in the dry season (Klingel 1969). Combined with a migration route of 100–150 km in each direction, Serengeti zebra cover at least 1,000 km^2 in a year, compared with annual home ranges of 49–566 km^2 in Kruger National Park (Smuts 1975). Plains zebra subpopulations cover areas from 28–136 km^2 in Zululand (Brooks 1982) to 1,530–1,560 km^2 in Kruger National Park (Smuts 1974).

Seasonal movements are associated with changes in food quality. Mountain zebra summer on the plateau, where they have a high selectivity of vegetation communities, and winter on hillsides, where they use a wider range of vegetation communities (Novellie *et al.* 1988; Winkler and Owen-Smith 1995). Home ranges of Cape mountain zebra breeding herds are 3–16 km^2 in Mountain Zebra National Park, and core areas change with seasonal movements (Penzhorn 1982*b*). Hartmann's zebra in southwest Africa and Namibia have winter grazing areas of 6–20 km^2 and smaller ranges in the summer; winter and summer areas at one site were separated by 120 km (Penzhorn 1988). Like migratory plains zebra that track resource abundance across seasons (Young *et al.* 2005), Hartmann's zebra do not appear to select for specific areas for a home range, as their distribution is associated with rainfall pattern (Penzhorn 1982*b*). In Ethiopia, Grevy's zebra have a larger area of optimal habitat in the dry season than in the wet season, whereas like other equids they use a smaller geographic area when forage quantity and quality are high (Kebede *et al.* 2012).

Home range size of feral horses varies considerably within and among populations (McCort 1984), and feral horse bands generally show high fidelity to home ranges (Berger 1986). In coniferous forest habitat, home ranges average 48 km^2 (Girard *et al.* 2013), which is 33 km^2 larger than that reported by Salter and Hudson (1982) in similar habitat. In Oregon, USA, home range size of feral horses averages 12–27 km^2, and in similar Great Basin habitat, their low-altitude (fall/winter) home ranges average 6.7 km^2, whereas high-altitude (summer) home ranges average 25.1 km^2 (Berger 1986). Home range of feral horse bands in the Red Desert of Wyoming, USA, varies from 73 to 303 km^2, and six bands overlapped home range areas during that study (Miller 1983). Band home ranges in New Zealand are smaller than those reported by Berger (1977) in Grand Canyon, USA, and Miller (1983) in the Red Desert but are comparable to those recorded by Berger (1986) in the Granite Range, USA, and Salter and Hudson (1982) in Alberta. Band sizes were correlated to home range size, indicating home range size is likely related to resource demand (Linklater *et al.* 2000). Information on

Fig. 4.2 Feral horses seek out a subalpine snow patch for hydration in the high elevations of the Pryor Mountains, Montana, USA. Photo by Jason I. Ransom

feral bachelor home range size is rarely reported, but Linklater *et al.* (2000) reported relatively small home ranges for bachelors compared to bands, whereas other studies have reported larger home ranges for bachelors than bands (Berger 1986). Almost no information on home range size is available for feral burros: Moehlman (1974) reported that female home ranges are 1.3–18.6 km^2 and male home ranges are 2.3–40.7 km^2.

Water

Equids are dependent on water sources in warm seasons; thus the distribution of water strongly influences distribution, space use, and movement patterns of equid species. In winter, equids are able to eat snow for hydration, allowing them to forage farther from water (Feh *et al.* 2002; Kaczensky *et al.* 2010) (fig. 4.2).

Przewalski's horses in Kalamaili drink twice a day, in the early afternoon and early evening (Wang *et al.* 2012). Free-ranging Przewalski's horses drank less frequently postrelease than when in captivity (reintroduced, mean 1.3 ± 0.5 times per day; captive, 4.8 ± 1.5; Chen 2008; Zhang 2010), suggesting that drinking frequency depends on availability. In Great Gobi B SPA, Przewalski's horses drink once daily in spring and autumn and every other day in summer, when vegetation is green and rich in water content (Ganbaatar 2003). Although water is not limiting in winter, forage availability can be problematic in deep snow.

Watering sites strongly determine summer ranges of equids, particularly wild asses. In spring, with high water content in plants, Asiatic wild asses can go without watering, but when water content drops below 50–55%, they remain within 10–19 km of water (Bannikov 1971). During summer, African wild asses subsist on dry and hard grass, but in winter there is enough moisture in forage that they can go two to three days without drinking (Mungall 1979). In Ethiopia, they are generally found within 30 km of known water sources (Moehlman 2002). Khulan will also dig holes as deep as 60 cm in dry riverbeds to access water in summer (Feh *et al.* 2002) (fig. 4.3). Kiang feed on green patches to fulfill their water requirements because they do not appear to drink frequently from running water (Schaller 1998; St-Louis and Côté 2009). Onager use areas farther from water and roads in the cold season, but in the hot season they use higher slopes and areas closer to water (Nowzari *et al.* 2012). Although more arid adapted than plains zebra (Rubenstein 2010), Grevy's zebra tend to remain <10 km from water and are not observed >18 km from nearest water source (de Leeuw *et al.* 2001; Hostens 2009; Mwazo 2012).

Frequency of drinking varies by sex, reproductive status, and species. Mountain zebra are physiologically dependent on almost daily access to drinking water (Joubert 1973; Joubert and Louw 1976). The species' range in the more desert areas of Namibia was able to expand when artificial water sources were established (Novellie *et al.* 2002). Lactating plains zebra females stay within 2 km of water, drinking every one or two days, twice as often as nonlactating females, which may move up to 4.5 km away and only drink every two to five days (Becker and Ginsberg 1990; Sundaresan *et al.* 2007*a*,*b*). In a study of a single khulan mare, she did not drink daily but rather every 1.5–2.2 days during summer and on average every 2.3–3.8 days in spring and autumn (Kaczensky *et al.* 2010). Khulan drink at least every third day during the hot season. They have maximal intervals of five to six days between drinking events during the cooler and dry months such as April, May, and September, and do not visit water points between the middle of December and the end of February because there is ample snow. Lone khulan or small groups visit water sites during all 24 hours of the day without an apparent diurnal pattern in the southwest Gobi (Kaczensky *et al.* 2010), but different diurnal patterns have been described from other areas (e.g., China; see chap. 3). Average distance to nearest water source for khulan in the southwest Gobi Mongolia is 9.0 ± 2.9 km, and there are seasonal differences (Kaczensky *et al.* 2008). Mean watering frequency of feral horses in the Australian outback is 2.67 days (range 1–4 days; Hampson *et al.* 2010). Feral horses have been recorded up to 55 km from their watering points, and some horses walk for 12 hours to water from feeding grounds (Hampson *et al.* 2010). Watering activities in a feral horse population in Oregon are most intense during the first and last periods of daylight (Ganskopp and Vavra 1986). Feral horses remain closer to water during hot, dry summer months compared to spring, when additional sources and seeps are available (Ganskopp and Vavra 1986). Feral burros drink once every 24 hours during hot, dry summers, and females with young foals drink several times per day in Death Valley, California, USA (Moehlman 1974).

There is evidence that equids compete for water with domestic livestock and sympatric wildlife. In arid and semiarid ecosystems, water is a scarce resource, and potential for competition is high. In Africa, zebra are increasingly constrained by the presence of livestock,

Fig. 4.3 A group of Asiatic wild asses digging for water in the arid southeast Gobi Desert, Mongolia. Photo by Petra Kaczensky

especially as their movements often take them outside protected areas (Hostens 2009). They avoid areas with high livestock density (Sundaresan *et al.* 2007*b*; Hostens 2009; Kebede *et al.* 2012) but are forced in proximity of livestock because of their need to access water sources (Low *et al.* 2009). In the western United States, equids compete and prevail over smaller-bodied ruminants. For example, feral horses in the Great Basin have little dietary overlap with pronghorn (*Antilocapra americana*) (Meeker 1979; McInnis and Vavra 1987), yet pronghorn spend more time in vigilance behavior when horses are present, and in pronghorn–feral horse interactions, pronghorn are excluded from water almost 50% of the time (Gooch 2014).

Diet

Equids are primarily grazers and typically select graminoids over other species when available (table 4.1). Diet of Przewalski's horses in Hustai National Park consists primarily of grasses, making up 75–83% even in winter (Siestes *et al.* 2009), and in Great Gobi B SPA, Przewalski's horses move from pastures with dry grasses to pastures with woody plants in winter (Ganbaatar 2003), likely resulting in decreased graminoids in winter diet compared to summer. Harris and Miller (1995) determined that *Stipa* spp. constitutes ~95% of kiang diet, and Schaller (1998) reported that 84.3–100% of the diet of kiangs is made up of graminoids. Kiang select for higher plant biomass and percentage of green foliage compared to random sites in the same habitats, while at the plant level, grasses are selected over forbs and shrubs, and sedges are used in proportion to their availability during all seasons (St-Louis and Côté 2014) (fig. 4.4).

Plains zebra diet consists almost entirely of grasses (Dekker 1997; Bodenstein *et al.* 2000), with occasional browse to maintain protein levels (Berry and Louw 1982). During the dry season, plains zebra spend more time foraging (Okello *et al.* 2002; Owen-Smith and Goodall 2014) and graze longer grasses than in the wet season (Stelfox *et al.* 1986; Awiti 1997; Arsenault and Owen-Smith 2011; Kleynhans *et al.* 2011), select sites with a lower level of fiber than wildebeest or cattle (Voeten and Prins 1999), and even eat dry leaves (McNaughton 1985). They eat a broader range of species during the dry season than in wet season, depending on availability (Owen-Smith *et al.* 2013). Mountain zebra are also tall-grass grazers (Novellie 1990; Novellie and Winkler 1993; Kraaij and Novellie 2010), only including browse in their diet as quality and quantity of grass decline in winter (Penzhorn 1982*a*; Novellie *et al.* 1988; Penzhorn and Novellie 1991). They select greener grass with a high leaf:stalk ratio at a height of 40–150 mm in Mountain Zebra National Park (Grobler 1983; Novellie 1990; Penzhorn and Novellie 1991; Novellie and Winkler 1993). Grevy's zebra are predominantly grazers, although browse can comprise up to 30% of their diet during drought, and there is significant variation in diet between individuals (Churcher 1993; Williams 2002; Kleine 2010).

Major constituents of feral horse diets are graminoids, while forbs and shrubs play a more limited—although sometimes still significant—role particularly in winter (Hansen 1976; Krysl *et al.* 1984*a*,*b*; Crane *et al.* 1997). Several studies report grass species remaining at or above 83% of feral horse diet in all seasons (Salter and Hudson 1979; McInnis and Vavra 1987; Smith *et al.* 1998), with sedges (*Carex* spp.) being important dietary components in bogs/meadows and stream sides (Salter and Hudson 1979; Crane *et al.* 1997). Smith *et al.* (1998) reported that feral horse diets consist of 91% grasses, 8% shrubs, and 1% forbs and unknowns. Salter and Hudson (1979) and Hubbard and Hansen (1976) indicate *Carices* as the major dietary constituent of feral horses. Grasses comprise 66–76% of summer feral burro diets in the Cottonwood Mountains of Death Valley National Park (five times more than predicted on the basis of their availability), but burros switch to shrubs (50–81%) from September to April (Ginnett and Douglas 1982). Jordan *et al.* (1979) reported that *Muhlenbergia porteri* comprises 24% of July burro diets but is a minor component of plant communities.

Equids likely do not have the capacity to detoxify plant phenolic compounds (Janis 1976; Duncan 1992); thus high amounts of phenolics in forbs and shrubs may deter equids from feeding on them. The minimum protein content required to meet metabolic functions in herbivores is around 6–7% (Owen-Smith 2002). Protein content is generally lower for grasses and sedges than forbs and shrubs, so in fall and winter, forbs and shrubs can meet metabolic requirements better than graminoids.

Seasonal differences in equid diets are due to changes in forage availability across seasons (Salter and Hudson 1979; Crane *et al.* 1997; Hanley and Hanley 1982; McInnis and Vavra 1987). Smith *et al.* (1998) reported that feral horse diet is consistent throughout the year; however, other authors have observed seasonal differences in specific grasses consumed or changes in use of forbs during spring. In Hustai National Park, Przewalski's horses select a greater variety of vegetation classes in spring and autumn, less in summer, and only few in winter (King and Gurnell 2005). Bannikov

Table 4.1. **Percentage of total diet of wild and feral equids**

Species	Location	Season	Grasses	Forbs	Browse	Forbs and Browse Combined
E. caballus (feral)	North America[1]	spring	91.03 ± 3.58 (85.6–99.0)	7.66 ± 4.41 (0–33.0)	4.50 ± 0.94 (trace–8.7)	3.0 ± 2.0 (1.0–5.0)
		summer	84.32 ± 2.76 (58.0–100)	9.18 ± 2.18 (0–27.0)	12.38 ± 3.36 (1.0–33.4)	7.4 ± 3.34 (trace–18.0)
		autumn	76.05 ± 8.55 (8.0–100)	11.28 ± 3.98 (0–37.0)	6.98 ± 2.05 (2.0–22.0)	0.50 ± 0.50 (0–1.0)
		winter	97.17 ± 5.53 (36.0–100)	16.75 ± 6.34 (0–52.0)	5.77 ± 1.65 (1.0–12.1)	11.25 ± 8.09 (0–23.0)
E. f. przewalskii	Mongolia,[2] China[3]	spring	~50 (—)			
		summer	72.0 ± 22 (50–94)	4.4 (—)	1.6 (—)	
		autumn		primary		
		winter	78.53 ± 2.07 (75.8–82.6)	15.87 ± 2.63 (10.7–primary)	5.60 ± 0.56 (4.9–6.7)	
E. asinus (feral)	India,[4] Arizona,[5] California[6]	spring	8.24 ± 5.62 (0.2–30.1)	64.0 ± 10.92 (34.5–98.0)	24.62 ± 6.67 (1.0–38.1)	
		summer	30.84 ± 9.49 (0–66.0)	19.13 ± 3.31 (11.2–37.2)	46.03 ± 9.77 (9.0–82.3)	
		autumn	12.73 ± 6.06 (2.3–23.3)	8.83 ± 0.64 (8.0–10.1)	74.00 ± 5.66 (64.2–83.8)	
		winter	17.70 ± 13.82 (0–86.0)	27.03 ± 8.09 (10.9–46.9)	50.90 ± 12.89 (0–82.9)	
E. hemionus	Mongolia[7]	spring		primary		
		summer	primary			
		autumn	35.0 (—)	65.0 (65–primary)	—	
		winter	primary			
E. kiang	India[8]	summer	92.88 ± 2.10 (86.8–98.0)	3.28 ± 0.65 (1.6–4.9)	1.02 ± 0.45 (0–2.5)	
		autumn	92.63 ± 1.56 (90.0–95.4)	3.37 ± 1.01 (1.7–5.2)	2.27 ± 1.50 (0–5.1)	
E. zebra	South Africa[9]	summer	primary			
		winter	primary		some use	
E. quagga	South Africa[10]	annual	81.92 ± 2.40 (76.0–100)	7.0 (—)	1.5 (—)	

Sources:

[1]Hansen 1976, Hubbard and Hansen 1976, Hansen and Clark 1977, Hansen *et al.* 1977, Vavra and Sneva 1978, Salter and Hudson 1979, Hanley and Hanley 1982, Krysl *et al.* 1984*b*, McInnis and Vavra 1987, Kissell *et al.* 1996, Crane *et al.* 1997, Smith *et al.* 1998, Nordquist 2011.

[2]Siestes *et al.* 2009.

[3]Meng 2007.

[4]Mishra *et al.* 2004.

[5]Jordan *et al.* 1979, Potter and Hansen 1979, Seegmiller and Ohmart 1981.

[6]Mohelman 1974, Woodward and Ohmart 1976.

[7]Lengger *et al.* 2007, Horacek *et al.* 2012.

[8]St-Louis and Côté 2014.

[9]Penzhorn 1982*a*, Novellie *et al.* 1988.

[10]Landman and Kerley 2001, Arsenault and Owen-Smith 2011.

Note: The mean value across studies per season ± standard error is shown. The range of individual study means is given in parentheses after the point estimate unless data are from only one study. The word "primary" indicates that these plants were the dominant food source. Jordan *et al.* (1979) used microhistological analysis of stomach contents; Horacek *et al.* (2012) used gas chromatography; and Arsenault and Owen-Smith (2011) used relative use of grass species. All other studies used microhistological analysis of fecal samples to determine diet constituents (box 4.1).

Fig. 4.4 Kiang persist on minimal forage at 4,550–6,000 m elevation in the Tso Kar basin of eastern Ladakh, India, where they seek out meadows with the highest forage biomass (St-Louis and Côté 2014). Photo by Antoine St-Louis

(1971) reported the diet of Asiatic wild asses across their range to include cereals and wormwood in summer, with proportions varying according to locale, conditions, and time of year, whereas in spring they feed on *Ephemerae*, such as *Poa* and *Bromus*. Ongoing work using isotope analyses (box 4.1) suggests khulan switch between the grass-dominated diet of a typical grazer in summer to a mixed grass / shrub diet in winter (M. Burnik-Sturm, personal communication).

Studies demonstrate that equids select and are also able to subsist on nongraminoids. Khur in India have been observed eating seedpods (Shah 1993) and use hooves to break up woody vegetation to obtain succulent forbs at the base of woody plants. They feed mostly on a mixed diet of forbs and grasses like *Cyperus capillaris*, *Andropogon* spp., *Dichanthium annulatum*, *Aristida alscansiouis*, and *Iseilema prostratum* (Jadhav 1979). Khur also feed on tree leaves of *Salvadora oleoides* and *Salvadora persica*, and leaves and pods of *Prosopis juliflora* (Smielowski and Raval 1988). Grevy's zebras are predominantly grazers, although browse can comprise up to 30% of their diet during drought, and there is significant variation in diet between individuals (Churcher 1993; Williams 2002; Kleine 2010). Feral horse diets in New Mexico contain a lower percentage of grasses (50%) than studies of feral horse diet in other habitat types have found (Hansen 1976), and feral burros are primarily browsers in Death Valley, California (Moehlman 1974). In a review of feral burro diets in the Mojave Desert, USA, Abella (2008) reported that a native annual forb constituted the greatest proportion (11%) of burro diets in three of seven annual diet studies. Other studies report shrubs as the major component (58–84%; Woodward and Omart 1976) of feral burro diets in all seasons except spring, when they focus on forbs (Moehlman 1974; Woodward and Omart 1976). In the Chemehuevi Mountains of California, USA, feral burros were reported to feed on 39 different plant species, with *Plantago insularis* and *Cercidium floridum* being most common. These two species combined with *Prosopis* spp. and *Pluchea sericea* make up over 50% of the annual diet of feral burros (Woodward and Ohmart 1976).

Although feral burros, like other equids, can forage on a variety of plant species, they forage selectively when given the opportunity (Smith 1969; Douglas and

BOX 4.1 TOOLS FOR DIET ANALYSES

Microhistology

The most common method to determine diet of ungulates has been microhistological analysis of feces (Hubbard and Hansen 1976; Hansen and Clark 1977; Hansen *et al.* 1977; Olsen and Hansen 1977; Salter and Hudson 1979; Hanley and Hanley 1982; Krysl *et al.* 1984*b*; McInnis and Vavra 1987; Crane *et al.* 1997; Smith *et al.* 1998). Fresh fecal samples are prepared with an alcohol solution and analyzed with a binocular compound microscope. The frequency of occurrence of individual plant species is recorded in 100 microscope fields. Histological characteristics are used to determine plant species. Size and shape of epidermal hairs, cell shapes, and crystals within epidermal cells are used to determine forbs species. Presence and position of cork cells, silica cells, silico-suberose couples/asperities, and size and shape of guard and subsidiary cells of the stomata are used to identify graminoids (Storr 1961; Sparks and Malechek 1968; Williams 1969).

Observational Studies

Observational studies are used to determine forages consumed (Krysl *et al.* 1984*a*; Mayes and Duncan 1986; Menard *et al.* 2002) by using scan sampling at a predetermined rate (e.g., every two minutes) for a discrete time period. Activity is recorded and plants consumed. The method allows the potential to directly measure nutrients consumed, but the presence of observers could alter behavior and influence forage consumption.

Stable Isotopic Analysis

Isotopic analysis has been used for several decades (Deniro and Epstein 1978; Tiezen *et al.* 1983; Ambrose and Deniro 1986). Hair has been used in to extract $\delta^{13}C$ and $\delta^{15}N$ signatures, which are used to characterize diet (Schoeninger *et al.* 1998; Macko *et al.* 1999; O'Connell and Hedges 1999; Chambers and Doucett 2008). Hair or tissue—such as muscle, blood, or feces—is collected in the field, cleaned, weighed, and placed in a mass spectrometer, which burns the sample, ionizes it, and separates the ionized particles by mass, quantifying atoms with different masses. Forage samples are also collected, dried, ground, and weighed. Once isotopic values are obtained for the tissue and forage samples, an analysis program (IsoSource) tests all possible combinations for the forage samples. If there is a possibility of the sources adding up to the tissue value, the possible combination value is stored. IsoSource gives ranges that indicate all possibilities of source contributions to the tissue sample (Nordquist 2011). With isotopic analysis, one tail hair can provide several years of chronological data; however, IsoSource is limited in the number of sources that can be put into the program.

Plant DNA Barcoding

A recently developed method is plant deoxyribonucleic acid (DNA) barcoding. When ungulates consume plants, not all of the plant DNA is digested before defecation. Thus plant DNA can be isolated from ungulate feces and used to reconstruct diet (Valentini *et al.* 2009; Kowalczyk *et al.* 2011). Total plant DNA is extracted from the sample, DNA is amplified using primers that capture all possible plant taxa, and the polymerase chain reaction–amplified DNA is sequenced. A reference database is then used to match resulting sequences to given plant taxa. Calibration studies have suggested the percent of all sequences identified to a given taxa is proportional to the relative amount of biomass in the diet (Willerslev *et al.* 2014), yet more research is needed to calibrate the technique for factors such as chloroplast density in tissues or preferential digestion. Over 1,000 sequences per sample can often be recovered, providing high-resolution information on the diversity of plants consumed.

Photo by M. Dumbleton

Hiatt 1987). Within a habitat type, mountain zebra are highly selective, using only 7 of 17 grass species at feeding sites and 26% of plants available (Grobler 1983; Penzhorn and Novellie 1991). Out of 20 plant communities available in Mountain Zebra National Park, breeding herds preferred 9 in summer, 7 in winter, and 3 in both seasons; bachelors had no preference (Penzhorn and Novellie 1991). In Kenya, plains zebra eat 6.2–7 species of grasses on average (Stewart and Stewart 1970), with zebra in South Africa eating 19 (Landman and Kerley 2001) to 27 (Boyers 2011) species, depending on the area, and displaying a seasonal change in species preference: 6 of 13 species preferred in the early dry season and 9 of 17 in the late dry season (Owen-Smith *et al.* 2013).

Conclusion

Resource selection by equids is driven mostly by forage availability. Studies have shown that Przewalski's horses select the most productive communities, kiangs select for forage abundance, Grevy's zebra and plains zebra track resource abundance across seasons selecting areas with high grass biomass, and mountain zebra select habitat with >50% grass cover. Some equids, such as Asiatic wild ass and some feral horse herds, use habitat relative to its availability. Feral horses and feral burros are found on almost all continents, their distribution assisted by the spread of human culture, but today they are mostly in habitats of low human value and use. All equids are dependent on water in warm months and can eat snow for hydration in winter. Wild asses in particular are highly arid adapted and survive several days without drinking water, instead obtaining hydration from plants. Equids can dig for water in dry seasons, and dig through snow for forage in winter if necessary. Pests influence habitat selection and daily habitat use in most equids, particularly Przewalski's horses, khulan, and some herds of feral horses. Equids are ubiquitous in their avoidance of people and human settlement, and most equid species compete with pastoralism or livestock for access to water and forage.

Equids are primarily grazers and select graminoids over other species when available, relying on browse as quality and quantity of grass declines. The exception is wild asses and feral burros, for which browsing is common. Equids are bulk or roughage feeders, selecting quantity over quality. Their digestion is less efficient than ruminants; they can re-ingest semidigested food via coprophagy. Equids consume large quantities of low-quality forage, and they survive and thrive in some of the harshest environmental conditions and extreme terrestrial habitats on earth. The breadth of ecosystems they inhabit is truly astounding.

ACKNOWLEDGMENTS

The authors thank the US Geological Survey for funding research on wild horses and burros in the western United States, and the Bureau of Land Management for their commitment to using science to improve management of wild horses and burros. They are grateful to P. Gogan and E. Beever for providing useful comments on early drafts of the chapter, and thank the anonymous reviewers and editors for helpful review comments. The authors extend their appreciation to J. Ransom and P. Kaczensky for organizing this book and for inviting our contributions on habitat and diet of equids.

REFERENCES

Abella, S.R. 2008. A systematic review of wild burro grazing effects on Mojave Desert vegetation, USA. Environmental Management 41:809–819.

Ambrose, S.H., and M.J. Deniro. 1986. The isotopic ecology of east African mammals. Oecologia 69:395–406.

Arsenault, R., and N. Owen-Smith. 2011. Competition and coexistence among short-grass grazers in the Hluhluwe-iMfolozi Park, South Africa. Canadian Journal of Zoology 89:900–907.

Averbeck, C., A. Apio, M. Plath, and T. Wronski. 2009. Environmental parameters and anthropogenic effects predicting the spatial distribution of wild ungulates in the Akagera savannah ecosystem. African Journal of Ecology 47:756–766.

Awiti, A.O. 1997. Habitat use patterns by Burchell's zebra (*Equus burchelli bohmi* Matschie) in Nairobi National Park, Kenya. Thesis, Moi University, Eldoret, Kenya.

Bahloul, K., O.B. Pereladova, N. Soldatova, G. Fisenko, E. Sidorenko, and A.J. Sempéré. 2001. Social organization and dispersion of introduced kulans (*Equus hemionus kulan*) and Przewalski horses (*Equus przewalskii*) in the Bukhara Reserve, Uzbekistan. Journal of Arid Environments 47:309–323.

Bannikov, A.G. 1971. The Asiatic wild ass: Neglected relative of the horse. Animals 13:580–585.

Bartlam-Brooks, H.L.A., M.C. Bonyongo, and S. Harris. 2011. Will reconnecting ecosystems allow long-distance mammal migrations to resume? A case study of a zebra *Equus burchelli* migration in Botswana. Oryx 45:210–216.

Bartlam-Brooks, H.L.A., P.S.A. Beck, G. Bohrer, and S. Harris. 2013. In search of greener pastures: Using satellite images to predict the effects of environmental change on zebra migration. Journal of Geophysical Research: Biogeosciences 118:1427–1437.

Becker, C., and J.R. Ginsberg. 1990. Mother-infant behaviour of wild Grevy's zebra: Adaptations for survival in semidesert East Africa. Animal Behaviour 40:1111–1118.

Bell, R. 1971. A grazing ecosystem in the Serengeti. Scientific American 225:86–93.

Berger, J. 1977. Organizational systems and dominance in feral horses in the Grand Canyon. Behavioral Ecology and Sociobiology 2:131–146.

Berger, J. 1986. Wild horses of the Great Basin: Social competition and population size. University of Chicago Press, Chicago, USA.
Berry, H.H., and G.N. Louw. 1982. Nutritional balance between grassland productivity and large herbivore demand in the Etosha National Park. Madoqua 13:141–150.
Bodenstein, V., H.H. Meissner, and W. van Hoven. 2000. Food selection by Burchell's zebra and blue wildebeest in the Timbavati area of the Northern Province Lowveld. South African Journal of Wildlife Research 30:63–72.
Boyers, M. 2011. Do zebra (*Equus quagga*) select for greener grass within the foraging area? University of the Witwatersrand, Johannesburg, South Africa.
Brooks, C.J. 2005. The foraging behaviour of Burchell's Zebra (*Equus burchelli antiquorum*). University of Bristol, Bristol, UK.
Brooks, P.M. 1982. Zebra, wildebeest and buffalo subpopulation areas in the Hluhluwe-Corridor-Umfolozi Complex, Zululand, and their application in management. South African Journal of Wildlife Research 12:140–146.
Burkepile, D.E., C.E. Burns, C.J. Tambling, E. Amendola, G.M. Buis, N. Govender, V. Nelson, D.I. Thompson, A.D. Zinn, and M.D. Smith. 2013. Habitat selection by large herbivores in a southern African savanna: The relative roles of bottom-up and top-down forces. Ecosphere 4:art139.
Chambers, C.L., and R.R. Doucett. 2008. Diet of the Mogollon vole as indicated by stable-isotope analysis (delta C-13 and delta N-15). Western North American Naturalist 68:153–160.
Chen, J. 2008. Utilization of food, water and space by released Przewalski horse (*Equus przewalski*) with reference to survival strategies analysis. [In Chinese.] Thesis, Beijing Forestry University, Beijing, China.
Churcher, C.S. 1993. *Equus grevyi*. Mammalian Species 453:1–9.
Crane, K.K., M.A. Smith, and D. Reynolds. 1997. Habitat selection patterns of feral horses in southcentral Wyoming. Journal of Range Management 35:152–158.
de Boer, W.F., and H.H.T. Prins. 1990. Large herbivores that strive mightily but eat and drink as friends. Oecologia 82:264–274.
de Leeuw, J., M.N. Waweru, O.O. Okello, M. Maloba, P. Nguru, M.Y. Said, H.M. Aligula, I.M.A. Heitkönig, and R.S. Reid. 2001. Distribution and diversity of wildlife in northern Kenya in relation to livestock and permanent water points. Biological Conservation 100:297–306.
Dekker, B. 1997. Calculating stocking rates for game ranches: Substitution ratios for use in the Mopani Veld. African Journal of Range and Forage Science 14:62–67.
Dekker, B., N. Van Rooyen, and J. Du P Bothma. 1996. Habitat partitioning by ungulates on a game ranch in the Mopani veld. South African Journal of Wildlife Research 26:117–122.
Deniro, M.J., and S. Epstein. 1978. Influence of diet on the distribution of carbon isotopes in animals. Geochimica et Cosmoschimica Acta 42:495–506.
Douglas, C.L., and H.D. Hiatt. 1987. Food habits of feral burros in Death Valley, California. No. 006/46. Cooperative National Park Resources Studies Unit, University of Nevada, Las Vegas, USA.
Duncan, P. 1992. Zebras, asses, and horses: An action plan for the conservation of wild equids. International Union for Conservation of Nature, Gland, Switzerland.
Feh, C., B. Munkhtuya, S. Enkhbold, and T. Sukhbaatar. 2001. Ecology and social structure of the Gobi khulan *Equus hemionus* subsp. in the Gobi B National Park, Mongolia. Biological Conservation 101:51–61.
Feh, C., N. Shah, M. Rowen, R. Reading, and S.P. Goyal. 2002. Status and action plan for the Asiatic wild ass (*Equus hemionus*). Pages 62–71 *in* P.D. Moehlman (Ed.), Equids: Zebras, asses and horses. Status survey and conservation action plan. International Union for Conservation of Nature, Gland, Switzerland.
Fischhoff, I.R., S.R. Sundaresan, J.E. Cordingley, and D.I. Rubenstein. 2007. Habitat use and movements of plains zebra (*Equus burchelli*) in response to predation danger from lions. Behavioral Ecology 18:725–729.
Ganbaatar, O. 2003. Takhi's (*Equus przewalskii* Polj., 1883) home range and water use. Thesis, National University of Mongolia, Ulaanbaatar, Mongolia.
Ganskopp, D., and M. Vavra. 1986. Habitat use of feral horses in the northern sagebrush steppe. Journal of Range Management 39:207–212.
Georgiadis, N.J., M.A. Hack, and K. Turpin. 2003. The influence of rainfall on zebra population dynamics: Implications for management. Journal of Applied Ecology 40:125–136.
Ginnett, T.F., and C.L. Douglas. 1982. Food habits of feral burros and desert bighorn sheep in Death Valley National Monument. Desert Bighorn Council Transactions 26:81–87.
Girard, T.L., E.W. Bork, S.E. Nielsen, and M.J. Alexander. 2013. Seasonal variation in habitat selection by free-ranging feral horses within Alberta's forest reserve. Rangeland Ecology and Management 66:428–437.
Gooch, A.M. 2014. The impacts of feral horses on the use of water by pronghorn on the Sheldon National Wildlife Refuge, Nevada. Thesis, Brigham Young University, Provo, UT, USA.
Greyling, T., S.S. Cilliers, and H. VanHamburg. 2007. Vegetation studies of feral horse habitat in the Namib Naukuluft Park, Namibia. South African Journal of Botany 73:328.
Grinder, M.I., P.R. Krausman, and R.S. Hoffmann. 2006. *Equus asinus*. Mammalian Species 794:1–9.
Grobler, J.H. 1983. Feeding habits of the Cape mountain zebra *Equus zebra zebra* Linn. 1758. Koedoe 26:159–168.
Groom, R., and S. Harris. 2010. Factors affecting the distribution patterns of zebra and wildebeest in a resource-stressed environment. African Journal of Ecology 48:159–168.
Groves, C. 1974. Horses, asses, and zebras in the wild. Ralph Curtis, Hollywood, FL, USA.
Hampson, B.A., M.A. de Laat, P.C. Mills, and C.C. Pollitt. 2010. Distances traveled by feral horses in "outback" Australia. Equine Veterinary Journal 42:582–586.
Hanley, T.A., and K.A. Hanley. 1982. Food resource partitioning by sympatric ungulates on Great Basin rangeland. Journal of Range Management 35:152–158.
Hansen, R.M. 1976. Foods of free-roaming horses in southern New Mexico. Journal of Range Management 29:347.
Hansen, R.M., and R.C. Clark. 1977. Foods of elk and other ungulates at low elevations in northwestern Colorado. Journal of Wildlife Management 41:76–80.
Hansen, R.M., R.C. Clark, and W. Lawhorn. 1977. Foods of wild horses, deer, and cattle in the Douglas Mountain area, Colorado. Journal of Range Management 30:116–118.

Harris, R., and D. Miller. 1995. Overlap in summer habitats and diets of Tibetan Plateau ungulates. Mammalia 59:197–212.

Henley, S.R., D. Ward, and I. Schmidt. 2007. Habitat selection by two desert-adapted ungulates. Journal of Arid Environments 70:39–48.

Hirst, S.M. 1975. Ungulate-habitat relationships in a South African woodland/savanna ecosystem. Wildlife Monographs 44:3–60.

Hopcraft, J.G.C., J.M. Morales, H.L. Beyer, D.T. Haydon, M. Borner, A.R.E. Sinclair, and H. Olff. 2010. Serengeti wildebeest and zebra migrations are affected differently by food resources and predation risks. Pages 143–166 *in* Ecological implications of food and predation risk for herbivores in the Serengeti. Rijksuniversiteit Groningen, Groningen, Netherlands.

Horacek, M., M. Burnik Sturm, and P. Kaczensky. 2012. First stable isotope analysis of Asiatic wild ass tail hair from the Mongolian Gobi. Exploration into the Biological Resources of Mongolia 2012:85–92.

Hostens, E. 2009. Modelling the migration of Grevy's zebra in function of habitat type using remote sensing. Thesis, Universitit Gent, Gent, Belgium.

Hubbard, R.E., and R.M. Hansen. 1976. Diets of wild horses, cattle, and mule deer in the Piceance Basin, Colorado. Journal of Range Management 29:389–392.

Jadhav, S.A. 1979. Wildlife management plan for wild ass sanctuary, Dhrangadhra, Gujurat State, India. 1979–1980 and 1983–1984. Unpublished report. Dhrangadhra, India.

Janis, C. 1976. The evolutionary strategy of the Equidae and the origins of rumen and caecal digestion. Evolution 30:757–774.

Jordan, J.W., G.A. Ruffner, S.W. Carothers, and A.M. Phillips III. 1979. Summer diets of feral burros (*Equus asinus*) in Grand Canyon National Park, Arizona. Pages 15–22 *in* R.H. Denniston (Ed.), Symposium on the ecology and behavior of wild and feral equids. University of Wyoming, Laramie, USA.

Joubert, E. 1973. Habitat preference, distribution and status of the Hartmann zebra *Equus zebra hartmannae* in South West Africa. Madoqua 1:5–15.

Joubert, E., and G.N. Louw. 1976. Preliminary observations on the digestive and renal efficiency of Hartmann's zebra *Equus zebra hartmannae*. Madoqua 10:119–121.

Kaczensky, P., O. Ganbaatar, H. Von Wehrden, and C. Walzer. 2008. Resource selection by sympatric wild equids in the Mongolian Gobi. Journal of Applied Ecology 45:1762–1769.

Kaczensky, P., V. Dreslez, D. Vetter, H. Otgonbayar, and C. Walzer. 2010. Water use of Asiatic wild asses in the Mongolian Gobi. Exploration into the Biological Resources of Mongolia 2010:291–298.

Kaczensky, P., O. Ganbaatar, N. Altansukh, N. Enkhsaikhan, C. Stauffer, and C. Walzer. 2011. The danger of having all your eggs in one basket—Winter crash of the re-introduced Przewalski's horses in the Mongolian Gobi. PLoS ONE 6:e28057.

Kebede, F., A. Bekele, P.D. Moehlman, and P.H. Evangelista. 2012. Endangered Grevy's zebra in the Alledeghi Wildlife Reserve, Ethiopia: Species distribution modeling for the determination of optimum habitat. Endangered Species Research 17:237–244.

Kgathi, D.K., and M.C. Kalikawe. 1993. Seasonal distribution of zebra and wildebeest in Makgadikgadi Pans Game Reserve, Botswana. African Journal of Ecology 31:210–219.

King, S.R.B. 2002. Home range and habitat use of free-ranging Przewalski horses at Hustai National Park, Mongolia. Applied Animal Behaviour Science 78:103–113.

King, S.R.B., and J. Gurnell. 2005. Habitat use and spatial dynamics of takhi introduced to Hustai National Park, Mongolia. Biological Conservation 124:277–290.

Kissell, R.E., L.R. Irby, and R.J. Mackie. 1996. Competitive interactions among bighorn sheep, feral horses, and mule deer in Bighorn Canyon National Recreation Area and Pryor Mountain Wild Horse Range. Completion Report CA-1268-1-9017. Montana State University, Bozeman, USA.

Kleine, L. 2010. Stable isotope ecology of the endangered Grevy's zebra (*Equus grevyi*) in Laikipia, Kenya. Thesis, University of Puget Sound, Puget Sound, WA, USA

Kleynhans, E.J., A.E. Jolles, M. Bos, and H. Olff. 2011. Resource partitioning along multiple niche dimensions in differently sized African savanna grazers. Oikos 120:591–600.

Klingel, H. 1969. The social organisation and population ecology of the plains zebra (*Equus quagga*). Zoologica Africana 4:249–263.

Kowalczyk, R., P. Taberlet, E. Coissac, A. Valentini, C. Miquel, T. Kamiński and J. M. Wójcik. 2011. Influence of management practices on large herbivore diet—Case of European bison in Białowieża Primeval Forest, Poland. Forest Ecology and Management 2614:821–828.

Kraaij, T., and P.A. Novellie. 2010. Habitat selection by large herbivores in relation to fire at the Bontebok National Park (1974–2009): The effects of management changes. African Journal of Range and Forage Science 27:21–27.

Krysl, L.J., B.F. Sowell, M.E. Hubbert, G.E. Plumb, T.I. Jewett, M.A. Smith, and J.W. Waggoner. 1984*a*. Horses and cattle grazing in the Wyoming Red Desert. 2. Dietary quality. Journal of Range Management 37:252–256.

Krysl, L.J., M.E. Hubbert, B.F. Sowell, G.E. Plumb, T.K. Jewett, M.A. Smith, and J.W. Waggoner. 1984*b*. Horses and cattle grazing in the Wyoming Red Desert. 1. Food habits and dietary overlap. Journal of Range Management 37:72–76.

Landman, M., and G.I.H. Kerley. 2001. Dietary shifts: Do grazers become browsers in the Thicket Biome? Koedoe 44:31–36.

Lengger, J., F. Tataruch, and C. Walzer. 2007. Feeding ecology of Asiatic wild ass *Equus hemionus*. Exploration into the Biological Resources of Mongolia 2007:93–97.

Linklater, W.L., E.Z. Cameron, K.J. Stafford, and C.J. Veltman. 2000. Social and spatial structure and range use by Kaimanawa wild horses (*Equus caballus*: Equidae). New Zealand Journal of Ecology 24:139–152.

Liu, Z. 2013. Study on habitat selectivities and community protection awareness of the reintroduced *Equus przewalskii* in Mt. Kalamaili Ungulate Nature Reserve. [In Chinese.] Thesis, Xinjiang University, China.

Low, B., S.R. Sundaresan, I. R. Fischhoff, and D.I. Rubenstein. 2009. Partnering with local communities to identify conservation priorities for endangered Grevy's zebra. Biological Conservation 142:1548–1555.

Lugauer, B. 2010. Differences of movement pattern between Asiatic wild ass (*Equus hemionus*) and Przewalski's horse (*Equus ferus przewalskii*). Thesis, University of Vienna, Vienna, Austria.

Macandza, V.A., N. Owen-Smith, and J.W. Cain III. 2012. Habitat and resource partitioning between abundant and relatively rare grazing ungulates. Journal of Zoology 287:175–185.

Macko, S.A., M.H. Engel, V. Andrusevich, G. Lubec, T.C. O'Connell, and R.E.M. Hedges. 1999. Documenting the diet in ancient human populations through stable isotope analysis of hair. Philosophical Transactions of the Royal Society of London B: Biology Sciences 354:65–76.

Mayes, E., and P. Duncan. 1986. Temporal patterns of feeding behavior in free-ranging horses. Behaviour 96:105–129.

McCort, W.D. 1984. Behavior of feral horses and ponies. Journal of Animal Science 58:493–499.

McInnis, M.L., and M. Vavra. 1987. Dietary relationships among feral horses, cattle, and pronghorn in southeastern Oregon. Journal of Range Management 40:60–66.

McNaughton, S.J. 1985. Ecology of a grazing ecosystem: The Serengeti. Ecological Monographs 55:259.

Meeker, J.O. 1979. Interactions between pronghorn antelope and feral horses in northwestern Nevada. Thesis, University of Nevada, Reno, USA.

Menard, C., P. Duncan, G. Fleurance, J. Georges, and M. Lila. 2002. Comparative foraging and nutrition of horses and cattle in European wetlands. Journal of Applied Ecology 39:120–133.

Meng, Y. 2007. Studies on the food plants, food preference and foraging strategy of released Przewalski horse. [In Chinese.] Thesis, Beijing Forestry University, Beijing, China.

Miller, R. 1983. Seasonal movements and home ranges of feral horse bands in Wyoming's Red Desert. Journal of Range Management 36:199–201.

Mishra, C., S.E. Van Wieren, P. Ketner, I.M.A. Heitkönig, and H.H.T. Prins. 2004. Competition between domestic livestock and wild bharal *Pseudois nayaur* in the Indian Trans-Himalaya. Journal of Applied Ecology 41:344–354.

Moehlman, P.D. 1974. Behavior and ecology of feral asses (*Equus asinus*). Thesis, University of Wisconsin–Madison, Madison, USA.

Moehlman, P.D. 2002. Status and action plan for the African wild ass (*Equus africanus*). Pages 2–10 *in* P.D. Moehlman (Ed.), Equids: Zebras, asses and horses. Status survey and conservation action plan. International Union for Conservation of Nature, Gland, Switzerland.

Moehlman, P.D., H. Yohannes, R. Teclai, and F. Kebede. 2008. *Equus africanus*. IUCN Red List of Threatened Species. Version 2013.2. International Union for Conservation of Nature, Gland, Switzerland. Available at www.iucnredlist.org.

Moehlman, P.D., D.I. Rubenstein, and F. Kebede. 2013. *Equus grevyi*. IUCN Red List of Threatened Species. Version 2013.2. International Union for Conservation of Nature, Gland, Switzerland. Available at www.iucnredlist.org.

Mungall, E.C. 1979. Habitat preferences of Africa's recent *Equidae*, with special reference to the extinct quagga. Pages 159–172 *in* R.H. Denniston (Ed.), Symposium on the ecology and behavior of wild and feral equids. University of Wyoming, Laramie, USA.

Mwangi, E.M., and D. Western. 1998. Habitat selection by large herbivores in Lake Nakuru National Park, Kenya. Biodiversity and Conservation 7:1–8.

Mwazo, A.G. 2012. Distribution and vegetation association of Grevy's zebra (*Equus grevyi*) in Tsavo East National Park and the surrounding ranchlands. Kenyatta University, Nairobi, Kenya.

Nandintsetseg, D., and B. Boldgiv, B. 2007. The home range and habitat use of takhi (*Equus przewalski* Poljakov, 1881) reintroduced to the Hustai National Park. Pages 1–30 *in* Hustai National Park Trust annual report. Hustai National Park Trust, Ulaanbaatar, Mongolia.

Nordquist, M.K. 2011. Stable isotope diet reconstruction of feral horses (*Equus caballus*) on the Sheldon National Wildlife Refuge, Nevada, USA. Thesis, Brigham Young University, Provo, Utah, USA.

Novellie, P. 1990. Habitat use by indigenous grazing ungulates in relation to sward structure and veld condition. Journal of the Grassland Society of Southern Africa 7:16–23.

Novellie, P., and A. Winkler. 1993. A simple index of habitat suitability for Cape mountain zebras. Koedoe 36:53–59.

Novellie, P., L.J. Fourie, O.B. Kok, and M.C. van der Westhuizen. 1988. Factors affecting the seasonal movements of Cape mountain zebras in the Mountain Zebra National Park. South African Journal of Zoology 23:13–19.

Novellie, P., M. Lindeque, P. Lindeque, P.H. Lloyd, and J. Koen. 2002. Status and action plan for the mountain zebra (*Equus zebra*). Pages 28–42 *in* P.D. Moehlman (Ed.), Equids: Zebras, asses and horses. Status survey and conservation action plan. International Union for Conservation of Nature, Gland, Switzerland.

Nowzari, H.M., M. Hemami, M. Karami, M.M.K. Zarkesh, B. Riazi, and D.I Rubenstein. 2012. Habitat associations of Persian wild ass (*Equus hemionus onager*) in Qatrouyeh National Park, Iran. Presented at the 4th International Wildlife Management Congress, 9–12 July 2012, Durban, South Africa.

Nowzari, H.M., M. Hemanmi, M.M. Karami, B. Kheirkhah Zarkesh, H. Riazi, and D.I. Rubenstein. 2013. Habitat use by the Persian onager, *Equus hemionus onager* (Perissodactyla: Equidae) in Qatrouyeh National Park, Fars, Iran. Journal of Natural History 47:2795–2814.

O'Connell, T.C., and R.E.M. Hedges. 1999. Investigation in the effect of diet on modern human hair isotopic values. American Journal of Physical Anthropology 108:409–425.

Okello, M.M., R.E.L. Wishitemi, and F. Muhoro. 2002. Forage intake rates and foraging efficiency of free-ranging zebra and impala. South African Journal of Wildlife Research 32:93–100.

Olsen, F.W., and R.M. Hansen. 1977. Food relations of wild free-roaming horses to livestock and big game, Red Desert, Wyoming. Journal of Range Management 30:17–20.

Orlando, L., A. Ginolhac, G. Zhang, D. Froese, A. Albrechtsen, *et al.* 2013. Recalibrating *Equus* evolution using the genome sequence of an early Middle Pleistocene horse. Nature 499:74–78.

Owen-Smith, N. 2002. Adaptive herbivore ecology. Cambridge University Press, New York, USA.

Owen-Smith, N. 2013. Daily movement responses by African savanna ungulates as an indicator of seasonal and annual food stress. Wildlife Research 40:232.

Owen-Smith, N., and V. Goodall. 2014. Coping with savanna seasonality: Comparative daily activity patterns of African ungulates as revealed by GPS telemetry. Journal of Zoology 293:181–191.

Owen-Smith, N., E. Le Roux, and V. Macandza. 2013. Are relatively rare antelope narrowly selective feeders? A sable antelope and zebra comparison. Journal of Zoology 291:163–170.

Penzhorn, B.L. 1982*a*. Habitat selection by Cape mountain zebras in the Mountain Zebra National Park. South African Journal of Wildlife Research 12:48–54.

Penzhorn, B.L. 1982*b*. Home range sizes of Cape mountain zebras *Equus zebra zebra* in the Mountain Zebra National Park. Koedoe 25:103–108.

Penzhorn, B.L. 1988. *Equus zebra*. Mammalian Species 314:1–7.

Penzhorn, B.L., and P.A. Novellie. 1991. Some behavioural traits of Cape mountain zebras (*Equus zebra zebra*) and their implications for the management of a small conservation area. Applied Animal Behaviour Science 29:293–299.

Potter, R.L., and R.M. Hansen. 1979. Feral burro food habits and habitat relations, Grand Canyon National Park, Arizona. Pages 143–157 *in* R.H. Denniston (Ed.), Symposium on the ecology and behavior of wild and feral equids. University of Wyoming, Laramie, USA.

Ransom, J.I., P. Kaczensky, B.C. Lubow, O. Ganbaatar, and N. Altansukh. 2012. A collaborative approach for estimating terrestrial wildlife abundance. Biological Conservation 153:219–226.

Rasool, G. 1992. Tibetan wild ass—Verging on extinction. Tiger Paper 19:16–17.

Rubenstein, D.I. 2010. Ecology, social behavior, and conservation in zebras. Advances in the Study of Behavior 42:231–258.

Salter, R.E. 1978. Ecology of feral horses in western Alberta. Thesis, University of Alberta, Edmonton, Canada.

Salter, R.E., and R.J. Hudson. 1979. Feeding ecology of feral horses in western Alberta. Journal of Range Management 32:221–225.

Salter, R.E., and R.J. Hudson. 1982. Social organization of feral horse in western Canada. Applied Animal Ethology 8:207.

Schaller, G.B. 1998. Wildlife of the Tibetan Steppe. University of Chicago Press, Chicago, USA.

Schoeninger, M.J., U.T. Iwaniec, and L.T. Nash. 1998. Ecological attributes recorded in stable isotope ratios of arboreal prosimian hair. Oecologia 113:222–230.

Seegmiller, R.F., and R. Ohmart. 1981. Ecological relationships of feral burros and desert bighorn sheep. Wildlife Monographs 78:3–58.

Shah, N.V. 1993. The ecology of the wild ass (*E. h. khur*) in the Little Rann of Kutch, Gujarat, India. Thesis, University of Baroda, Gujarat, India.

Shah, N.V. 1994. Status survey of southern kiang (*Equus kiang polyodon*) in Sikkim, India. Report. University of Baroda, Baroda, Gujasrat, India, and Zoological Society for the Conservation of Species and Populations, Munich, Germany.

Shah, N.V. 1996. Status and distribution of western kiang (*Equus kiang kiang*) in Changthang Plateau, Ladakh, India. Report. University of Baroda, Baroda, Gujasrat, India, and Gujarat Nature Conservation Society, Gujarat, India.

Sharma, B.D., J. Clevers, R. De Graaf, and N.R. Chapagain. 2004. Mapping *Equus kiang* (Tibetan wild ass) habitat in Surkhang, Upper Mustang, Nepal. Mountain Research and Development 24:149–156.

Shem, M.M., S.E. Van Wieren, I.M.A. Heitkönig, and H.H.T. Prins. 2013. Seasonal resource use and niche breadth in an assemblage of coexisting grazers in a fenced park. Open Journal of Ecology 3:383–388.

Siestes, D.J., G. Faupin, W.F. de Boer, C.B. de Jong, R.H.G. Henkens, D. Usukhjargal, and T. Batbaatar. 2009. Resource partitioning between large herbivores in Hustai National Park, Mongolia. Mammalian Biology 74:381–393.

Smielowski, J.M., and P.P. Raval. 1988. The Indian wild ass—Wild and captive populations. Oryx 22:85–88.

Smith, C. 1969. Burro problems in the Southwest. Desert Bighorn Council Transactions 13:91–97.

Smith, C., R. Valdez, J.L. Holechek, P.J. Zwank, and M. Cardenas. 1998. Diets of native and non-native ungulates in southcentral New Mexico. Southwestern Naturalist 43:163–169.

Smuts, G.L. 1974. Game movements in the Kruger National Park and their relationship to the segregation of subpopulations and the allocation of culling compartments. Journal of the South African Wildlife Management Association 4:51–58.

Smuts, G.L. 1975. Home range sizes for Burchell's zebra *Equus burchelli antiquorum* from the Kruger National Park. Koedoe 18:139–146.

Sparks, D.R., and J.C. Malechek. 1968. Estimating percentage dry weights in diets using a microscope technique. Journal of Rangeland Management 21:264–265.

Stelfox, J.G., D.G. Peden, H. Epp, R.J. Hudson, S.W. Mbugua, J.L. Agatsiva, and C.L. Amuyunzu. 1986. Herbivore dynamics in southern Narok, Kenya. Journal of Wildlife Management 50:339.

Stewart, D.R.M., and J. Stewart. 1970. Food preference data by faecal analysis for African plains ungulates. Zoologica Africana 15:115–129.

St-Louis, A., and S.D. Côté. 2009. *Equus kiang* (Perrisodactyla: Equidae). Mammalian Species 835:1–11.

St-Louis, A., and S.D. Côté. 2014. Resource selection in a high-altitude rangeland equid, the kiang (*Equus kiang*): Influence of forage abundance and quality at multiple scales. Canadian Journal of Zoology 92:239–249.

Storr, G.M. 1961. Microscopic analysis of faeces: A technique for ascertaining the diet of herbivorous mammals. Australian Journal of Biological Science 14:157–164.

Sundaresan, S.R., I.R. Fischhoff, and D.I. Rubenstein. 2007*a*. Male harassment influences female movements and associations in Grevy's zebra (*Equus grevyi*). Behavioral Ecology 18:860–865.

Sundaresan, S.R., I.R. Fischhoff, H.M. Hartung, P. Akilong, and D.I. Rubenstein. 2007*b*. Habitat choice of Grevy's zebras (*Equus grevyi*) in Laikipia, Kenya. African Journal of Ecology 46:359–364.

Thaker, M., A.T. Vanak, C.R. Owen, M.B. Ogden, and R. Slotow. 2010. Group dynamics of zebra and wildebeest in a woodland savanna: Effects of predation risk and habitat density. PLoS ONE 5:e12758.

Tiezen, L.L., T.W. Boutton, K.G. Tesdahl, and N.A. Slade. 1983. Fractination and turnover of stable carbon isotopes in animal tissues: Implications for delta carbon-13 analysis of diet. Oecologia 57:32–37.

Traill, L.W., and R.C. Bigalke. 2006. A presence-only habitat suitability model for large grazing African ungulates and its utility for wildlife management. African Journal of Ecology 45:347–354.

Valeix, M., A.J. Loveridge, S. Chamaillé-Jammes, Z. Davidson, F. Murindagomo, H. Fritz, and D.W. Macdonald. 2009. Behavioral adjustments of African herbivores to predation risk by lions: Spatiotemporal variations influence habitat use. Ecology 90:23–30.

Valentini, A., F. Pompanon, and P. Taberlet. 2009. DNA barcoding for ecologists. Trends in Ecology and Evolution 242:110–117.

Van Soest, P.J. 1994. Nutritional ecology of the ruminant, 2nd ed. Cornell University Press, Ithaca, NY, USA.

Vavra, M., and F. Sneva. 1978. Seasonal diets of five ungulates grazing the cold desert biome. Pages 435–437 *in* D.N. Hyder (Ed.), Proceedings of the first international rangeland congress. Denver, CO, USA.

Voeten, M.M., and H.H.T. Prins. 1999. Resource partitioning between sympatric wild and domestic herbivores in the Tarangire region of Tanzania. Oecologia 120:287–294.

Walter, M. 2002. The population ecology of wild horses in the Australian Alps. Thesis, University of Canberra, ACT, Australia.

Wang, H., Z.Q. He, H. Wang, and Y. Niu. 2012. Study on survival status of reintroduced equus Przewalskii in Dunhuang West Lake National Nature Reserve. [In Chinese.] Journal of Gansu Forestry Science and Technolgy 37:45–46, 65.

Watson, L.H., and P. Chadwick. 2007. Management of Cape mountain zebra in the Kammanassie Nature Reserve, South Africa. South African Journal of Wildlife Research 37:31–39.

Watson, L.H., H.E. Odendaal, T.J. Barry, and J. Pietersen. 2005. Population viability of Cape mountain zebra in Gamka Mountain Nature Reserve, South Africa: The influence of habitat and fire. Biological Conservation 122:173–180.

Watson, L.H., T. Kraaij, and P. Novellie. 2011. Management of rare ungulates in a small park: Habitat use of bontebok and Cape mountain zebra in Bontebok National Park assessed by counts of dung groups. South African Journal of Wildlife Research 41:158–166.

Willerslev, E., J. Davison, M. Moora, M. Zobel, E. Coissac, *et al.* 2014. Fifty thousand years of Arctic vegetation and megafaunal diet. Nature 506:47–51.

Williams, O.B. 1969. An improved technique for identification of plant fragments in herbivore feces. Journal of Range Management 22:51–52.

Williams, S.D. 1998. Grevy's zebra: Ecology in a heterogeneous environment. Thesis, University College London, University of London, London, UK.

Williams, S.D. 2002. Status and action plan for Grevy's zebra (*Equus grevyi*). Pages 11–27 *in* P.D. Moehlman (Ed.), Equids: Zebras, asses and horses. Status survey and conservation action plan. International Union for Conservation of Nature, Gland, Switzerland.

Winkler, A., and N. Owen-Smith. 1995. Habitat utilisation by Cape mountain zebras in the Mountain Zebra National Park, South Africa. Koedoe 38:83–93.

Wockner, G., F.J. Singer, and K.A. Schoenecker. 2004. An animal location-based habitat suitability model for bighorn sheep and wild horses in Bighorn Canyon National Recreation Area and the Pryor Mountain Wild Horse Range, Montana, and Wyoming. *In* K.A. Schoenecker (Compiler), Bighorn sheep habitat studies, population dynamics, and population modeling in Bighorn Canyon National Recreation Area, 2000–2003. Open File Report 2004-1337. US Geological Survey, Reston, VA, USA.

Woodward, S.L., and R.D. Ohmart. 1976. Habitat use and fecal analysis of feral burros (*Equus asinus*), Chemehuevi Mountains, California, 1974. Journal of Range Management 29:482–485.

Young, T.P., T.M. Palmer, and M.E. Gadd. 2005. Competition and compensation among cattle, zebras, and elephants in a semi-arid savanna in Laikipia, Kenya. Biological Conservation 122:351–359.

Zhang, F. 2010. The study on behavior time budget and its influencial factors of Przewalski's horse (*Equus przewalskii*). [In Chinese.] Thesis, Beijing Forestry University, Beijing, China.

5

Equids and Ecological Niches: Behavioral and Life History Variations on a Common Theme

DANIEL I. RUBENSTEIN,
QING CAO, AND JUSTINE CHIU

All species are limited in where they can live. This is certainly the case for equids. Yet, over historical time, they have collectively occupied an astonishing array of habitats ranging from the mesic grasslands of the Asia and North America, to the more arid savannas of Africa, to the hot and cold deserts of Asia and Africa. Horses, zebras, and asses are evolutionarily closely related, comprising the same genus, *Equus*. Thus, not surprisingly, they show only minor variations on a common body plan. On the one hand, it is not too surprising that today they collectively continue to cover so much of the earth's surface and occupy so many different niches. On the other hand, with such a common body plan, it is somewhat surprising that each species only occupies a small and distinct part of the genus's overall range. Understanding why the genus is widespread, but with the range of each species being circumscribed, requires examining what factors shape the niche of each species, the degree to which niches overlap among the seven equids, and the physiological processes and behavioral actions that help shape these patterns.

The Niche Concept

Joseph Grinnell (1917) was the first ecologist to develop the concept of the niche. He viewed the niche as a species's "place" or "role" in an ecological community. Features such as what foods an animal ate or how it escaped from predators helped characterize its place among other species. But this concept was hard to apply to a wide range of species. Hutchinson (1957) instead focused on identifying environments and their characteristics—such as ranges of humidity and temperature—within which populations of a species could persist. By identifying the range of environmental conditions under which growth rates were equal to or greater than death rates, Hutchinson bounded a species's place on earth and created the concept of the n-dimensional hyperspace. Hutchinson's niche concept was quantitative and measurable and emphasized what a species does and how it survives over identifying its place in a community. Such "persistence niches" were divided into two types: those within which a species could survive in the absence of any other species—the fundamental niche—and those where it could persist in the presence of other species populating the landscape—the realized niche. In essence, the realized niche is often bounded by physiological limits on one edge and by interactions with other species at the other edge. If the fundamental niches of two species overlap extensively, and one species is competitively superior to the other, then the superior species will likely dominate and exclude the other from the region of niche overlap.

Hutchinson's niche was relatively easy to define when the variables were associated with physical properties of the environment. Range limits associated with temperature and humidity were easy to identify and measure, but when it came to features with discrete properties such as diet items, it was much harder to demarcate niche boundaries. To overcome this constraint, Robert MacArthur, Hutchinson's student, along with Richard Levins, formulated the concept of the "resource-utilization" niche (MacArthur and Levins 1967). While it still maintained a quantitative and operational framework, it measured what individuals within species actually did, and what resources they actually used. In essence, the utilization niche precisely defined the natural history of a species, often on the basis of its behavior. The ratio of the width of the niche utilization curve (w) and the distance between the midpoint of each species utilization curve (d) determined whether species could coexist. So long as the ratio $d/w > 1$, overlap among fundamental niches would be small enough to allow species to coexist. If the inequality was reversed and was much smaller than 1, then one species was likely to outcompete the other and reduce the realized niche breadth of the inferior competitor.

When the elements of Hutchinson's persistence niche are combined with MacArthur's resource-utilization niche, it becomes possible to examine how the physiology and behavior of a species shape the boundaries of its fundamental and realized niches, thus revealing how a species copes with its physical and biotic environments. To understand how equids distribute themselves across landscapes and how they cope with the needs and actions of other herbivorous species—both wild and those under human control—we must first characterize each equid species's fundamental niche, then its overlap with those of other equids and other herbivores, and finally explore which behavioral and physiological mechanisms enable equid species to persist, and at times coexist, when the overlap with other equids and more distantly related herbivores is large.

Equid Commonalities and Specificities

All equids share a common body plan consisting of a long body, four long legs tipped with a single hoof, a long neck with a large head, and an elongated face with large, flexible ears. A long mane covers the neck, and a long tail is attached at the rear. All equids ferment vegetation in a cecum, located after the stomach. Because there is no temporal bottleneck when passing food to the stomach or the cecum, equids can subsist on low-quality forage that they quickly pass through their guts and only partially process (see chap. 4). As a result, they can survive in some of the least productive habitats on earth, but they have to spend significantly more time foraging to sustain themselves than their equivalent-sized ruminant counterparts.

All equid species live in groups, but the groups vary in composition and cohesion (Klingel 1975; Rubenstein 1986, 1994; see chap. 2). Horses (*Equus caballus*, *E. ferus przewalskii*), plains zebra (*E. quagga*), and mountain zebra (*E. zebra*) live in closed-membership family groups consisting of a male, a group of females, and their immature young. Grevy's zebra (*E. grevyi*) and the wild asses (*E. asinus*, *E. africanus*, *E. hemionus*), however, live in more open societies where individual associations, apart from those of mothers and their young, are much more transitory. In such fission-fusion societies, males forsake associating with particular females, instead establishing resource-based territories that attract females. Typically, females show affinities for females of similar reproductive states. Those with young foals need to drink daily, aggregating around water and forming temporary associations with particular territorial males, while those that do not need to drink as often wander more widely in search of large quantities of food. For those equids living in family groups, females are loyal to particular males so long as they remain vigilant and reduce harassment from marauding males, especially the bachelors. In plains zebra, when cuckolding pressure from bachelors becomes intense, stallions from family groups come together to form coalitions to keep bachelors away from their females. As a result, herds composed of many family groups form (Rubenstein and Hack 2004). Females in fission-fusion societies would also benefit by such male protection, but searching widely for food forces all but those females with young foals to wander through the territories of many males each day (Sundaresan *et al.* 2007; Rubenstein 2010).

Equid Niches

In general, all equids live in relatively open habitats and subsist mostly on grasses, although some species include a moderate amount of browse in their diets. In order to characterize Hutchinson's persistence niches of the seven equids, principal component analysis (PCA) was used to identify clusters of environmental features

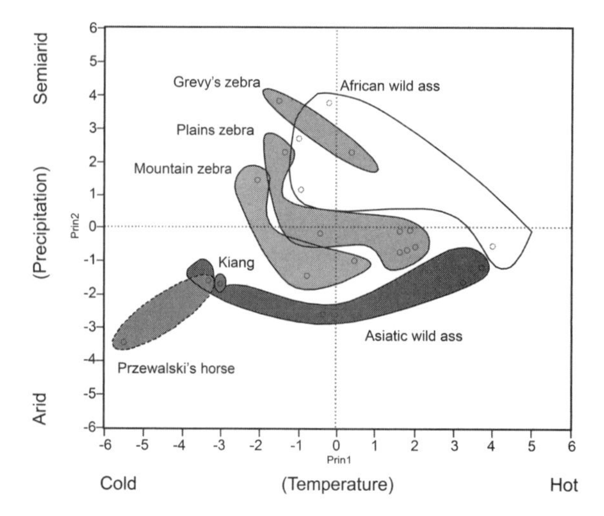

Fig. 5.1 Visualization of equid niches based on principal component analysis of monthly temperature, precipitation, and number of rainy days for January, May, August, and October from a variety of wild populations of each species. Minimum convex polygons depict the niche boundaries for each species.

that demarcated niches boundaries. Minimum and maximum temperatures in January, May, August, and October, along with amount of precipitation and total number of rainfall days in the same months at known equid locations, were the environmental features considered. Three principal components emerged that explained a total of 81% of the overall variation among the variables. The first principal component explained 34%, the second 27%, and the third 20% of the total variance. What is most interesting is the fact that of the original variables, temperature (minimum and maximum temperatures in January, May, and October) loaded the strongest on the first component, while the precipitation (amount of rainfall in May, August, and October as well as the number of rainy days per month) loaded strongest on the second component. When these points are arrayed on a bivariate plot of PCA1 (temperatures) and PCA2 (precipitation), the seven equids show strikingly different fundamental niches (fig. 5.1). In this two-dimensional niche space, five of the seven species showed fundamental niches that were unique and showed no overlap with those of any other equid. Only the wild asses showed any overlap with other equids: African wild asses with Grevy's zebra, and Asiatic wild asses with Przewalski's horses. For both Grevy's zebra and Przewalski's horses, regions of overlap occurred in the hottest part of niche space. For the Przewalski's horses, overlap also occurred in the driest part of its niche. Details about ranging, drinking, and foraging behavior help explain why some equids show no niche overlap while others do.

The African Equids

Precipitation and temperature shape the abundance and quality of vegetation available to grazers. The Normalized Difference Vegetation Index (NDVI) that measures the reflectance of light in the red and near-infrared regions provides an estimate of the amount of photosynthetic tissue in a defined area. NDVI values vary across months comprising dry and wet seasons at the locations inhabited by the different African equids (fig. 5.2). During the May rains, the three zebra species occupy habitats that show higher levels of photosynthesis than asses do (fig. 5.2). During the August dry season, however, these differences vanish, in part because variation among sites is so large and as a result the niches of the four equids show no significant differences (fig. 5.2). Because the geographical ranges of African asses rarely overlap those of mountain, plains, and Grevy's zebras, convergence of their utilization niches is of little consequence. But the convergence does show that even the more mesic-adapted species—plains and mountain zebra—can survive at least temporarily under conditions faced year-round by the wild asses, and to a lesser extent the Grevy's zebras.

The Grevy's and plains zebras show both niche convergence and geographic range overlap, yet they manage to coexist despite showing strong co-occurrence in numbers within similar microhabitats (Rubenstein 2010). Both species prefer habitats with light bush, densely vegetated grasslands dominated by tall grasses, and areas of high biomass (table 5.1). They also prefer

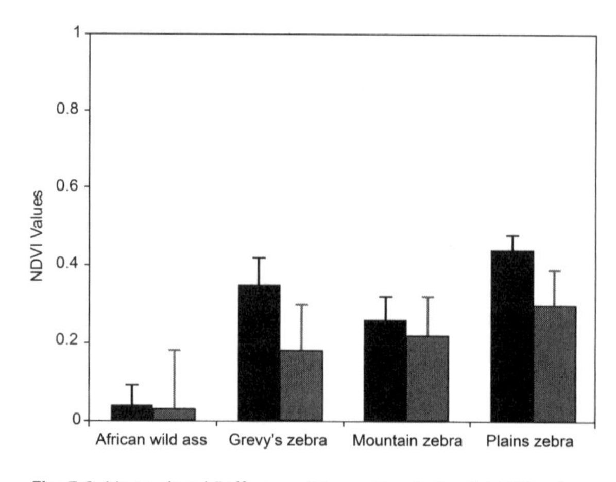

Fig. 5.2 Normalized Difference Vegetation Index (NDVI) values of African equids during the wet (black bars) and dry seasons (gray bars). Means and standard errors are shown.

Table 5.1. **Characteristics of foraging and drinking behavior of Grevy's and plains zebras**

Feature	Grevy's Zebra	Plains Zebra	Statistic (*t*)	df	*P* Value
Visibility index	4.3 ± 0.02	4.0 ± 0.01	1.13	125	NS
Ground cover (%)	88.2 ± 0.05	90.1 ± 0.07	0.69	105	NS
Height of the leaf table (cm)	11.9 ± 1	12.04 ± 1.1	2.1	122	NS
Biomass (g m^{-2})	6,010 ± 400	6,324 ± 463	0.56	105	NS
Green leaf (%)	58.9 ± 1.8	56.4 ± 3.6	0.6	122	NS
Peak time to drink	9:00–10:00 a.m.	9:00–10:00 a.m.			
Distance to water (m)	810 ± 55	633 ± 64	3.5	602	<0.0005
Digestible organic matter	63.39 ± 0.07	61.34 ± 0.41	2.68	24	<0.02
Potassium (%)	2.03 ± 0.025	1.54 ± 0.021	2.38	49	<0.008
Water (%)	65.2 ± 0.05	50.3 ± 0.07	2.1	85	<0.02

Note: All values are means ± standard error.

to drink at the same time of day, although Grevy's zebra range farther from water on average than plains zebra. This is a recipe for competition, and when plains zebra outnumber Grevy's zebra, plains zebra suppress the intake rate of Grevy's zebra, whereas when Grevy's zebra outnumber plains zebra, Grevy's zebra have no impact on the intake rate of plains zebra (Rubenstein 2010). Yet competition is minimized because niche overlap is avoided for three reasons.

First, where both zebras coexist, Grevy's zebra avoid black cotton soils, whereas plains zebra regularly spend time on them (fig. 5.3). Second, the two zebra species show little dietary overlap. Based on the frequency of occurrence of deoxyribonucleic acid (DNA) in the scat of these species, Grevy's zebra have broader fundamental niches than plains zebra (0.396 vs. 0.298, respectively) with a niche overlap value of 0.434. Other pairwise overlap values for a variety pairs of ruminant grazers range from 0.577 to 0.987, revealing that overlap among zebra species is relatively low (Kartzinel *et al.* 2015). Species-specific analyses of diet items also show that plains and Grevy's zebra mostly consume different foods (table 5.2) and these foods differ in nutritional quality. The grasses of plains zebra contain significantly more digestible organic matter (DOM), more potassium (K), and less water than those of Grevy's zebra (Reo 2012). Based on DNA barcoding of the undigested vegetation in zebra scat, both species tend to consume large quantities of *Pennisitum* grasses. Apart from this similarity, plains zebra eat *Brachiaria, Bothriochola*, and *Setaria*, which are common on black cotton soils, as well as *Triandra*, which is found on black cotton as well as red lateritic soils. Grevy's zebra, however, focus on *Cynodos*, *Harpacne*, *Aristids*, and *Chloris*, which are mostly red soil specialists (Kartzinel *et al.* 2015). And third, although both species prefer to drink between 9:00 a.m. and noon, at least half the population of Grevy's zebra—females that are not lactating—need only drink every three to five days (Becker and Ginsberg 1990). Therefore a sizeable population of Grevy's zebra can graze in plains zebra–free zones far from water.

Although different aridity and temperature tolerances determine differences in Hutchinsonian persistence niches, differences in dietary and drinking patterns reveal differences in MacArthurian utilization niches. Responses and susceptibility to predation add yet another dimension to niche differentiation. In areas where both plains and Grevy's zebras coexist and where lion densities are relatively high, Grevy's zebra are more susceptible to predation by lions than plains zebra (Rubenstein 2010). By using light microscopy to quantify the frequency of each zebra species' hairs in lion scat, those of Grevy's zebra are disproportionately more abundant than those of plains zebra. Although Grevy's zebra are larger than plains zebra and could provide more food per kill, their greater vulnerability is likely the result of two factors. First, they use closed habitats more frequently than plains zebra, and that is where lions tend to reside during daytime. And second, Grevy's zebra females are often left on their own because territorial males are patrolling their territories in search of receptive females or cuckolding males.

The Asian Equids

Although plains and Grevy's zebras often share geographical areas, their niches remain separate and distinct because their diets, watering needs, and antipredatory abilities differ (fig. 5.1). This does not appear to be the situation for Asiatic wild asses and Przewalski's horses. In Asia, Przewalski's horses and the asses show both niche and range overlap (fig. 5.1), especially in the Gobi and Dzungarian Deserts. Przewalski's horses thrive on the cold mesic steppes of Mongolia (Usukh-

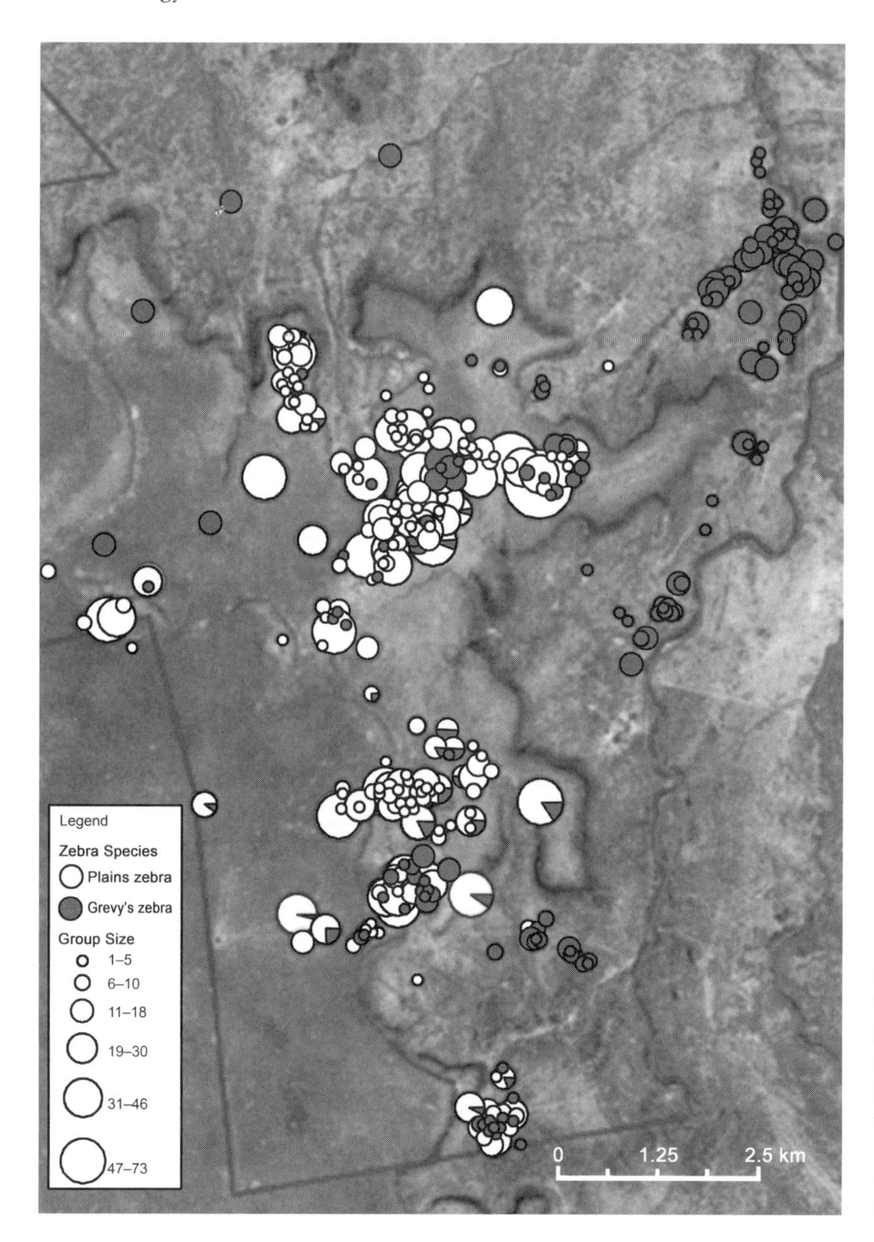

Fig. 5.3 Sightings of Grevy's zebra and plains zebra during June and July 2013 on the Mpala conservancy, Laikipia, Kenya. The size of the circle is proportional to the number of zebras seen. The lighter areas depict red soil, and the darker areas depict black cotton soil.

jargal and Bandi 2013), as do wild asses in Iran (Nowzari *et al.* 2013) and the Gobi, which also thrive in hot deserts (Feh *et al.* 2002; Kaczensky *et al.* 2008). In both the Kalamaili Nature Reserve, China, and the Great Gobi B Strictly Protected Area, Mongolia, populations of Przewalski's horses were reintroduced because the last survivors were seen there before becoming extinct in the wild at the end of the 1960s. Whether they ever flourished in these locales with their persistence niches, stretching to encompass these environmental states, remains unknown. Recent studies on movements, diet choice, drinking habits, and interspecific aggression, however, are providing important insights on whether these two equids can coexist when their ranges overlap in this region of their niche spaces.

In the Gobi Desert, Przewalski's horses inhabit small, restricted home ranges centered around permanent watering points where they forage disproportionately on highly productive grass communities. Wild asses exhibit just the opposite patterns. Their ranges are much larger than those of the horses, and they reveal no preference for particular plant communities. Instead, they utilize them in proportion to their abundance (see chap. 4). They range far from water, and instead of living in small, closed membership family groups as do horses, they form large herds whose

Table 5.2. **Dietary comparisons of Grevy's and plains zebra**

Vegetation Species Name	Grevy's Zebra	Plains Zebra	*P* Value
Pennisetum stramineum			NS
Pennisetum mezianum			NS
Cynodon dactylon			0.002
Cynodon plectostachyus			0.004
Bothriochola insuculpta			0.002
Brachiaria eruchiformis			0.008
Setaria sp.			0.008
Digitaria milanjiana			NS
Themeda triandra			0.002
Eragrostis sp.			0.04
Harpacne schimperi			NS
Aristida congesta			0.002
Chloris sp.			0.002
Asteracea			NS
Fabaceae			NS

Note: Shadings represent frequency of occurrence (FOO) of species-specific DNA sequences based on comparisons of the P6 region of the chloroplast from field samples compared to published sequences. Dark gray signifies FOO values ranging from 1.0 to 0.75; medium gray from 0.74 to 0.50; light gray from 0.50 to 0.10; and white <0.10. *P* values <0.05 signify significant differences. NS, not significant.

composition tends to vary over time (Kaczensky *et al.* 2008). These patterns make it difficult to know whether competition for productive foraging patches separates the realized niches of these two equids. Is it that the smaller asses voluntarily avoid or are actively repulsed by the larger horses? Or do the niche differences result from size-dependent reductions in energy needs that allow asses to economically track and utilize widely dispersed patches of asynchronously appearing "green-up" patches of vegetation?

Based on the movement tracks of Przewalski's horses fitted with GPS tags and sightings of horses and asses at fixed watering points captured by motion-triggered cameras in the Kalamaili Nature Reserve in China's Dzungarian Desert, answers to these questions are emerging (table 5.3). As a mesic-adapted species, it is not surprising to see Przewalski's horses drinking on average about two times per day under desert conditions (Zhang *et al.* 2015). What is surprising, though, is that horses and asses are rarely seen drinking together; camera traps show that horses and asses drink concurrently from the same watering point <1% of the time. That the Przewalski's horses drank almost exclusively from one water hole, whereas the asses drank mostly from the reserve's three other watering points—all of which were significantly more saline than the one used by the horses—reinforces the notion that the two equids actively avoid each other (Zhang *et al.* 2015). In addition, the daily drinking rhythms of the two species are nonoverlapping: horses confine 80% of their drinking to daylight hours, peaking at noon, while asses confine 85% of their drinking bouts to twilight or nighttime hours (Zhang *et al.* 2015).

But is the shift by asses to nighttime drinking voluntary, or is it imposed upon them by Przewalski's horses? On the one hand, asses continued to drink at night at a water point rarely visited by horses, thus suggesting that asses prefer nighttime drinking. In the few instances when both were together at watering points, however, the asses became highly vigilant, often fleeing when Przewalski's horses approached (Zhang *et al.* 2015). Furthermore, allopatric populations of asses at the northern edge of the reserve spent more time drinking during the day than those sharing habitats with horses (Q. Cao, personal communication). At least in the Kalamaili Nature Reserve, niche overlap between horses and asses is large, and when populations co-occur, horses outcompete asses, reducing the size of their realized utilization niche.

In the Gobi Desert, water has been shown to be a major factor constraining space use in Przewalski's horses, especially in the dry season (Kaczensky *et al.* 2008). The same applies in the Dzungarian Desert. Normally, horses drink at permanent watering points. But during summers, short pulses of rain are common. For up to two days after these storms, daily temperatures drop and small pools of water persist. During these pulsed rain events, the rate of daily drinking is significantly reduced (table 5.3). In addition, the day range of horses searching for forage also increases dramatically. During dry periods, Przewalski's horses on average travel 2–3 km from water in search of food.

Table 5.3. **Changes in Przewalski's horse behavior associated with pulsed rain events**

Behavior	Prior to Rain Event	After Rain Event	Statistic	*P* Value
Daily drinking rate (times per day)	1.9 ± 0.03	0.98 ± 0.02	*F* (1, 57) = 15.23	<0.0001
Distance traveled to feeding site (km)	2.47 ± 0.52	4.2 ± 0.61	*F* (1, 57) =9.43	<0.005
Green plant parts (%)	42.2 ± 3.7	61.6 ± 7.2	*F* (1, 34) = 17.76	<0.0001

Note: All values are means ± standard errors.

But immediately after pulsed rains, distances traveled can reach 6 km, with family groups led by subordinate males ranging the farthest ($F_{3,57} = 6.85$; $p < 0.005$). Clearly, increasing the dispersion of water frees Przewalski's horses to range more widely, and they do so because they are likely to be able to drink wherever they find food. And because high-quality food quickly appears after rains (table 5.3), horses attempt to take advantage of this high-quality but ephemeral resource. Again, groups led by subordinate males foraged in the greenest patches ($F_{3,57} = 11.10$; $p < .0001$). When forced to drink at fixed watering points, however, male dominance determines where groups drink and which groups drink the most (Chui 2012). Thus, when conditions are hot, local rainstorms both lower daytime temperatures and create a mosaic of short-lived water puddles that enable the weakest competitors to reach ungrazed and productive green-up patches of vegetation without having to sacrifice their ability to drink.

Social Flexibility

Equids exhibit broad fundamental niches that are rarely overlapping. Behavioral flexibility, especially with respect to fluidity of social relationships, fosters adaptability. The asses and Grevy's zebra live in open membership groups in which individuals come together and break apart with ease. Consequently, their societies are highly flexible. But when the number, strength, and diversity of social relationships are quantified, the connections among individuals in wild ass societies tend to be more numerous and diverse, but weaker than those in the societies of Grevy's zebra (Sundareson *et al.* 2007). Given that the likelihood of individual asses finding transitory patches of greening vegetation after rain pulses is essential for their survival and that the spread of information is greatest in networks that are highly connected (Rubenstein 2015), selection for diffuse but highly connected networks should be stronger in the asses than Grevy's zebra. This suggests that social flexibility may enable asses to live in the harshest and least predictable of habitats, thus making the niches of the asses the broadest of the equids (fig. 5.1).

If this is indeed true, then the fact that plains zebra, but not horses, exhibit a higher level of sociality by forming herds that fuse and fission depending upon social context again suggests that a need for social flexibility may be a force shaping their niche as well. Unlike in horse societies where bachelor groups are small, those in plains zebra societies can be as large—and at times larger—than family groups. Such large bachelor aggregations likely form to reduce the per capita risk of bachelors being preyed upon by lions. When faced with such high levels of cuckolding pressure, plains zebra stallions could benefit by rapidly assembling coalitions with other stallions to fend off these heightened threats (Rubenstein and Hack 2004). By constructing large and dense networks composed of other stallions and bachelors, individual stallions reinforce both dominance over a particular segment of the bachelor community and reciprocal relationships with a large but circumscribed set of stallions. Because the persistence niches of plains zebra are much larger than those of Przewalski's horses (fig. 5.1), well-connected networks among stallions may potentiate social flexibility, which in turn widens the fundamental niches of plains zebras relative to those of horses (Rubenstein 2015).

Equids and Bovids: Niches and the Dynamics of Human–Equid Conflict

Most herders view wildlife—especially horses, zebras, and asses—as pests or even vermin because they assume that their niches overlap those of bovids. But do they? Differences in fermentation systems suggest that zebra should be able to eat coarser vegetation than cattle (Duncan *et al.* 1990). Barcoding of DNA sequences of vegetation in the scat of Boran zebu cattle and plains zebra on a common rangeland reveals that they do. While cattle have broader niches than plains zebra (0.404 vs. 0.297, respectively) dietary overlap between the species is moderate (0.577), surpassing the overlap of plains and Grevy's zebras (0.434) but lower than that of cattle and Cape buffalo (0.563). Yet experiments in which herds of only cattle or donkeys—surrogates for plains zebra because they have the same body plan, basic physiology, and are more amenable to following and observing what they eat—or mixtures of the two species showed that when grazing together over a six-month period encompassing seasons of rainy as well as dry conditions, both cattle and donkey grew faster than equivalent-sized herds composed only of conspecifics (Odadi *et al.* 2011). Because the donkeys ate the stems and stalks that ordinarily reduce cattle ingestion rates, donkeys actually changed the structure of the vegetation, increasing the availability of food items preferred by cattle. In this way, facilitation by surrogate zebras reduced niche overlap, as each species could specialize on preferred items. As a result, niche overlap need not always lead to negative interactions among species. Depending on morphology and physiology, interactions can reduce competition and under certain conditions enhance coexistence via facilitation.

Conclusion

Collectively, equids are distributed over much of the earth's surface, exhibiting a broad niche encompassing wide ranges of precipitation and temperature. Perhaps what is most striking, however, is that the individual species' niches hardly overlap. Instead, they abut each other in niche space, suggesting an orderly replacement of species at particular environmental tipping points. African wild asses push to the extremes of temperature and aridity, with plains and mountain zebras regressing toward the mean with respect to both environmental factors. Kiang (*E. kiang*) and Przewalski's horses persist under cold and wet conditions, with Asiatic wild asses being the most catholic of the Asian equids and exhibiting the broadest niche. Of all these species, only two pairs exhibit overlapping niches—the African wild ass and the Grevy's zebra, and the Asiatic wild ass and the Przewalski's horse—and only the pair of Asian species shows overlapping geographical ranges. Thus, even though there should be no competition among any of the other pairs, their fundamental niches have diverged over evolutionary time, suggesting that physiological and behavioral adaptations drive niche diversification.

Physiological tolerances of environmental conditions shape the boundaries of Hutchinson's persistence niches, while behavioral interactions between species often shape MacArthur's utilization niches. Because so few equids show both niche and range overlap, one niche boundary is typically determined by environmental conditions while the other boundary tends to be shaped by biotic interactions. For the pair of zebras sharing geographic ranges in east Africa, similarities in habitat choices and watering strategies suggest that their fundamental utilization niches should overlap. But avoidance of certain soil types by Grevy's zebra leads to significant differences in the quality and species of grasses consumed by the two species. Similarly, their different social and mating systems enable them to respond to environmental fluctuations in different ways. These differences help separate the realized niches. And although plains zebra, when abundant, can reduce the foraging abilities of Grevy's zebra in areas where they co-occur, Grevy's zebra can thrive in areas where water is localized, abundant, and free of predators.

A variation on this theme plays out in the deserts of China and Mongolia where the Asiatic wild ass and the reintroduced Przewalski's horse co-occur. Tolerance of the extremely cold temperatures provides the horses with an exclusive refuge, but they can also persevere in extremely arid conditions. Studies in the Gobi left undetermined whether asses voluntarily avoided areas where horses lived or whether asses, because of their small size, developed physiological and behavioral strategies involving a flexible fission-fusion social system composed of a large information network that increased the likelihood of individual asses finding transitory and unpredictably appearing patches of greening vegetation. Studies in the Dzungarian Desert of China suggest that asses do not voluntarily cede access to the areas or resources that Przewalski's horses use. Instead, they cede access to the horses because horses are larger and dominate them, keeping asses away from the best watering points and the forage growing near them. But control of these resources only works for horses so long as there is enough low-saline water for them to drink repeatedly throughout the day. To compensate for these constraints, asses must either drink at night from high-quality water points or during the day at watering sites that are highly saline that horses avoid. Drinking and feeding trials on onagers (*E. hemionus onager*) and Przewalski's horses in a zoo show that onagers consume significantly less water and energy per day than Przewalski's horses (Owusu-Akyaw 2012), supporting the idea that they are physiologically well adapted for traveling economically over long distances without being constrained by access to water. In addition, the flexible nature of ass sociality and the benefits of information exchange allow them to range freely and widely in search of greening vegetation patches. But with increasing pressure during the winter from domestic sheep when they also seek these resources, being subordinate to horses during the summer may be creating a year-round realized niche that is narrowing and becoming too circumscribed to sustain the asses.

At least among the co-evolved living equids, even when fundamental niches and geographic ranges overlap, a combination of behavioral and physiological adaptations allows each species to reduce overlap so that coexistence is possible.

In today's world, niche overlap with human-managed grazers presents strong challenges for equids (see chap. 12). At least in Kenya, overlap in the utilization niche is stronger between plains zebra and cattle than between plains and Grevy's zebras. Fortunately, fundamental differences in ingestion and fermentation processes generate a grazing succession that facilitates growth and health of the plains zebra when associating with cattle. In turn, by consuming plant parts of mutually preferred species that cattle find difficult to ingest, donkeys as surrogate plains zebra change the canopy structure of grasslands in ways that increase the foraging success of cattle. In this case, the large overlap in the fundamental niches of the two species leads to interactions that benefit each species more than when only foraging with conspecifics.

From these patterns it is clear that equid niches allow the genus to span a wide range of environmental conditions, but that this range emerges from the systematic subdivision of niche space into largely non-overlapping niches. When niches actually overlap, either with other equids or human-managed herbivores, flexibility in behavior is the key to reducing effective niche overlap, thus fostering coexistence.

ACKNOWLEDGMENTS

The ideas in this chapter emerge from extensive fieldwork in the United States, Israel, Africa, and China, along with many discussions with colleagues, postdocs, and graduate and undergraduate students from all over the world. Without the permission and permits from many government agencies, the authors could not have accomplished any of the fieldwork. Over the years, many grants from the US National Science Foundation have funded the authors' studies.

REFERENCES

Becker, C.D., and J.R. Ginsberg. 1990. Mother-infant behaviour of wild Grevy's zebra: Adaptations for survival in semi-desert East Africa. Animal Behaviour 40:1111–1118.

Chui, J. 2012. Time budgeting and social behavior of reintroduced Przewalski's horses in Kalamaili Nature Reserve, China. Thesis, Princeton University, Princeton, NJ, USA.

Duncan, P., T.J. Foose, I.J. Gordon, C.G. Gakahu, and M. Lloyd. 1990. Comparative nutrient extraction from forages by grazing bovids and equids: A test of the nutritional model of equid/bovid competition and coexistence. Oecologia 84:411–418.

Feh, C., N. Shah, M. Rowen, R.P. Reading, and S.P. Goyal. 2002. Status and action plan for the Asiatic wild ass (*Equus hemionus*). Pages 62–71 *in* P.D. Moehlman (Ed.), Equids: Zebras, asses and horses. Status survey and conservation action plan. International Union for Conservation of Nature, Gland, Switzerland.

Grinnell, J. 1917. The niche-relationships of the California thrasher. The Auk 34:427–433.

Hutchinson G.E. 1957. Concluding remarks. Proceedings of the Cold Spring Harbor Symposium 22:415–427.

Kaczensky, P., O. Ganbaatar, H. Von Wehrden, and C. Walzer. 2008. Resource selection by sympatric wild equids in the Mongolian Gobi. Journal of Applied Ecology 45:1762–1769.

Kartzinel, T.R., P.A. Chen, T.C. Coverdall, D.L. Erickson, W.J. Kress, M.L. Kuzmina, D.I. Rubenstein, W. Wang, and R.M. Pringle. 2015. DNA metabarcoding illuminates dietary niche partitioning by large African herbivores. Proceedings of the National Academy of Sciences of the USA 112:8019–8024.

Klingel, H. 1975. Social organization and reproduction in equids. Journal of Reproduction and Fertility Supplement 23:7–11.

MacArthur, R., and R. Levins. 1967. The limiting similarity, convergence, and divergence of coexisting species. American Naturalist 101:377–385.

Nowzari, H., M. Hemami, M. Karami, M.M.K. Zarkesh, B. Riazi, and D. I. Rubenstein. 2013. Habitat use by the Persian onager *Equus hemionus onager* (Perissodactyla: Equidae) in Qatrouyeh National Park, Fars, Iran. Journal of Natural History 47:2795–2814.

Odadi, W.O., M.K. Karachi, S.A. Adulrazak, and T.P. Young. 2011. African wild ungulates compete with or facilitate cattle depending on season. Science 333:1753–1755.

Owusu-Akyaw, A. 2012. A comparative nutritional analysis of Przewalski's horses and Asiatic wild asses. Thesis, Princeton University, Princeton, NJ, USA.

Reo, B. 2012. A multidisciplinary approach to the nutrition and land-use patterns of the plains and Grevy's zebra. Thesis, Princeton University, Princeton, NJ, USA.

Rubenstein, D.I. 1986. Ecology and sociality in horses and zebras. Pages 282–302 *in* D.I. Rubenstein and R.W. Wrangham (Eds.), Ecological aspects of social evolution. Princeton University Press, Princeton, NJ, USA.

Rubenstein, D.I. 1994. The ecology of female social behavior in horses, zebras, and asses. Pages 13–28 *in* P. Jarman and A. Rossiter (Eds.), Animal societies: Individuals, interactions, and organization. Kyoto University Press, Kyoto, Japan.

Rubenstein, D.I. 2010. Ecology, social behavior, and conservation in zebras. Pages 231–258 *in* R. Macedo (Ed), Advances in the study of behavior, vol. 42: Behavioral ecology of tropical animals. Elsevier, Oxford, UK.

Rubenstein, D.I. 2015. Networks of terrestrial ungulates: Linking form and function. Pages 184–196 *in* J. Krause, R. James, D.W. Franks, and D.P. Croft (Eds.), Animal social networks. Oxford University Press, Oxford, UK.

Rubenstein, D.I., and M. Hack. 2004. Natural and sexual selection and the evolution of multi-level societies: Insights from zebras with comparisons to primates. Pages 266–279 *in* P. Kappeler and C.P. van Schaik (Eds.), Sexual selection

in primates: New and comparative perspectives. Cambridge University Press, Cambridge, UK.
Sundaresan, S.R., I.R. Fischhoff, J. Dushoff, and D.I. Rubenstein. 2007. Network metrics reveal differences in social organization between two fission-fusion species, Grevy's zebra and onager. Oecologia 151:140–149.
Usukhjargal, U., and N. Bandi. 2013. Reproduction and mortality of re-introduced Przewalski's horse *Equus przewalskii* in Hustai National Park, Mongolia. Journal of Life Sciences 7:624–631.
Zhang, Y., Q. Cao, D.I. Rubenstein, S. Zang, M. Songer, P. Leimgruber, H. Chu, J. Cao, K. Li, and D. Hu. 2015. Water use patterns of sympatric Przewalski's horse and Khulan: Interspecific comparison reveals niche differences. PLoS ONE 10:e0132094.

Wild and Feral Equid Population Dynamics

JASON I. RANSOM, LAURA LAGOS, HALSZKA HRABAR, HANIYEH NOWZARI, DORJ USUKHJARGAL, AND NATALIA SPASSKAYA

Wild equid populations are shaped by births, deaths, emigration, and immigration, each of which is subject to a suite of biological and ecological inputs across space and time. How these inputs influence population dynamics varies between populations as well as among species. In our rapidly changing world, human actions also increasingly affect wildlife populations (Sanderson *et al.* 2002). Fragmented landscapes, hunting of predators, reintroduction of species, invasive species, competition with humans and their livestock for resources, direct killing, and climate change all can influence population dynamics (see chap. 12). Most wild equids occupy arid or semiarid steppe, grassland, and desert environments, but some feral populations also thrive in forests, montane, and coastal environments, and across hot and cold, xeric and mesic, conditions. The array of factors potentially influencing these populations is great, but some fundamental elements influencing birth, survival, and movement are common to all.

Food can be one of the most important natural factors limiting populations of large herbivores (Fowler 1987; Gaillard *et al.* 1998, 2000; Gaidet and Gaillard 2008). As such, some populations may exhibit density dependence and be limited because of their own impact on forage resources, or because population density itself influences age-specific natality, behavior, or survival (Choquenot 1991; Gaillard *et al.* 1998, 2000; Flux 2001; Linklater *et al.* 2004; Bonenfant *et al.* 2009). When such populations lack resources, body condition of individuals can diminish and disease and parasitism can increase, all with the cumulative effect of reducing fecundity and survival, and ultimately limiting population growth (Eberhardt 1977). Some populations may persist without exhibiting density dependence, changing through time without realized self-imposed resource limitation. Those populations may still be limited by predation or climate events (Ellis and Swift 1988; Georgiadis *et al.* 2003). Over time, any given population may move between exhibiting density dependence and apparent density independence as the influencing factors change. Regardless of the specific paradigm, both deterministic and stochastic mechanisms play important roles in shaping populations (Björnstad and Grenfell 2001).

Attempting to draw broad inferences across equid population states, processes, and inputs is a challenging task, but the wild and feral members of family Equidae are perhaps more similar than not in their equation of life on earth. Some species, notably African wild ass (*Equus af-*

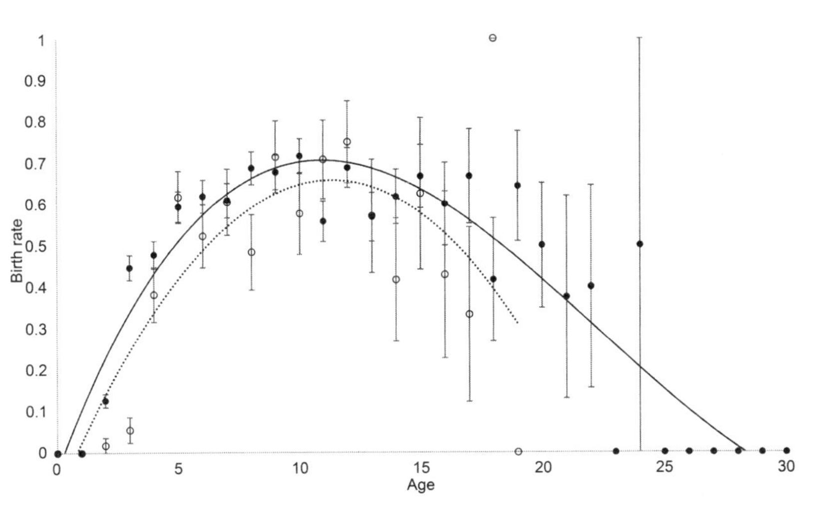

Fig. 6.1 Mean age-specific birth rates of horses ± standard error, as derived from *E. caballus* (feral) in the western United States (solid dots and solid trend line, *n* = 1,013 births, excluding females that received fertility control treatments) (Ransom *et al.* 2011; J. Ransom, unpublished data), and *E. f. przewalskii* (open dots and dashed trend line, *n* = 217 births, excluding females in the year of their introduction) (Great Gobi B Strictly Protected Area, Mongolia, unpublished data).

ricanus) and kiang (*E. kiang*), are vastly understudied, and considerable gaps persist in our knowledge of even their basic ecology. Other species, such as the feral horse (*E. caballus*), provide rich data sets that can inform us with in-depth ecological perspectives. In this chapter, we have synthesized data from studies on equids across the globe toward understanding the common and diverging natural influences acting to shape their populations. The populations in these studies ranged on 6 continents across >91 degrees of latitude, from savannah to mountains and deserts to islands, and thus were influenced by a diversity of biotic and abiotic factors. We focus here on natality, survival, and population growth. We briefly discuss movement into and out of populations, but largely defer that topic to chapter 13. The influence of humans on these populations is a broad topic that can profoundly affect survival, and we must defer that topic to chapter 12.

Natality

All species of wild equid appear to follow the same general distribution of age-specific natality that is observed in horses, with primiparous births most typically occurring from females age 2–5 years, and peak birth rates occurring roughly for females age 6–17 (fig. 6.1). There is some evidence that when forage is abundant, female equids are more likely to give birth at an earlier age (Berger 1986; Saltz and Rubenstein 1995; Tatin *et al.* 2009). When forage and water becomes exhausted, age of primiparous birth can increase and birth rates for all females can decline quickly (Ginsberg 1989). The majority of feral asses (*E. asinus*) in Death Valley, USA, for example, were observed giving primiparous birth at age ≥4 years while persisting in an extremely harsh environment (Moehlman 1974). (Note that sexual maturity is reached at 1.5 years; Grinder *et al.* 2006.) In the wild, female plains zebra (*E. quagga*) have been documented giving birth as old as age 18 (Klingel 1969), mountain zebra (*E. zebra*) at age >21 (Joubert 1974*a*), Przewalski's horses (*E. ferus przewalskii*) at age 20 (O. Ganbaatar, unpublished data), and feral horses as old as age 24 (fig. 6.1).

Horses and zebras typically give birth to one offspring annually, though extraordinarily rare cases of twins have been reported (Roelle *et al.* 2010; Vodička 2010). Feral asses have been observed giving birth to either one or two offspring (Grinder *et al.* 2006), as have Asiatic wild asses (*E. hemionus*) (Schook *et al.* 2013). The cases of twins born to Asiatic wild asses all resulted in early mortality of at least one sibling, and it is unknown how frequently twin births occur in the wild. Wild females of all asses are most typically observed with only one offspring of the year. The mean birth rate for equid females ≥2 years old across all species and studies we reviewed ranged from 0.20 in mountain zebras to 0.68 in feral asses, with a mean across species of 0.47 (95% confidence interval, or CI = 0.35–0.59) (table 6.1). Note that the lowest value from individual studies in this range arose from Przewalski's horses in the Gobi Desert following a catastrophic winter climate event (birth rate = 0.04) (Kaczensky *et al.* 2011), and the maximum value occurred in a small Asiatic wild ass (onager) population that started to increase rapidly six years after reintroduction (birth rate = 1.00) (Saltz and Rubenstein 1995). Despite these extremes, however, birth rates among wild equid species are remarkably similar when put in context of the pressures selecting for fitness.

Birth rates in wild populations reflect offspring born and observed, rather than pregnancy rate, which is presumably higher in most cases. Pregnancy rates

Table 6.1. **Mean birth, survival, and population growth rates of wild and feral equids**

Species	Populations	Mean Adult Birth Rate	Mean Foal Survival Rate	Mean Adult Survival Rate	Mean Population Growth Rate (λ)
E. caballus (feral)[1]	Tornquist Park, Argentina; Garden Station, Australia; Alberta, Canada; Sable Island, Canada; La Camargue, France; Namib Desert, Namibia; Kaimanawa Ranges, New Zealand; Peneda-Gerês, Portugal; Vodnyi Island, Russia; Galicia, Spain; Exmoor, UK; New Forest, UK; Snowdonia National Park, UK; Assateague Island, USA; Garfield Flat, USA; Granite Range, USA; Great Basin, USA; Jicarilla Wild Horse Territory, USA; Little Book Cliffs, USA; McCullough Peaks, USA; Montgomery Pass, USA; Pine Nut and Pah Rah, USA; Pryor Mountain, USA; Red Desert, USA; Salmon, USA; multiple locations across western USA	0.56 ± 0.001 (0.23–0.92)	0.84 ± 0.002 (0.09–0.97)	0.90 ± 0.001 (0.79–0.98)	1.18 ± 0.001 (0.84–1.39)
E. f. przewalskii[2]	Xinjiang, China; Le Villaret, France; Great Gobi B, Mongolia; Hustai National Park, Mongolia	0.36 ± 0.011 (0.27–0.75)	0.61 ± 0.003 (0.59–0.83)	0.86 ± 0.003 (0.83–0.98)	1.13 ± 0.001 (1.12–1.18)
E. asinus (feral)[3]	Dorisvale, Australia; Victoria River, Australia; Bandelier National Monument, USA; Death Valley, USA; Grand Canyon, USA; Mojave Desert, USA	0.68 ± 0.003 (0.41–0.89)	—	—	1.19 (—)
E. africanus[4]	Denkelia Desert, Eritrea	0.30 ± 0.066 (0.00–0.71)	—	—	—
E. hemionus[5]	Qatrouyeh National Park, Iran; Makhtesh Ramon, Israel; Great Gobi B, Mongolia; Badghyz Nature Reserve, Turkmenistan	0.48 ± 0.004 (0.15–0.76)	0.57 ± 0.001 (0.40–0.64)	0.86 (0.85–0.86)	1.04 ± 0.001 (0.95–1.15)
E. kiang[6]	Tibetan Plateau	0.50 (—)	—	—	—
E. zebra[7]	Ai-Ais/Fish River Canyon National Park, Namibia; BüllsPort Guest Farm, Namibia; Gondwana Cañon Park, Namibia; NamibRand Nature Reserve, Namibia; Camdeboo National Park, South Africa; De Hoop Nature Reserve, South Africa; Karoo National Park, South Africa; Mountain Zebra National Park, South Africa	0.20 ± 0.002 (0.16–0.45)	0.98 ± 0.001 (0.87– 0.98)	0.92 ± 0.001 (0.91– 0.93)	1.13 ± 0.002 (0.85–1.23)
E. grevyi[8]	Barsalinga, Kenya; Dvur Kralove Zoological Garden (translocated from Kenya); Laikipia, Kenya; Lewa, Kenya; Ngare Ndare, Kenya; Samburu Buffalo Springs Game Reserve, Kenya	0.64 (—)	0.57 ± 0.084 (0.27– 0.90)	0.85 ± 0.002 (0.77– 0.88)	1.02 —

***Table 6.1.* continued**

Species	Populations	Mean Adult Birth Rate	Mean Foal Survival Rate	Mean Adult Survival Rate	Mean Population Growth Rate (λ)
E. quagga [9]	Laikipia, Kenya; Athi-Kapiti Plains, Kenya; Serengeti, Tanzania; Ngorongoro, Tanzania; Hwange National Park, Zimbabwe; iMfolozi-Hluhluwe, South Africa; multiple private reserves, South Africa; Kruger National Park, South Africa; Boland Lanbou, South Africa; Elandsberg Private Reserve, South Africa	0.48 ± 0.056 (0.23–0.75)	0.67 ± 0.120 (0.13–0.96)	0.84 ± 0.025 (0.79–0.88)	0.98 ± 0.028 (0.92–1.01)

Sources:

[1]Pellegrini 1971, Tyler 1972, Feist and McCullough 1975, Nelson 1978, Perkins *et al.* 1979, Boyd 1980, Wolfe 1980, Eberhardt *et al.* 1982, Salter and Hudson 1982, Seal and Plotka 1983, Keiper and Houpt 1984, Berger 1986 Siniff *et al.* 1986, Wolfe *et al.* 1989, Garrott and Taylor 1990, Berman 1991, Turner *et al.* 1992, Baker 1993, Monard *et al.* 1997, Greger and Romney 1999, Ashley 2000, Gomes and Oom 2000, Goodloe *et al.* 2000, Ashley and Holcombe 2001, Turner and Morrison 2001, Linklater *et al.* 2004, Greyling 2005, Grange *et al.* 2009, Scorolli and Lopez Cazorla 2010, Contasti *et al.* 2012, Dawson and Hone 2012, Ransom 2012, Spasskaya *et al.* 2012, Stanley and Shultz 2012, Lagos 2013.

[2]Chen *et al.* 2008, Tatin *et al.* 2009, Kaczensky *et al.* 2011, Usukhjargal and Bandi 2013.

[3]Moehlman 1974, Norment and Douglas 1977, Mogart 1978, White 1980, Ruffner and Carothers 1982, Johnson *et al.* 1987, Perryman and Muchlinski 1987, Choquenot 1990, 1991.

[4]P. Moehlman, unpublished data.

[5]Solomatin 1973, 1977, Wolfe 1979, Zhirnov and Ilyinsky 1986, Saltz and Rubenstein 1995, Feh *et al.* 2001, Ziaee 2008, Nowzari 2011, Hemami and Momeni 2013, Lkhagvasuren *et al.* in review.

[6]Schaller 1998.

[7]Joubert 1971, Penzhorn 1985, Lloyd and Rasa 1989, Novellie *et al.* 1992, 1996, Smith *et al.* 2007, Gosling 2013, Eastern Cape Parks, unpublished data, H. Hrabar, unpublished data, South African National Parks, unpublished data.

[8]Dobroruka *et al.* 1987, Rubenstein 1989, 2010, Rowen 1992, Williams 1998.

[9]Klingel 1965, 1969, Petersen and Casebeer 1972, Spinage 1972, Smuts 1976*a,b*, Ohsawa 1982, Rubenstein 1989, Hack *et al.* 2002, Georgiadis *et al.* 2003, Grange *et al.* 2004, Barnier *et al.* 2012, Grange *et al.* 2012, 2015, H. Hrabar and G. Kerley, unpublished data.

Note: Table includes some semiwild populations in managed reserves. Mean rates from each source study were weighted by sample size (number of individuals sampled) and length of study (years) to generate the mean values shown for each species ± standard error. The range of individual study means is given in parentheses after the point estimate. Birth rate is total number of foals divided by all females at least 2 years of age, and survival rate for adults represents individuals ≥2 years of age. Foal survival represents individuals surviving from birth to age 1.

have not been rigorously measured across equid species in the wild, but the feral horses of Kaimanawa, New Zealand, exhibited pregnancy rates of 0.79–0.94, with an average birth rate observed at only 0.49 (Linklater *et al.* 2004). Similarly, feral horses in Idaho, USA, exhibited an average pregnancy rate (across all ages) of 0.68, with only 8% of 1- to 3-year-olds becoming pregnant, increasing to 93% of 10- to 12-year-olds becoming pregnant (Seal and Plotka 1983). Plains zebra pregnancy rates in the Serengeti were reported ranging from 0.69 in 3- to 5-year-olds to 0.88 in 13- to 16-year-olds (Grange *et al.* 2004), but birth rates in similar systems of east Africa were reported as only 0.26–0.51 (Klingel 1969; Georgiadis *et al.* 2003). It is possible that in many studies the reported birth rates may be negatively biased because offspring aborted, stillborn, or depredated immediately after birth were undetected by observers. Abortion is extraordinarily unlikely after three months of gestation in horses, which implies that the environmental conditions early in pregnancy are perhaps most influential in the successful birth of a foal (Roberts 1986).

Interbirth interval (the time between subsequent offspring) among equids can vary depending on individual-level reproductive history and resource availability. Primiparous feral horses, for example, were less likely to foal in the year following their first foal, whereas multiparous females were more likely to foal if they had foaled the previous year (Roelle *et al.* 2010). The interbirth interval among plains zebra, on the other hand, appeared to be longer if the first foal was male (Barnier *et al.* 2012). Feral horses also demonstrated a protracted interbirth interval after producing a male foal under stressed resource conditions (Monard *et al.* 1997).

Both strong physiological and environmental inputs influence the rate and phenology of conception and subsequent year births in most wild equid populations. Hormones associated with reproduction (such as estradiol, progesterone, and testosterone) fluctuate cyclically through time in both males and females (Kirkpatrick *et al.* 1977; Nuñez *et al.* 2011; Schook *et al.* 2013), reaching an annual apex of reproductive potential when enough sunlight and warm temperatures have stimulated the hormonal cascade (Ransom *et al.* 2013). This mechanism can help regulate birth cycles, and ultimately produces birth pulses that are often synchronized with forage availability. Estrus, birth rates, and birth phenology have been correlated with rainfall in some populations of zebras (Klingel 1969; Joubert 1974*b*; Williams 1998; Sinclair *et al.* 2000; Rubenstein 2010), feral horses (Siniff *et al.* 1986; Berman 1991; Scorolli and Lopez Cazorla 2010; Ransom *et al.* 2013), and Asiatic wild asses (Saltz *et al.* 2006). This synchrony with forage resources may increase female and neonate survival, especially because nutritional requirements of lactating females are elevated and higher-quality forage can result in increased body condition and nutritional support for neonates (Ginsberg 1989; National Research Council 2007). In a study of three temperate populations of feral horses where birth phenology was perturbed, estimated survival of foals decreased 1.4% for every 10 days after peak forage that parturition occurred (Ransom *et al.* 2013).

Environmental pressures can contribute greatly to birth phenology, and in some cases can truncate the birth season to a short, synchronized pulse. Predation, for example, has been posited to influence the timing of births under the premise that neonates are most vulnerable in their first hours or days of life, and as such predators can only harvest a finite number of young in a short period of time; consequently, if all young are born together, more should survive (Darling 1938; Sinclair *et al.* 2000). Parturition season for Asiatic wild asses in the Mongolian Gobi (where wolves prey on them), for example, typically begins mid-June and concludes around mid-July (P. Kaczensky, personal communication), and similarly the same species in Badghyz State Reserve, Turkmenistan (13° latitude farther south), was reported to foal primarily between 14 April and 14 May, but some foals were recorded as early as late as 28 July (Wolfe 1979). In contrast, a Sri Lanka feral ass population located almost on the equator and with no predators present exhibited no birth seasonality, and offspring were born uniformly across every month of the year (Santiapillai *et al.* 1999). Predation pressure may not be the primary driver of length of birthing season, however; length of gestation, birth interval, and duration of resource availability can all be important factors (Williams 1998). Wildebeest (*Connochaetes taurinus*) sympatric with Grevy's zebra (*E. grevyi*) in Kenya, for example, exhibited a birth season constrained to only two weeks, whereas the Grevy's zebra present in the same environmental conditions gave birth over the course of months (Williams 1998). In that system, rainfall was highly stochastic, and thus seasonal forage availability did not exhibit a pronounced peak.

Gestation period is similar among equid species, though environmental inputs can create plasticity (Howell and Rollins 1951). Average gestation length ranges from 327 to 357 days in horses (Howell and Rollins 1951; Card and Hillman 1993; Monfort *et al.* 1994), 321 to 404 days in asses (Grinder *et al.* 2006; Pagan *et al.* 2009; Schook *et al.* 2013), and 364 to 390 days in zebras (Grubb 1981; Penzhorn 1988; Churcher 1993). Gestation period in kiang is the least quantified, with reports ranging from 210 to 365 days (St-Louis and Côté 2009). Given the nearly yearlong gestation across all equid species, we could expect estrus, reproductive behavior, and conception to peak between the spring equinox and the summer solstice, waning thereafter with subsequent parturition occurring roughly at the same time the following year. Where such empirical data were present in the literature, this pattern was evident across equid species, though it is more variable in populations near the equator (fig. 6.2).

While body condition, resource availability, and biological inputs all contribute to conception and birth success, these factors may also influence sex ratio of young born into the population. Female feral horses in poor body condition at the time of conception have been observed producing more female foals (Cameron *et al.* 1999), for example, aligning with the much-debated Trivers-Willard hypothesis, which posits that when populations are experiencing more stress (and thus individuals have decreased body condition), more female offspring will be produced because they can increase fitness more quickly than inputs of male offspring (Trivers and Willard 1973). Sex ratio of foals across all equid species and populations we reviewed is of course quite variable, precisely because of these environmental inputs, but on average we found that the ratio of males to females born did not differ from parity (1.1 males to 1 female; 95% CI = 0.93–1.29:1). The extreme situations were 2.2 males to 1 female in a population of mountain zebra ($n = 16$ foals; H. Hrabar, unpublished data), and 2.2 males to 1 female in an Asiatic wild ass population that was experiencing a high growth rate ($n = 53$ foals; Saltz and Rubenstein 1995). In the latter case, it is interesting that while more males were being produced in a density-independent environment

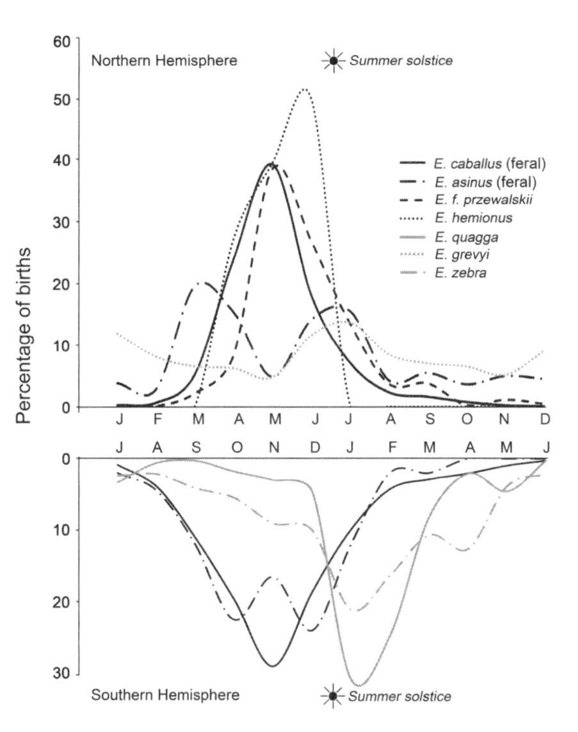

Fig 6.2 Percentage of births by month across equid species. Data are smoothed and shown as the mean across populations. Data for *E. caballus* derived from Tyler 1972, Boyd 1980, Keiper and Houpt 1984, Berger 1986, Berman 1991, Lucas *et al.* 1991, Ashley 2000, Gomes and Oom 2000, Linklater *et al.* 2004, Greyling 2005, Nuñez *et al.* 2010, Scorolli and Lopez Cazorla 2010, Lagos 2013, Ransom *et al.* 2013, and N. Spasskaya, unpublished data; data for *E. asinus* derived from Moehlman 1974, McCool *et al.* 1981, Ruffner and Carothers 1982, and Santiapillai *et al.* 1999; data for *E. f. przewalskii* derived from Chen *et al.* 2008, Prague Zoo 2010 (wild and semiwild populations only), and Usukhjargal and Bandi 2013; data for *E. hemionus* derived from Wolfe 1979 and O. Ganbaatar, unpublished data; data for *E. quagga* derived from Klingel 1965, Smuts 1976*b*, and Sinclair *et al.* 2000; data for *E. grevyi* derived from Dobroruka *et al.* 1987 and Rowen 1992; data for *E. zebra* derived from Joubert 1974*b*, Penzhorn 1985, Westlin-van Aarde *et al.* 1988, and H. Hrabar, unpublished data.

(in concurrence with Cameron and Linklater 2007), the investigators noted that more female foals were being produced by primiparous mothers of age 2 or 3 years as compared to multiparous mothers of age 4 or 5 years. This may reflect an evolutionary disparity in effective maternal investment, but that hypothesis was not specifically tested. The opposite extreme was observed in a resource-stressed population of plains zebra where the sex ratio was 0.27 males to 1 female (n = 19 foals; Hrabar and Kerley 2013) and a feral horse population where the sex ratio of 0.47 males to 1 female was observed (n = 25 foals; Stanley and Shultz 2012).

Survival

Most feral horses, feral asses, and some Przewalski's horses and mountain zebra live in environments without significant predation pressure and exhibit relatively high survival rates across all age classes. Most other wild equid populations are subject to predation from a variety of large predators, and thus experience locally reduced survival rates. Despite this dichotomy of population paradigms, age-specific survival is surprisingly comparable among equid species, with some elevated mortality in foals, relative security in the adult cohort, and then a gradual decrease as senescence prevails (fig. 6.3). The resulting age structure of wild populations is also relatively consistent among species and populations, with foals typically comprising 8–15% of a population, juveniles 13–28% of a population, and adults 71–78% of a population (Berman 1991; Tatin *et al.* 2003; Grange *et al.* 2004; Smith *et al.* 2007). Equids are long-lived animals, and where individuals were followed from birth, known ages for individual wild Przewalski's horses exceeded 25 years old (Prague Zoo 2010) and for feral horses exceeded 29 years old (Ransom 2012). Equids generally live longer in captivity owing to more controlled environmental conditions and applied veterinary care; as such, captive Asiatic wild asses >27 years old have been reported (Volf 2010), as have Przewalski's horses >30 years old (Ballou 1994). Mortalities in wild equid populations arise from a variety of causes. Deaths of adult Przewalski's horses (n = 145) at Hustai National Park, Mongolia, for example, were attributed to blood parasites (2.1%), endoparasites (1.4%), disease (7.6%), stillbirth (9.7%), injury (12.4%), predation (22.1%), starvation (31.1%), and other/unknown reasons (13.8%) (Usukhjargal and Bandi 2013). Nine percent of foal deaths (25 of 278 foals) in the same study were attributed to infanticide by stallions (box 6.1).

Annual foal survival rates reported across the studies we reviewed averaged 0.71 (95% CI = 0.53–0.88), and adult survival rates averaged 0.88 (95% CI = 0.85–0.90) (table 6.1). Much of the variation in annual survival rates of both foals and adults by population may be attributed to localized predation. A large number of predators have been documented killing both young and adult equids, including African lion (*Panthera leo*), leopard (*P. pardus*), cheetah (*Acinonyx jubatus*), and spotted hyena (*Crocuta crocuta*) for feral horses, zebras, and asses in Africa; wolf (*Canis lupus*) and snow leopard (*Uncia uncia*) for Asiatic asses, Przewalski's horses, and kiang in the Middle East and Asia; wolf and brown bear (*Ursus arctos*) for feral horses in Europe; and puma

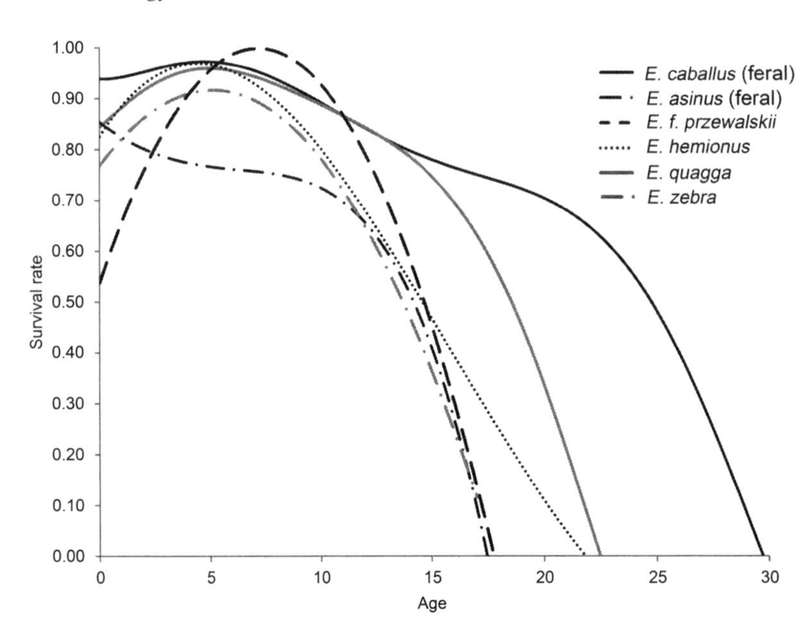

Fig. 6.3 Trend in age-specific survival rates for equid species. Data for *E. caballus* derived from Ransom 2012; data for *E. asinus* derived from McCool *et al.* 1981; data for *E. f. przewalskii* derived from Hustai National Park, unpublished data; data for *E. hemionus* derived from Saltz and Rubenstein 1995 and Lkhagvasuren *et al.* in review; data for *E. quagga* derived from Spinage 1972; data for *E. zebra* derived from Joubert 1974*a*.

(*Puma concolor*) and jaguar (*P. onca*) for feral horses and feral asses in the Americas. In one feral horse population in the western United States, foal survival was reported to be only 27%, and 82% of foal carcasses found showed evidence of being killed by puma (Turner *et al.* 1992); however, most feral horse populations coexist with pumas and experience limited depredation. In the Gobi Desert of Mongolia, 23% of wolf scats analyzed contained remains of Asiatic ass (Feh *et al.* 2001), but it is unknown how many of those samples arose from predation as opposed to scavenging.

Analysis of predator diets has provided some evidence of selection for equids, with a wide range of selection to nonselection across species and seasons. Wolves in northwest Spain clearly selected for feral horses (Ivlev index = 0.99–1.00) (Lagos 2013), but a similarly pack-organized canine, African wild dog (*Lycaon pictus*), specifically avoided hunting zebra even when they were relatively abundant (Hayward *et al.* 2006). In Lewa, Kenya, African lions strongly selected Grevy's zebra over plains zebra, even though plains zebra outnumbered Grevy's zebra four to one (Rubenstein 2010). In the Serengeti, Tanzania, 90% of all spotted hyena kills were adult zebra when the available ungulate assemblage was quite diverse (Kruuk 1972). In systems where feral horses and predators most typically coexist, a high percentage of all mortality is attributed to a sole predator as opposed to disease, accident, or other natural causes. Wolves in Peneda-Gerês National Park, Portugal, for example, have been responsible for 89% of horse mortalities, and in nearby Galicia, Spain, that figure rises to 96% (Gomes and Oom 2000; Lagos 2013). In systems with multiple predators or sympatric prey species, the percentage of equid mortality attributed to depredation can be somewhat ameliorated, such as plains zebras in the Serengeti, Tanzania, where 59–74% of all mortality was attributed to African lion and spotted hyena (Schaller 1972), or Przewalski's horses in Hustai National Park, where 13–50% of horse mortalities were attributed to wolves (Van Duyne *et al.* 2009).

At a finer scale, predator species may also select for one sex of prey over the other; thus they may affect prey populations differently at long timescales. Spotted hyenas at Ngorongoro, Tanzania, killed 2.25 female plains zebra for every one male, but lions at Serengeti, Tanzania, killed only 0.83 female plains zebra for every one male (Kruuk 1972). Longitudinal studies indicate that shifts in diet can also occur over time, as in the Serengeti, Tanzania, where 19–20% of spotted hyena diet was composed of plains zebra before the 1970s, but after that time it diminished to <9% (Kruuk 1972). The zebra population remained relatively stable over that time period (Hack *et al.* 2002), but changes in other prey populations or competing predator populations could have contributed to the apparent dietary shift.

Some behavioral traits may also influence survival. Group formation in ungulates has been posited to be a function of open environments where foraging can be optimized because groups of animals are more vigilant for predators than isolated individuals can be (Kie 1999). Equids form groups of various sizes and composition, which may influence predation risk (see chap. 2). Schaller (1972) reported that lions were 13% more successful when plains zebras were alone rather than

BOX 6.1 INFANTICIDE AMONG EQUIDS

Infanticide is not common in equids but has been reported in both captive and wild populations. This unique source of mortality arises when a stallion intentionally kills a foal, an action hypothesized to be a fitness-driven evolutionary behavior (Hrdy 1979). The concept posits that if a novel stallion kills an unrelated foal, then the foal's mother may more quickly return to estrus and be fertilized by the assailant male, thus increasing his fitness faster through reduction of interbirth interval. This concept has been discussed most broadly in the classic case of the African lion (Packer and Pusy 1982), but it has drawn the attention of equine ethologists because of their polyandrous mating systems.

Infanticide among the equids appears to be most prevalent in the harem-forming species and has been reported most frequently in captive conditions where artificial social constructs may elevate frequency (Pluháček *et al.* 2006). In the wild, infanticide has been documented at the highest rates in Przewalski's horses, where 9% (25 of 278 foals born 1993–2011) of all foal mortality in Hustai National Park, Mongolia, was attributed to this phenomenon (Usukhjargal and Bandi 2013). It was also reported in the same species at Kalamaili Ungulate Protected Area, China, where 83.3% of all foal mortalities (7 mortalities out of 24 births) was attributed to infanticide in the 5-year period following reintroduction (Chen *et al.* 2008). Nearly half of those deaths occurred in the first year, lending support to the proximate social stresses presented when novel stallions suddenly become available to social groups.

Infanticide was reported in domestic horses (Duncan 1982) and New Forest ponies (Tyler 1972), and attempted infanticide among feral horses has been documented across populations (Tyler 1972; Berger 1986; Gray 2009). When more than one adult male has been present in a band of feral horses, extraordinary maternal protection has also been reported, potentially minimizing the incidents of infanticide (Cameron *et al.* 2003). While the sample size is limited, there is some evidence that intruding stallions may selectively target male foals (all cases reported by Duncan 1982, with additional cases reported by Berger 1986), which could reflect the potential for reducing future competition from unrelated males, but that hypothesis has never been explicitly tested. Induced abortion has also been documented when novel stallions forcibly copulate with a pregnant female. At the Granite Range, USA, such feticide was reported for 78% of forced copulations by novel feral horse males (Berger 1986).

Harem-forming zebra have also demonstrated infanticide, with three cases being reported in wild mountain zebra (Penzhorn 1975), and multiple cases reported in captive plains zebra (of which several individuals originated from wild populations) (Pluháček *et al.* 2006). In a captive mountain zebra enclosure in Namibia, Joubert (1971) reported an attempted infanticide that failed when the female actively defended her foal, but the stallion then proceeded to kill two sympatric antelope instead. This event illustrates the uncertainties underlying our understanding of mechanisms and motivations related to infanticide (Pluháček *et al.* 2006). It is possible that some cases of infanticide are motivated by a more immediate need: survival. At Hluhluwe Reserve, South Africa, a severely injured plains zebra foal was observed in the center of a band vocalizing continuously and loudly after escaping from a predator attack. All adult members of the band were expressing elevated vigilance behaviors and nervousness, and the band stallion eventually approached and killed the foal with strikes to the head. While the genetic relationship of the stallion and foal was unknown, the silencing of the foal instantly reduced the exposure of the band to local predators (J. Ransom, unpublished data). Infanticide has not been reported for Grevy's zebra, but the territorial organization of this species can result in females not remaining with specific males long enough for the males to benefit from such behavior.

Photo by J. Ransom

in a group. Unlike most equids, where foals closely accompany their dams most or all of the time, Grevy's zebra sometimes leave their foals in aggregated juvenile groups or "kindergartens" (Williams 1998). The source of this behavior may arise from energetic constraints of young, but ultimately it may also place foals at risk of depredation, especially during nocturnal aggregations (Klingel 1974).

Because most equids inhabit arid environments, considerable interannual variance in precipitation and related productivity may also influence survival (Archer *et al.* 1999). One model simulation of zebra populations suggested that survival decreases more rapidly during dry seasons than reproduction increases during wet seasons, which would constrain population growth over time (Georgiadis *et al.* 2003). In contrast, neither wet season nor dry season rainfall at Kruger National Park, South Africa, influenced plains zebra survival in a longitudinal empirical study, though it did affect other ungulates in the system (Owen-Smith *et al.* 2005). Across ungulate species, it appears that interannual variance in precipitation influences juvenile survival more profoundly than it does adult survival (Gaillard *et al.* 1998, 2000). In feral horses, each centimeter of spring rainfall increased foal survival by 1.5%, did not measurably influence survival for animals up to age 19, and then significantly increased survival rates for animals ≥20 years old (Ransom 2012).

Population Growth

Quantifying population dynamics is generally not as simple as adding births and subtracting deaths. Understanding population growth and decline is further complicated by movement into and out of populations, and by catastrophic stochastic events that change rates both instantaneously and longitudinally. In addition to that volume of uncertainty, the size of most wild equid populations is not absolutely known or even estimated robustly (Moehlman 2002). We have reviewed studies where some measure of population growth was reported or could be reasonably derived. To discuss this more clearly, we refer to the theta-logistic model because it accommodates the nonlinear density dependence more indicative of equid populations (Gilpin and Ayala 1973):

$$\frac{dN}{dt} = rN\left(1 - \frac{N}{K}\right)^{\theta}$$

In this model, r is the intrinsic rate of increase, N is the population size, K is the carrying capacity, t is time, and θ is the nonlinear modifying parameter for the ratio of N:K. Conceptually, when $\theta < 1$, r decreases rapidly with population size; likewise, the population will quickly approach K when $\theta > 1$. For equids, this is a steeper curve than the classic logistic sigmoid curve, and the inflection point is much closer to K (fig. 6.4). In this example, a simplified deterministic population trajectory is used to illustrate growth of hypothetical equid population. To add a degree of reality, K is randomized (K') to illustrate how carrying capacity can vary annually with the amount and quality of resources available over time, which in turn can influence the maximum number of individuals that could survive at a given time (fig. 6.4, gray lines). Conceptually, when rapid growth is reported for a population, the theta-logistic model suggests that the population is most likely below α (the inflection point) on the sigmoid curve, and when slow growth is observed, the population might be closer to K at that point in time. Many studies report an instantaneous rate of increase (λ) rather than r. This simply reflects the growth rate of a population from one year to the next as if it had instantly occurred, or $\lambda_t = (N_t / N_{t-1})$ where t represents one year. For synthesis and discussion purposes, we have converted r values from individual studies to λ, using $\lambda = e^r$.

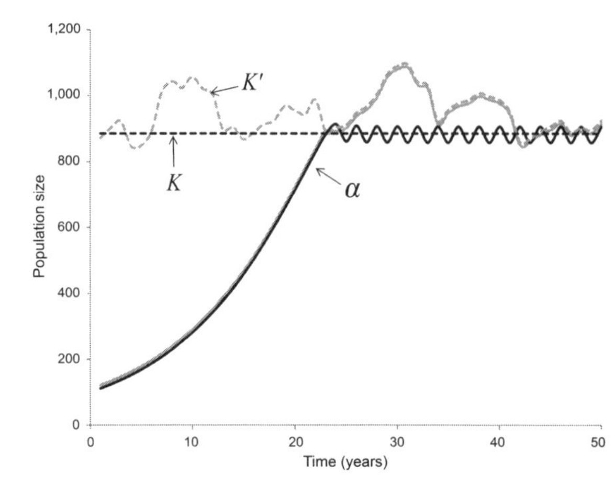

Fig. 6.4 An example of theta-logistic population growth for a simulated equid population over time (black line), where growth rate (r) is fixed and reaches the fixed carrying capacity of K. The same population is simulated with a randomized carrying capacity (K') in gray, illustrating the stochastic effects of resource availability on a single parameter. Point α represents the inflection point where population growth begins to moderate as the population approaches K.

Mean annual population growth (λ) in the studies we reviewed averaged 1.12 (95% CI = 1.04–1.19) across all species (table 6.1). While the reported range of growth rates was similar between species, the average growth rates for Asiatic wild ass, Grevy's zebra, and plains zebra were notably lower than the other species (1.04, 1.02, and 0.98, respectively). It is unknown whether this finding simply reflects the paucity of data for these species, is an artifact of their different life history strategies, or is indicative of a more complex set of factors. Robust population dynamics metrics for African wild ass and kiang are completely missing from the literature, illustrating critical science needs for better understanding equid ecology.

Some variability in population growth rates over time and between studies and species may be explained by the relationship between abundance and resources. Equids are generally well equipped to persist near *K* because their longevity and high survival rates, low recruitment rates, and low sensitivity to moderate environmental stressors result in relatively durable populations, barring catastrophic mortality events (Saltz 2002). When mass mortality events occur, equids are quite slow to recover because of their inherently low recruitment rates, but at the longitudinal scale, this is when equids should be most productive. For example, a small, reintroduced Asiatic wild ass (onager) population exhibited its maximum growth rate as abundance expanded to the resources available (Saltz and Rubenstein 1995). In feral horses, heavily managed populations that have persisted near *K* but then experience a large culling event can be suddenly reduced far below α, prompting a compensatory response in increased growth rate (Choquenot 1990; National Research Council 2013).

Focused predation can impart nonrandom influence on population growth, such as at Pryor Mountain, USA, where predation from specific pumas was increasingly limiting the horse population from 1993 to 2004 to a nadir of $\lambda = 0.97$ before the pumas were killed and the predation pressure was dramatically released (Ransom 2012). In a plains zebra population in South Africa, models of population growth including lion predation were far more supported than models of climate factors, though population growth of only one of the two populations was affected by number of lions present (Grange *et al.* 2012). While food, water, and predation have been widely investigated as agents of population constraint, behavior has also been posited to limit equid populations by reduced fecundity arising from social stress (Linklater 2000; Tatin *et al.* 2009).

Perhaps the most unpredictable natural influence on population growth of equids also carries the greatest magnitude of effect: catastrophic climate events such as extreme drought and storms. Such events can have lasting influence on population dynamics and in some cases can threaten extirpation of populations. The dzud winter (extreme cold with heavy snow) of 2009/2010 in the Mongolian Gobi, for example, suddenly reduced the Przewalski's horse population by 60% in a single season (Kaczensky *et al.* 2011). The population effect from direct mortality was compounded by a severe reduction in per capita births the following year, with only one female foaling. Though not well quantified, the extirpation of Asiatic wild asses from Kazakhstan at the end of the nineteenth century was documented following a similar dzud winter, though hunting by humans and competition with their livestock certainly contributed to the decline (Zhirnov and Ilyinsky 1986). Winter mass mortalities of feral horses in the American West have been documented with a 51% decline in a single winter at Pryor Mountain (Garrott and Taylor 1990) and 12% in a winter at the Granite Range (Berger 1983). The latter study indicated that severe snowstorm mortality might be more common than assumed and also referenced an earlier unpublished event in the Granite Range when 300 horses died (roughly 50% of the population).

Despite the extraordinary capacity of equids to persist in harsh arid environments, drought can still promulgate catastrophe. One severe mass mortality event related to drought was reported for plains zebra at the Tuli Game Reserve in southern Africa in 1983, where 199 carcasses and only 58 live zebra were detected following two years of drought (Walker *et al.* 1987). Hundreds of mountain zebra were reported dead in the Namibian Khomas Hochland during the 1968/1969 drought, with mortalities attributed to starvation, entrapment in mud at waterholes, and behavioral disorientation leading to increased human depredation (Joubert 1971). Large-scale drought-related mortalities are relatively common in the Australian outback, where tens of thousands of feral horses perish during extended drought. Similar events are generally precluded in the United States only by artificial water supplementation.

Severe thunderstorms have also had population-level effects on small equid populations. For example, a thunderstorm overtook Tornquist Provincial Park in Argentina, killing 193 feral horses in 48 hours (28% of the population) (Scorolli *et al.* 2006). Likewise, electrical storms at the high-altitude area of Pryor Mountain have produced fatal lightning strikes to individuals and entire feral horse bands, with an event in 1999 killing 4% of the population with a single lightning strike (US Bureau of Land Management, unpublished data).

Little has been quantified regarding movement of individuals into and out of equid populations in the context of population growth rate. But one only need consider the migrant case of the plains zebra to understand the ramifications of how migration can quickly influence access to resources and changes in body condition that may influence birth rates, or how such migrations attract predators that might influence survival rates (Naidoo *et al.* 2014). Localized movements of plains zebra in Kruger National Park, South Africa, required travel of 10–15 km every 1–2 days during the dry season for individuals to drink, thus reducing time available for foraging and resting (Cain *et al.* 2012). Feral horses in central Australia also traveled up to 15 km during drought in order to access water, and one population declined 51% in density during a severe drought; it is unknown how many of those animals died as opposed to emigrated from the area (Berman 1991). In a similar xeric environment, feral asses of Death Valley became more prevalent within 1.6 km of a water source in summer (Moehlman 1974). Anthropogenic disturbance at water holes has been implicated in movement of Grevy's zebra, with the energetic costs arising from long-distance movement to and from water being reflected in reduced foal survival (Williams 1998). Whether any of these traveling individuals truly moved from one population to another, affecting population demography, is unknown, but the catalysts for such movement clearly provide an opportunity for immigration and emigration, which ultimately can influence both population dynamics and genetic flow.

Conclusion

The wild and feral equid populations of the world share many common features, and they function similarly for a variety of reasons. Equids can be quite long-lived ungulates and demonstrate relatively low fecundity, meaning that, in the absence of extreme stressors, their populations should be rather stable over time. Unfortunately, this is seldom realized in a changing world where anthropogenic influence permeates nearly all populations (see chap. 12) and climate events can impart severe catastrophes. Recovering from such population decimations requires long timescales for equids. Ultimately, most wild equids persist in a state where extensive conservation measures are needed to prevent extirpation and most feral equids persist at the other end of the curve, where extensive management is necessary to mitigate resource damage.

We aimed in this synthesis to gain a better understanding of equid ecology and what naturally shapes their populations, but such findings are not functionally separate from the effects that humans have on these populations. Direct killing, encroachment, and resource competition remain the largest threats to wild equids and profoundly influence the dynamics of most populations. The scale of human impacts thus confounds attempts to understand the natural influences on these species. What we can conclude, however, is that precipitation and phenological rhythms can strongly influence births, whereas survival is more strongly influenced by predation and mass catastrophe. Equids also exhibit complex social behavior that may drive population dynamics in ways we simply have not quantified or fully understand (see chap. 3). Such uncertainties in both the changing environment and our shortfalls in understanding the depth of equid societies demand forward thinking conservation and careful management for a group of species that nearly all persist on the world's threatened and endangered list.

ACKNOWLEDGMENTS

The authors wish to thank the peer reviewers, who provided helpful and constructive comments that ultimately improved this synthesis. Special thanks are owed to the many colleagues that shared unpublished data and responded promptly to the authors' queries, all toward an open effort to better understand equids.

REFERENCES

Archer, S., W. MacKay, J. Mott, S.E. Nicholson, M. Pando Moreno, M.L. Rosenzweig, N.G. Seligman, N.E. West, and J. Williams. 1999. Arid and semi-arid land community dynamics in a management context. Pages 48–74 *in* T. W. Hoekstra, and M. Shachak (Eds.), Arid lands management: Toward ecological sustainability. University of Illinois Press, Chicago, USA.

Ashley, M.C. 2000. Feral horses in the desert: Population genetics, demography, mating, and management. Thesis, University of Nevada, Reno, USA.

Ashley, M.C., and D.W. Holcombe. 2001. Effect of stress induced by gathers and removals on reproductive success of feral horses. Wildlife Society Bulletin 29:248–254.

Baker, S. 1993. Survival of the fittest: A natural history of the Exmoor pony. Exmoor Books, Exeter, UK.

Ballou, J.D. 1994. Population biology. Pages 93–113 *in* L. Boyd and K. Houpt (Eds.), Przewalski's horse: The history and biology on an endangered species. State University of New York Press, Albany, USA.

Barnier, F., S. Grange, A. Ganswindt, H. Ncube, and P. Duncan. 2012. Inter-birth interval in zebras is longer following the birth of male foals than after female foals. Acta Oecologica 42:11–15.

Berger, J. 1983. Ecology and catastrophic mortality in wild horses: Implications for interpreting fossil assemblages. Science 220:1403–1404.

Berger, J. 1986. Wild horses of the Great Basin. University of Chicago Press, Chicago, USA.
Berman, D.M. 1991. The ecology of feral horses in central Australia. Thesis, University of New England, Armidale, NSW, Australia.
Björnstad, O.N., and B.T. Grenfell. 2001. Noisy clockwork: Time series analysis of population fluctuations in animals. Science 293:638–643.
Bonenfant, C., J.-M. Gaillard, T. Coulson, M. Festa-Bianchet, A. Loison, *et al.* 2009. Empirical evidence of density-dependence in populations of large herbivores. Advances in Ecological Research 41:313–357.
Boyd, L.E. 1980. The natality, foal survivorship, and mare-foal behavior of feral horses in Wyoming's Red Desert. Thesis, University of Wyoming, Laramie, USA.
Cain, J.W., III, N. Owen-Smith, and V.A. Macandza. 2012. The costs of drinking: Comparative water dependency of sable antelope and zebra. Journal of Zoology 286:58–67.
Cameron, E.Z., and W.L. Linklater. 2007. Extreme sex ratio variation in relation to change in condition around conception. Biology Letters 3:395–397.
Cameron, E.Z., W.L. Linklater, K.J. Stafford, and E.O. Minot. 2003. Social grouping and maternal behaviour in feral horses (*Equus caballus*): The influence of males on maternal protectiveness. Behavioural Ecology and Sociobiology 53:92–101.
Cameron, E.Z., W.L. Linklater, K.J. Stafford, and C.J. Veltman. 1999. Birth sex ratios relate to mare condition at conception in Kaimanawa horses. Behavioral Ecology 10:472–475.
Card, C.E., and R.B. Hillman. 1993. Parturition. Pages 567–573 *in* A.O. McKinnon and J.L. Voss (Eds.), Equine reproduction. Williams and Wilkins, Philadelphia, USA.
Chen, J., Q. Weng, J. Chao, D. Hu, and K. Taya. 2008. Reproduction and development of the released Przewalski's horses (*Equus przewalskii*) in Xinjiang, China. Journal of Equine Science 19:1–7.
Choquenot, D. 1990. Rate of increase for populations of feral donkeys in northern Australia. Journal of Mammalogy 71:151–155.
Choquenot, D. 1991. Density-dependent growth, body condition, and demography in feral donkeys: Testing the food hypothesis. Ecology 72:805–813.
Churcher, C.S. 1993. *Equus grevyi*. Mammalian Species 453:1–9.
Contasti, A.L., E.J. Tissier, J.F. Johnstone, and P.D. McLoughlin. 2012. Explaining spatial heterogeneity in population dynamics and genetics from spatial variation in resources for a large herbivore. PLoS ONE 7:e47858.
Darling, F.F. 1938. Bird flocks and the breeding cycle. Cambridge University Press, Cambridge, UK.
Dawson, M.J., and J. Hone. 2012. Demography and dynamics of three wild horse populations in the Australian Alps. Austral Ecology 37:97–109.
Dobroruka, L.J., A. Holejsovska, I. Maslova, and V. Novotny. 1987. An analysis of the population of Grevy's zebra. International Zoo Yearbook 26:290–293.
Duncan, P.B. 1982. Foal killing by stallions. Applied Animal Ethology 8:567–570.
Eberhardt, L.L. 1977. Optimal management strategies for marine mammals. Wildlife Society Bulletin 5:162–169.
Eberhardt, L.L., A.K. Majorowicz, and J.A. Wilcox. 1982. Apparent rates of increase for two feral horse herds. Journal of Wildlife Management 46:367–374.
Ellis, J.E., and D.M. Swift. 1988. Stability of African pastoral ecosystems: Alternate paradigms and implications for development. Journal of Range Management 41:450–459.
Feh, C., B. Munkhtuya, S. Enkhbold, and T. Sukhbaatar. 2001. Ecology and social structure of the Gobi khulan *Equus hemionus* subsp. in the Gobi B National Park, Mongolia. Biological Conservation 101:51–61.
Feist, J.D., and D.R. McCullough. 1975. Reproduction in feral horses. Journal of Reproduction and Fertility Supplement 23:13–18.
Flux, J.E.C. 2001. Evidence of self-limitation in wild vertebrate populations. Oikos 92:555–557.
Fowler, C.W. 1987. A review of density dependence in populations of large mammals. Pages 401–441 *in* H.H. Genoways (Ed.), Current mammalogy. Plenum, New York, USA.
Gaidet, N., and J.M. Gaillard. 2008. Density-dependent body condition and recruitment in a tropical ungulate. Canadian Journal of Zoology 86:24–32.
Gaillard, J.-M., M. Fest-Bianchet, and N.G. Yoccuz. 1998. Population dynamics of large herbivores: Variable recruitment with constant adult survival. Trends in Ecology and Evolution 13:58–63.
Gaillard, J.-M., M. Festa-Bianchet, N.G. Yoccoz, A. Loison, and C. Toïgo. 2000. Temporal variation in fitness components and population dynamics of large herbivores. Annual Review of Ecology and Systematics 31:367–393.
Garrott, R.A., and L. Taylor. 1990. Dynamics of a feral horse population in Montana. Journal of Wildlife Management 54:603–612.
Georgiadis, N., M. Hack, and K. Turpin. 2003. The influence of rainfall on zebra population dynamics: Implications for management. Journal of Applied Ecology 40:125–136.
Gilpin, M.E., and F.J. Ayala. 1973. Global models of growth and competition. Proceedings of the National Academies of Science 70:3590–3593.
Ginsberg, J.R. 1989. The ecology of female behaviour and male mating success in the Grevy's zebra, *Equus grevyi*. Symposium of the Zoological Society of London 61:89–110.
Gomes, J.C., and M.D.M. Oom. 2000. Demographic data on the garrano's feral population from the Peneda-Gerês National Park. [In Portugese.] Revista Portuguesa de Zootecnia 7:67–68.
Goodloe, R.B., R.J. Warren, D.A. Osborn, and C. Hall. 2000. Population characteristics of feral horses on Cumberland Island, Georgia and their management implications. Journal of Wildlife Management 64:114–121.
Gosling, L.M. 2013. Population ecology of Hartmann's mountain zebra: Comparisons between protected areas in southern Namibia. Fifth Progress Report. Namibia Nature Foundation, Windhoek, Namibia.
Grange, S., P. Duncan, J.-M. Gaillard, A.R.E. Sinclair, P.J.P. Gogan, C. Packer, H. Hofer, and M. East. 2004. What limits the Serengeti zebra population? Oecologia 140:523–532.
Grange, S., P. Duncan, and J.-M. Gaillard. 2009. Poor horse traders: Large mammals trade survival for reproduction during the process of feralization. Proceedings of the Royal Society of London B: Biological Sciences 276:1911–1919.

Grange, S., N. Owen-Smith, J.-M. Gaillard, D.J. Druce, M. Moleón, and M. Mgobozi. 2012. Changes of population trends and mortality patterns in response to the reintroduction of large predators: The case study of African ungulates. Acta Oecologica 42:16–29.

Grange, S., F. Barnier, P. Duncan, J.-M. Gaillard, M. Valeix, H. Ncube, S. Périquet, and H. Fritz. 2015. Demography of plains zebras (*Equus quagga*) under heavy predation. Population Ecology 57:201–214.

Gray, M.E. 2009. An infanticide attempt by a free-roaming feral stallion (*Equus caballus*). Biology Letters 5:23–25

Greger, P.D., and E.M. Romney. 1999. High foal mortality limits growth of a desert feral horse population in Nevada. Great Basin Naturalist 59:374–379.

Greyling, T. 2005. Factors affecting possible management strategies for the Namib feral horses. Thesis, North-West University, Potchesfstroom, South Africa.

Grinder, M.I., P.R. Krausman, and R.S. Hoffman. 2006. *Equus asinus*. Mammalian Species 794:1–9.

Grubb, P. 1981. *Equus burchelli*. Mammalian Species 157:1–9.

Hack, M.A., R. East, and D.I. Rubenstein. 2002. Status and action plan for the plains zebra (*Equus burchellii*). Pages 43–60 *in* P. Moehlman (Ed.), Equids: Zebras, asses and horses. Status survey and conservation action plan. International Union for Conservation of Nature, Gland, Switerland.

Hayward, M.W., J. O'Brien, M. Hofmeyer, and G.I.H. Kerley. 2006. Prey preferences of the African wild dog *Lycaon pictus*: Ecological requirements for their conservation. Journal of Mammalogy 87:1122–1131.

Hemami, M.-R., and M. Momeni. 2013. Estimating abundance of the endangered onager *Equus hemionus onager* in Qatruiyeh National Park, Iran. Oryx 47:266–272.

Howell, C.E., and W.C. Rollins. 1951. Environmental sources of variation in the gestation length of the horse. Journal of Animal Science 10:789–796.

Hrabar, H., and G. Kerley. 2013. Selective breeding in the Quagga Breeding Program—The effect of translocations and inbreeding on plains zebra reproduction. Report C127. Nelson Mandela Metropolitan University, Port Elizabeth, South Africa.

Hrdy, S.B. 1979. Infanticide among animals: A review, classification, and examination of the implications for the reproductive strategies of females. Ethology and Sociobiology 1:13–40.

Johnson, R.A., S.W. Carothers, and T.J. McGill. 1987. Demography of feral burros in the Mohave Desert. Journal of Wildlife Management 51:916–920.

Joubert, E. 1971. Ecology, behaviour, and population dynamics of the Hartmann zebra *Equus zebra hartmannae* Matchie, 1989 in south west Africa. Thesis, University of Pretoria, Pretoria, South Africa.

Joubert, E. 1974*a*. Notes on the reproduction in Hartmann zebra *Equus zebra hartmannae* in south west Africa. Madoqua 1:31–35.

Joubert, E. 1974*b*. Composition and limiting factors of a Khomas Hochland population of Hartmann zebra *Equus zebra hartmannae*. Madoqua 1:49–53.

Kaczensky, P., O. Ganbataar, N. Altansukh, N. Enkhsaikhan, C. Stauffer, and C. Walzer. 2011. The danger of having all your eggs in one basket—Winter crash of the re-introduced przewalski's horses in the Mongolian Gobi. PLoS ONE 6:e28057.

Keiper, R., and K. Houpt. 1984. Reproduction in feral horses: An eight-year study. American Journal of Veterinary Research 45:991–995.

Kie, J.G. 1999. Optimal foraging and risk predation: Effects on behavior and social structure in ungulates. Journal of Mammalogy 80:1114–1129.

Kirkpatrick, J.F., L. Wiesner, R.M. Kenney, V.K. Ganjam, and J.W. Turner. 1977. Seasonal variation in plasma androgens and testosterone in the North American wild horse. Journal of Endocrinology 72:237–238.

Klingel, H. 1965. Notes on the biology of the plains zebra (*Equus quagga boehmi* Matschie). East African Wildlife Journal 3:86–88.

Klingel, H. 1969. Reproduction in the plains zebra, *Equus burchelli boehmi*: Behaviour and ecological factors. Journal of Reproduction and Fertility Supplement 6:339–345.

Klingel, H. 1974. Social organisation and behaviour of the Grevy's zebra. [In German.] Zeitschrift für Tierpsychologie 36:36–70.

Kruuk, H. 1972. The spotted hyena: A study of predation and social behaviour. University of Chicago Press, Chicago, USA.

Lagos, L. 2013. Ecology of the wolf (*Canis lupus*), the wild pony (*Equus ferus atlanticus*) and the semi-extensive cattle (*Bos taurus*) in Galicia: Predator-prey interactions. [In Spanish.] Thesis, Universidad de Santiago de Compostela, Santiago de Compostela, Spain.

Linklater, W.L. 2000. Adaptive explanation in socio-ecology: Lessons from the Equidae. Biological Reviews 75:1–20.

Linklater, W.L., E.Z. Cameron, E.O. Minot, and K.J. Stafford. 2004. Feral horse demography and population growth in the Kaimanawa Ranges, New Zealand. Wildlife Research 31:119–128.

Lkhagvasuren, D., N. Batsaikhan, W.F. Fagan, E.C. Ghandakly, P. Kaczensky, *et al.* In review. Population structure of the Asiatic wild ass in Mongolia—Extracting age and mortality estimates from skull samples.

Lloyd, P.H., and O.A.E. Rasa. 1989. Status, reproductive success and fitness in Cape mountain zebra (*Equus zebra zebra*). Behavioral Ecology and Sociobiology 25:411–420.

Lucas, Z., J.I. Raeside, and K.J. Betteridge. 1991. Non-invasive assessment of the incidences of pregnancy and pregnancy loss in the feral horses of Sable Island. Journal of Reproduction and Fertility Supplement 44:479–488.

McCool, C.J., C.C. Pollitt, G.R. Fallon, and A.F. Turner. 1981. Studies of feral donkeys in the Victoria River-Kimberleys area: Observations on behaviour, reproduction and habitat and some possible control strategies. Australian Veterinary Journal 57:444–449.

Moehlman, P.D. 1974. Behavior and ecology of feral asses (*Equus asinus*). Thesis, University of Wisconsin–Madison, Madison, USA.

Moehlman, P.D. (Ed.). 2002. Equids: Zebras, asses and horses. Status survey and conservation action plan. International Union for Conservation of Nature, Gland, Switerland.

Mogart, J.R. 1978. Burro behavior and population dynamics, Bandelier National Monument, New Mexico. Thesis, Arizona State University, Tempe, USA.

Monard, A.M., P. Duncan, H. Fritz, and C. Feh. 1997. Variations in the birth sex ratio and neonatal mortality in a natural herd of horses. Behavioral Ecology and Sociobiology 41:243–249.

Monfort, S.L., N.P. Arthur, and D.E. Wildt. 1994. Reproduction in the Przewalski's horse. Pages 173–193 *in* L. Boyd and K. Houpt (Eds.), Przewalski's horse: The history and biology on an endangered species. State University of New York Press, Albany, USA.

Naidoo, R., M.J. Chase, P. Beytell, P. Du Preez, K. Landen, G. Stuart-Hill, and R. Taylor. 2014. A newly discovered wildlife migration in Namibia and Botswana is the longest in Africa. Oryx doi:10.1017/S0030605314000222.

National Research Council. 2007. Nutrient requirements of horses, 6th rev. ed. National Academies Press, Washington, DC, USA.

National Research Council. 2013. Using science to improve the BLM Wild Horse and Burro Program: A way forward. National Academies Press, Washington, DC, USA.

Nelson, K.J. 1978. On the question of male-limited population growth in feral horses (*Equus caballus*). Thesis, New Mexico State University, Las Cruces, USA.

Norment, C., and C.L. Douglas. 1977. Ecological studies of feral burros in Death Valley. National Park Service, Report No. 17. Cooperative Park Studies Unit / University of Nevada, Las Vegas, USA.

Novellie, P.A., P.H. Lloyd, and E. Joubert. 1992. Mountain zebra. Pages 6–9 *in* P. Duncan (Ed.), Zebras, asses, and horses: An action plan for the conservation of wild equids. International Union for Conservation of Nature, Gland, Switzerland.

Novellie, P.A., P.S. Millar, and P.H. Lloyd. 1996. The use of VORTEX simulation models in a long term programme of re-introduction of an endangered large mammal, the Cape mountain zebra (*Equus zebra zebra*). Acta Oecologica 17:657–671.

Nowzari, H. 2011. Population size, population structure and habitat use of the Persian onager (*Equus hemionus onager*) in Qatrouyeh National Park, Iran. [In Persian.] Thesis, Islamic Azad University, Tehran, Iran.

Nuñez, C.M.V., J.S. Adelman, and D.I. Rubenstein. 2010. Immunocontraception in wild horses (*Equus caballus*) extends reproductive cycling beyond the normal breeding season. PLoS ONE 5:e13635.

Nuñez, C.M.V., C.S. Asa, and D.I. Rubenstein. 2011. Zebra reproduction. Pages 2851–2865 *in* A.O. McKinnon, E.L. Squires, W.E. Vaala, and D.D. Varner (Eds.), Equine reproduction. Wiley-Blackwell, West Sussex, UK.

Ohsawa, H. 1982. Transfer of group members in plains zebra (*Equus burchelli*) in relation to social organization. African Study Monographs 2:53–71.

Owen-Smith, N., D.R. Mason, and J.O. Ogutu. 2005. Correlates of survival rates for 10 African ungulate populations: Density, rainfall and predation. Journal of Animal Ecology 74:774–788.

Packer, C., and A.E. Pusey. 1982. Cooperation and competition within coalitions of male lions: Kin selection or game theory? Nature 296:740–742.

Pagan, O., F. Von Houwald, C. Wenker, and B.L. Steck. 2009. Husbandry and breeding of Somali wild ass *Equus africanus somalicus* at Basel Zoo, Switzerland. International Zoo Yearbook 43:198–211.

Pellegrini, S.W. 1971. Home range, territoriality and movement patterns of wild horses in the Wassuk Range of western Nevada. Thesis, University of Nevada, Reno, USA.

Penzhorn, B.L. 1975. Behaviour and population ecology of the Cape mountain zebra, *Equus zebra zebra* L., 1758, in the Mountain Zebra National Park. Thesis, University of Pretoria, Pretoria, South Africa.

Penzhorn, B.L. 1985. Reproductive characteristics of a free-ranging population of Cape mountain zebra (*Equus zebra zebra*). Journal of Reproduction and Fertility 73:51–57.

Penzhorn, B.L. 1988. *Equus zebra*. Mammalian Species 314:1–7.

Perkins, A., E. Gevers, J.W. Turner Jr., and J.F. Kirkpatrick. 1979. Age characteristics of feral horses in Montana. Pages 51–58 *in* R.H. Denniston (Ed.), Symposium on the ecology and behavior of feral equids. University of Wyoming, Laramie, USA.

Perryman, P., and A. Muchlinski. 1987. Population dynamics of feral burros at the Naval Weapons Center, China Lake, California. Journal of Mammalogy 68:435–438.

Petersen, J.C.B., and R.L. Casebeer. 1972. Distribution, population status and group composition of wildebeest (*Connochaetes taurinus burchell*) and zebra (*Equus burchelli* Gray) on the Athi-Kapiti Plains, Kenya. Report KEN-Y1/526. United Nations Development Programme, New York, USA.

Pluháček, J., L. Bartos, and J. Vichova. 2006. Variation in incidence of male infanticide within species of plains zebra (*Equus burchelli*). Journal of Mammology 87:35–40.

Prague Zoo. 2010. International Studbook of the Przewalski Horse. Prague, Czech Republic.

Ransom, J.I. 2012. Population ecology of feral horses in an era of fertility control management. Thesis, Colorado State University, Fort Collins, USA.

Ransom, J.I., J.E. Roelle, B.S. Cade, L. Coates-Markle, and A.J. Kane. 2011. Foaling rates in feral horses treated with the immunocontraceptive porcine zona pellucida. Wildlife Society Bulletin 35:343–352.

Ransom, J.I., N.T. Hobbs, and J. Bruemmer. 2013. Contraception can lead to trophic asynchrony between birth pulse and resources. PLoS ONE 8:e54972.

Roberts, S.J. 1986. Abortion and other gestational diseases in mares. Pages 705–710 *in* D.A. Morrow (Ed.), Current therapy in theriogenology. W.B. Saunders, Philadelphia, USA.

Roelle, J.E., F.J. Singer, L.C. Zeigenfuss, J.I. Ransom, L. Coates-Markle, and K.A. Schoenecker. 2010. Demography of the Pryor Mountain wild horses, 1993–2007. Scientific Investigations Report 2010-5125. US Geological Survey, Reston, VA, USA.

Rowen, M. 1992. Mother-infant behavior and ecology of Grevy's zebra, *Equus grevyi*. Thesis, Yale University, New Haven, CT, USA.

Rubenstein, D.I. 1989. Life history and social organization in arid adpated ungulates. Journal of Arid Environments 17:145–156.

Rubenstein, D.I.. 2010. Ecology, social behavior, and conservation in zebras. Advances in the Study of Behavior 42:231–258.

Ruffner, G.A., and S.W. Carothers. 1982. Age structure, condition and reproduction of two *Equus asinus* (Equidae) populations from Grand Canyon National Park, Arizona. Southwestern Naturalist 27:403–411.

Salter, R.E., and R.J. Hudson. 1982. Social organization of feral horses in western Canada. Applied Animal Ethology 8:207–223.

Saltz, D. 2002. The dynamics of equid populations. Pages 118–123 *in* P. Moehlman (Ed.), Equids: Zebras, asses and horses. Status survey and conservation action plan. International Union for Conservation of Nature, Gland, Switzerland.

Saltz, D., and D.I. Rubenstein. 1995. Population dynamics of a reinstroduced Asiatic wild ass (*Equus hemionus*) herd. Ecological Applications 5:327–335.

Saltz, D., D.I. Rubenstein, and G.C. White. 2006. The impact of increased environmental stochasticity due to climate change on the dynamics of Asiatic wild ass. Conservation Biology 20:1402–1409.

Sanderson, E.W., M. Jaiteh, M.A. Levy, K.H. Redford, A.V. Wannebo, and G. Woolmer. 2002. The human footprint and the last of the wild. BioScience 52:891–904.

Santiapillai, C., W. Wijeyamohan, and K.R. Ashby. 1999. The ecology of a free-living population of the ass (*Equus africanus*) at Kalpitiya, Sri Lanka. Biological Conservation 91:43–53.

Schaller, G.B. 1972. The Serengeti lion: A study of predator-prey relations. University of Chicago Press, Chicago, USA.

Schaller, G.B.. 1998. Wildlife of the Tibetan Steppe. University of Chicago Press, Chicago, USA.

Schook, M.W., D.E. Wildt, R.B. Weiss, B.A. Wolfe, K.E. Archibald, and B.S. Pukazhenthi. 2013. Fundamental studies of the reproductive biology of the edangered Persian onager (*Equus hemionus onager*) result in first wild equid offspring from artificial insemination. Biology of Reproduction 89:41.

Scorolli, A., and A.C. Lopez Cazorla. 2010. Demography of feral horses (*Equus caballus*): A long-term study in Tornquist Park, Argentina. Wildlife Research 37:207–214.

Scorolli, A., A.C.L. Cazorla, and L.A. Tejera. 2006. Unusual mass mortality of feral horses during a violent rainstorm in Parque Provincial Tornquist, Argentina. Mastozoología Neotropical 13:255–258.

Seal, U.S., and E.D. Plotka. 1983. Age-specific pregnancy rates in feral horses. Journal of Wildlife Management 47:422–429.

Sinclair, A.R.E., S.A.R. Mduma, and P. Arcese. 2000. What determines phenology and synchrony of ungulate breeding in the Serengeti? Ecology 81:2100–2111.

Siniff, D.B., J.R. Tester, and G.L. McMahon. 1986. Foaling rate and survival of feral horses in western Nevada. Journal of Range Management 39:296–297.

Smith, R.K., A. Marais, P. Chadwick, P.H. Lloyd, and R.A. Hill. 2007. Monitoring and management of the endangered Cape mountain zebra *Equus zebra zebra* in the western Cape, South Africa. African Journal of Ecology 46:207–213.

Smuts, G.L. 1976*a*. Population characteristics of Burchell's zebra (*Equus burchelli antiquorum*, H. Smith, 1841) in the Kruger National Park. South African Journal of Wildlife Research 6:99–112.

Smuts, G.L. 1976*b*. Reproduction in the zebra mare *Equus burchelli antiquorum* from the Kruger National Park. Koedoe 19:89–132.

Solomatin, A.O. 1973. Kulan. [In Russian.] Akademiya Nauk, Moscow, Russia.

Solomatin, A.O. 1977. Kulan. Pages 249–276 *in* A.A. Kaletskii (Ed.), Hoofed wild animals. [In Russian.] Lesnaya Promyshlennost, Moscow, Russia.

Spasskaya, N.N., N.V. Shcherbakova, J.A. Ermilina, K.A. Mahotkina, D.S. Pchelkina, and A.E. Swinarenko. 2012. The results of a comprehensive monitoring populations of feral horses Vodny Island State Nature Biosphere Reserve "Rostovsky." [In Russian.] Proceedings of the Federal State Nature Biosphere Reserve "Rostovsky" 4:197–211.

Spinage, C.A. 1972. African ungulate life tables. Ecology 53:645–652.

Stanley, C.R., and S. Shultz. 2012. Mummy's boys: Sex differential maternal–offspring bonds in semi-feral horses. Behaviour 149:251–274.

St-Louis, A., and S.D. Côté. 2009. *Equus kiang* (Perissodactyla: Equidae). Mammalian Species 835:1–11.

Tatin, L., B.F. Darreh-Shoori, C. Tourenq, D. Tatin, and B. Azmayesh. 2003. The last populations of the critically endangered onager *Equus hemionus onager* in Iran: Urgent requirements for protection and study. Oryx 37:488–491.

Tatin, L., S.R.B. King, B. Munkhtuya, A.J.M. Hewison, and C. Feh. 2009. Demography of a socially natural herd of Przewalski's horses: An example of a small, closed population. Journal of Zoology 277:134–140.

Trivers, R.L., and D.E. Willard. 1973. Natural selection of parental ability to vary the sex ratio of offspring. Science 179:90–92.

Turner, J.W., Jr., and M.L. Morrison. 2001. Influence of predation by mountain lions on numbers and survivorship of a feral horse population. Southwestern Naturalist 46:183–190.

Turner, J.W., Jr., M.L. Wolfe, and J.F. Kirkpatrick. 1992. Seasonal mountain lion predation on a feral horse population. Canadian Journal of Zoology 70:929–934.

Tyler, S.J. (Ed.). 1972. The behaviour and social organization of the New Forest ponies. Part 1, vol. 5. Bailliere Tindall, London, UK.

Usukhjargal, D., and N. Bandi. 2013. Reproduction and mortality of re-introduced Przewalski's horse *Equus przewalskii* in Hustai National Park, Mongolia. Journal of Life Sciences 7:624–631.

Van Duyne, C., E. Ras, A.E.W. De Vos, W.F. De Boer, R.J.H. Henkens, and D. Usukhjargal. 2009. Wolf predation among reintroduced Przewalski horses in Hustai National Park, Mongolia. Journal of Wildlife Management 73:836–843.

Vodička, R. 2010. An abort of twins in a Przewalski horse mare. Pages 23–29 *in* M. Bobek (Ed.), Equus. Prague Zoo, Prague, Czech Republic.

Volf, J. 2010. Sixty years of kulan, *Equus hemionus kulan* (Groves et Mazák, 1967) breeding at Prague Zoo. Pages 31–55 *in* M. Bobek (Ed.), Equus. Prague Zoo, Prague, Czech Republic.

Walker, B.H., R.H. Emslie, N. Owen-Smith, and R.J. Scholes. 1987. To cull or not to cull: Lessons from a southern African drought. Journal of Applied Ecology 24:381–401.

Westlin-van Aarde, L.M., R.J. van Aarde, and J.D. Skinner. 1988. Reproduction in female Hartmann's zebra *Equus zebra hartmannae*. Journal of Reproductive Fertility 84:505–511.

White, M.L. 1980. A study of feral burros in Butte Valley, Death Valley National Monument. Thesis, University of Nevada, Las Vegas, USA.

Williams, S.D. 1998. Grevy's zebra: Ecology in a heterogeneous environment. Thesis, University College London, London, UK.

Wolfe, M.L. 1979. Population ecology of the kulan. Pages 205–218 *in* R.H. Denniston (Ed.), Symposium on the ecology and behavior of feral equids. University of Wyoming, Laramie, USA.

Wolfe, M.L. 1980. Feral horse demography: A preliminary report. Journal of Range Management 33:354–359.

Wolfe, M.L., L.C. Ellis, and R. Macmullen. 1989. Reproductive rates of feral horses and burros. Journal of Wildlife Management 53:916–924.

Zhirnov, L.V., and V.O. Ilyinsky. 1986. Wild ass. Pages 68–79 *in* V.Y. Sokolov (Ed.), The Great Gobi National Park: A refuge for rare animals of the central Asian deserts. United Nations Environment Programme, Moscow, Russia.

Ziaee, H. 2008. A field guide to the mammals of Iran, 2nd ed. [In Persian.] Wildlife Center Publications, Tehran, Iran.

PART II HISTORY AND MANAGEMENT

7

Genetics and Paleogenetics of Equids

EVA-MARIA GEIGL, SHIRLI BAR-DAVID, ALBANO BEJA-PEREIRA, E. GUS COTHRAN, ELENA GIULOTTO, HALSZKA HRABAR, TSENDSUREN OYUNSUREN, AND MÉLANIE PRUVOST

DNA, or deoxyribonucleic acid, is the molecule of life that determines morphology, function, and development of all living things. The study of both the information contained in these molecules and their transmission (i.e., genetics) allows us to describe phenotypic traits of a species as well as reconstruct its evolution, diversity, and relationship to other species. Genetic analyses are an invaluable and complementary part of the research necessary to understand the biology of a species and to aid in its preservation.

The genus *Equus* is the only remaining branch of a bushy evolutionary tree covering 55 million years. Within this genus, only the domesticated forms (the caballine horses and the donkeys) are spread over the planet and not endangered. The plains zebra (*Equus quagga*) is locally common in Africa, as is the kiang (*E. kiang*) in parts of Asia, but all the other equid species (African and Asiatic wild ass [*E. africanus* and *E. hemionus*, respectively], Przewalski's horse [*E. przewalskii*, also known as *E. ferus przewalskii*], mountain and Grevy's zebra [*E. zebra* and *E grevyi*, respectively]) are more or less at the edge of extinction. A deeper understanding of the phylogeny, genome evolution, and genetic diversity of these species is important for their conservation. In this chapter, we present an overview of the present-day knowledge concerning these topics. The mitochondrial DNA diversity and phylogeny of these species are shown as a phylogenetic tree based on whole mitochondrial genomes (fig. 7.1) (Vilstrup *et al.* 2013).

Despite a long-standing interest of paleontologists in the complex evolutionary history of the equids, the taxonomic relationships of the fossil forms are still not clear. Studies of equid chromosomes show a high rate of evolution within this genus, which has led to a large variety of different karyotypes that do not prevent gene flow (Jónsson *et al.* 2014). These karyotypic changes are not reflected in the morphology because extant equids rather strongly resemble each other, confirming the accelerated evolution that is also seen at the sequence level underlying the rapid radiation found among equid populations (Vilstrup *et al.* 2013).

The application of the extreme power of the latest DNA sequencing methods to the analysis of genetic material preserved in fossils has made it possible to revisit the most recent part of the phylogenetic relationships and the taxonomy of the equids with relatively high precision, leading to a taxonomic reappraisal and to the demise of several species present in

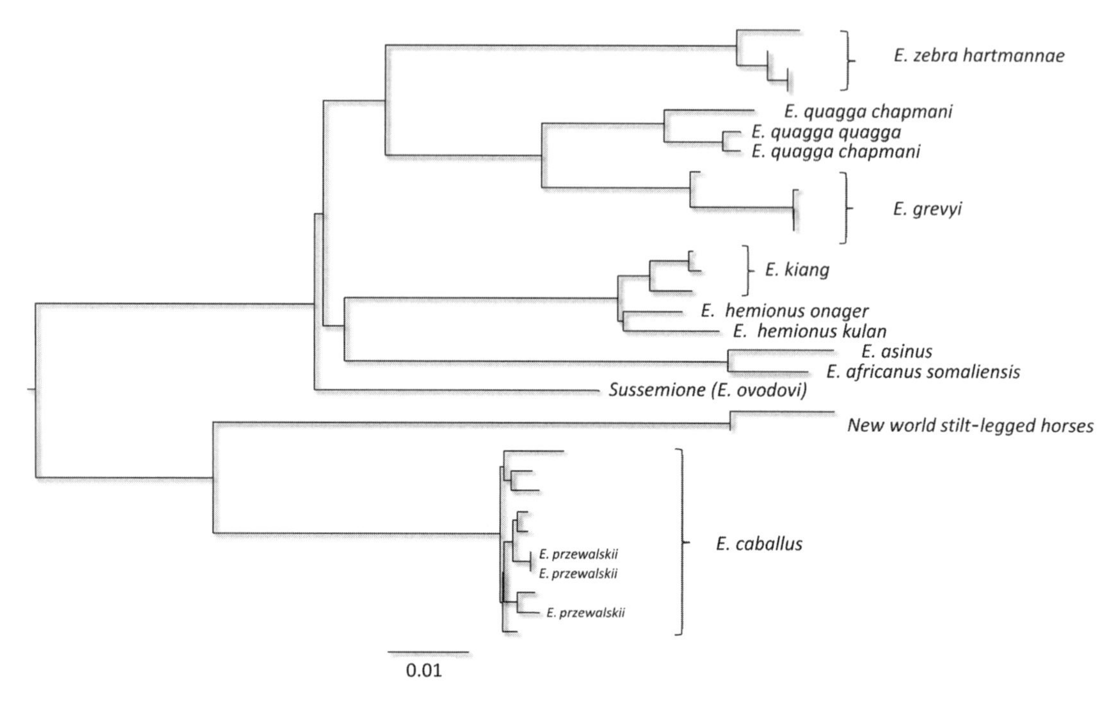

Fig. 7.1 Phylogenetic tree of whole mitochondrial genomes (after Vilstrup *et al.* 2013). The scale bar represents 0.01 nucleotide substitutions per site.

the Pleistocene (Orlando *et al.* 2003, 2008, 2009; Geigl and Grange 2012; Vilstrup *et al.* 2013). The most recent sequencing of whole genomes of several extinct and extant equid species allowed for the recalibration of the equid phylogeny, which now date the most common recent ancestor to around four million years (Orlando *et al.* 2013). In addition, the analysis in ancient equid remains of phenotypic traits—such as coat color—have increased our knowledge of the physical appearance of ancient horse populations beyond body size and shape. Because coat color is also one of the phenotypic traits that changed during the domestication process, when humans started to select for certain behavioral or physical characteristics, its analysis allowed the dating of the onset of the domestication of horses.

Whole genome sequence data from Pleistocene fossils now unequivocally assign domestic (*E. caballus*) and Przewalski's horses to two distinct evolutionary lineages (Orlando *et al.* 2013). This distinction was not clear before, as the Przewalski's horse, whose populations went through a dramatic bottleneck that almost led to its extinction, was sometimes considered to be the ancestor of the domestic horse.

The increasing power of paleogenetic and paleogenomic studies contributes considerably to a deeper understanding of the evolution and taxonomy of equids, which is also important in designing efficient strategies for the conservation of the relic populations, such as the Asiatic wild asses. If combined with the analysis of extant populations living in their natural habitat or in reserves where they have been (re)introduced, these phylogeographic studies can lead to the revision of conservation strategies so that they account for the past geographical range of the species. This type of genetic study is also advisable to better understand the evolution of reintroduced populations in a given area, such as the Asiatic wild ass in Israel, and will be invaluable for conservation biologists seeking to optimize strategies to rescue these threatened populations.

Genetic analyses of small populations of equids whose habitats are difficult or dangerous to access, such as in the case of the Somali wild ass (*E. a. somaliensis*), can be performed at present in a noninvasive manner using feces that have been collected in the wild. Such an approach can be used to better characterize the African wild ass and show that the Somali wild ass was probably not the ancestor of the domestic donkey. Microsatellite markers can also help demonstrate that until recently gene flow occurred between populations from Eritrea and Ethiopia, which by now are separated.

It is important to draw the attention of those who decide conservation policies to the results of these studies, since conservation strategies, which often neglect the genetics of the populations they aim to rescue, risk being ineffective, as in the example of the Cape mountain zebra. This is even true for the populations of feral horses (*E. caballus*) in North America. Although currently not endangered, a recently noticed reduction of the genetic variation of these horses could indicate a threat to their long-term survival.

This chapter aims to shed light from various angles on the genetics of various extant and extinct equid species, their evolution, their past and present distribution, and the consequences of population reduction on genetic diversity.

Karyotype Evolution and Genome Plasticity in the Genus *Equus*

With Contributions by Elena Giulotto and Elena Raimondi

In spite of their recent evolutionary divergence, morphological similarity, and capacity to interbreed (although infertile offspring are produced), equid species differ greatly at the karyotype level (Ryder 1978; Yang *et al.* 2003, 2004). The variation involves both the structure and the number of chromosomes, which ranges from 32 in *Equus zebra* to 66 in *E. przewalskii*. Using cross-species chromosome painting, Trifonov *et al.* (2008) calculated the rate of evolutionary chromosomal rearrangements, showing that it was extremely low in all ceratomorphs (<0.3 rearrangements per million years), while it was very rapid in the lineage of equids, ranging from 2.92 to 22.2 rearrangements per million years. According to these calculations, the equid genome evolution is one of the most rapid observed in nonrodent mammals (Trifonov *et al.* 2008).

In total, 53 chromosome fusion events occurred during the evolution of extant equids. In this genus, the majority of fusions are of the centromere-centromere type (33/53). Most chromosomes originating via centric fusions maintained the centromere at the site of fusion, although some underwent "centromere repositioning" or pericentric inversion. Centromere repositioning consists of the shift, during evolution, of centromere position on the chromosome without DNA rearrangement (Montefalcone *et al.* 1999; Rocchi *et al.* 2012). We discovered that, in the genus *Equus*, centromere repositioning played a crucial role in the extensive and rapid karyotype evolution. Using a comparative cytogenetic approach, we demonstrated that at least nine centromere repositioning events took place during the evolution of equids, six of which occurred in *E. asinus* (donkey) and one in horse chromosome 11 (ECA 11) (Carbone *et al.* 2006; Piras *et al.* 2009). These observations demonstrate that this phenomenon was exceptionally frequent in equids because such a high frequency of centromere repositioning has never been reported in any other evolutionary lineage.

Centromeres are essential for the proper chromosomal segregation in meiosis and mitosis. They are the sites of kinetochore assembly and spindle fiber attachment, and consist of protein-DNA complexes, in which the DNA component is typically characterized by the presence of extended arrays of tandem repeats (satellite DNA), but these do not specify centromere function or identity. There is general agreement that the centromere's genetic identity is determined epigenetically by proteins rather than the DNA they contain (Allshire and Karpen 2008). We performed an extensive analysis of the chromosomal distribution of centromeric satellite tandem repeats in *E. caballus* (ECA), *E. asinus* (EAS), *E. grevyi* (EGR), and *E. quagga* (EBU). Our results demonstrated that the organization of such sequences is atypical: several centromeres lack satellite DNA, at least at the fluorescence in situ hybridization (FISH)-resolution level, while satellite repeats are often present at noncentromeric termini, probably corresponding to relics of ancestral, now inactive, centromeres (Piras *et al.* 2010; Raimondi *et al.* 2011). In one instance, the centromere of horse chromosome 11, we proved the total absence of satellite repeats at the sequence level (Wade *et al.* 2009). Taking advantage of the availability of the horse genome sequence, we carried out a chromatin immunoprecipitation (ChIP)-on-chip assay: an array of the ECA11 centromeric region was hybridized with horse chromatin, immunoprecipitated with an antibody against the kinetochore proteins CENP-A or CENP-C. Using this approach, we demonstrated that the centromeric function resides within a DNA sequence totally devoid of satellite DNA (Wade *et al.* 2009).

The rapidly evolving *Equus* species gave us the opportunity to catch snapshots of evolutionarily new centromeres in different stages of "immaturity." On the basis of these observations, we propose a model for the birth, evolution, and complete maturation of centromeres (Piras *et al.* 2010). The first step would consist in the shift of the centromeric function to a new position lacking satellite DNA, while the satellite DNA from the old centromere remains in the ancestral position, as we observed in a number of donkey and zebra chromosomes; a subsequent step would be the loss of the leftover satellite DNA; and finally, at a later stage, satellite repeats would colonize the new centromere, giving rise to "mature" centromeres. Thus, in the case of ECA 11, we may surmise that it is a "young," evolutionarily new centromere, as suggested by the complete lack of satellite DNA.

The satellite-less equid centromeres represent a new and powerful model system offering a clear advantage with respect to other artificial or clinical model centromeres: they are natural centromeres, stably present in all individuals of a given species, and can therefore be used as an ideal tool to study the normal mammalian centromere. The nonrepetitive nature of these centromeres and the availability of the complete sequence of the horse genome provide the chance to analyze centromere architecture and plasticity at the molecular level with particular reference to protein-binding domains (Purgato *et al.* 2014), nucleosome organization, and DNA methylation; in fact, while the roles of many of these components in mitosis have been documented, relatively little is known about the detailed structure of the chromatin fiber that underlies these functions.

The uncommon evolutionary plasticity of the equid genome is also proved by the high mobility of insertion DNA elements, such as numts (nuclear sequences of mitochondrial origin). Insertion of mitochondrial DNA fragments into nuclear chromosomes—together with the insertion of transposons, retroviruses, and telomeric-like sequences—is a driving force in evolution. Indeed, numts, similarly to transposons and interstitial telomeres (Salem *et al.* 2003; Nergadze *et al.* 2004, 2007; Hazkani-Covo *et al.* 2010), have been successfully used in several studies as evolutionary markers. We analyzed mitochondrial DNA insertions in the horse reference genome sequence (Nergadze *et al.* 2010) and found that insertion polymorphisms are exceedingly high in the horse with respect to the human genome. Our results support the hypothesis that the genome of equid species is in an ongoing evolutionary state.

Depicting Wild Horses of the Past with Ancient DNA

With Contributions by Mélanie Pruvost

The study of the evolution of horses, and in particular of their domestication process, is complicated by the fact that the mitochondrial DNA (mtDNA) and nuclear microsatellite diversity of present-day horse populations is high and lacks significant geographic structure and that the diversity of the pseudoautosomal region of the Y chromosome is extremely low (Lippold *et al.* 2011; Achilli *et al.* 2012). This finding indicates that several maternal lineages had been domesticated, but a higher level of resolution of the domestication event could not be achieved. The recent publication of the sequence of the genome of a Middle Pleistocene horse, the oldest full genome sequence so far (560,000–780,000 years before present; Orlando *et al.* 2013), allowed a reappraisal of the evolution of equids in the Pleistocene. Such studies are so far limited to few samples owing to the high costs and the difficulties of the sequencing of ancient genomes, restraining it to well-preserved fossils and often preventing insights into the diversity of past populations. Therefore, until these approaches are less costly and challenging, well-chosen genetic markers give a useful picture of certain aspects of the evolution and of the migration of populations.

Shifts in phenotypic trait markers can be valuable indicators for alterations in both natural selection (climatic shifts causing changes of forest coverage or nutrition) and artificial selection (domestication). Domestication boosted major phenotypic changes in domesticates compared to their wild ancestors (Cieslak *et al.* 2011), for example, and coat color is likely to have been selected by humans during the domestication of horses. To identify a change in the phenotypes occurring during domestication, one needs to know the situation before domestication. The phenotype of undomesticated horses during the Late Pleistocene could be deduced from parietal art because horses are the most represented animals in caves during the Upper Paleolithic in Europe. They are depicted most of the time in natural, everyday postures such as standing, feeding, fighting, running, and trotting, which indicates that humans had a good knowledge of the behavior of horses in their natural environment. Horses represented in cave paintings show various coat colors, in particular, brown, black, and spotted (dark spots on white coat). Moreover, by analyzing basic coat color, pelage length, and seasonal patterns in horses represented in Paleolithic art throughout Europe, a regional cline in pelage change was reconstructed and concluded that living wild horses (*E. przewalskii*) have similar pelage characteristics (Guthrie 2006). Using a paleogenetic approach, we analyzed whether these coat colors corresponded to the coats of Pleistocene horses and whether there was a change in colors over time from the Late Pleistocene to the domestication of the horse during the Bronze Age.

The genetic systems underlying the large amount of phenotypic variations in coat color observed in modern horses has been well characterized, and >300 genes and associated loci have been identified thus far (Rieder 2009). The skin and hair of horses contain only two different kinds of pigments called

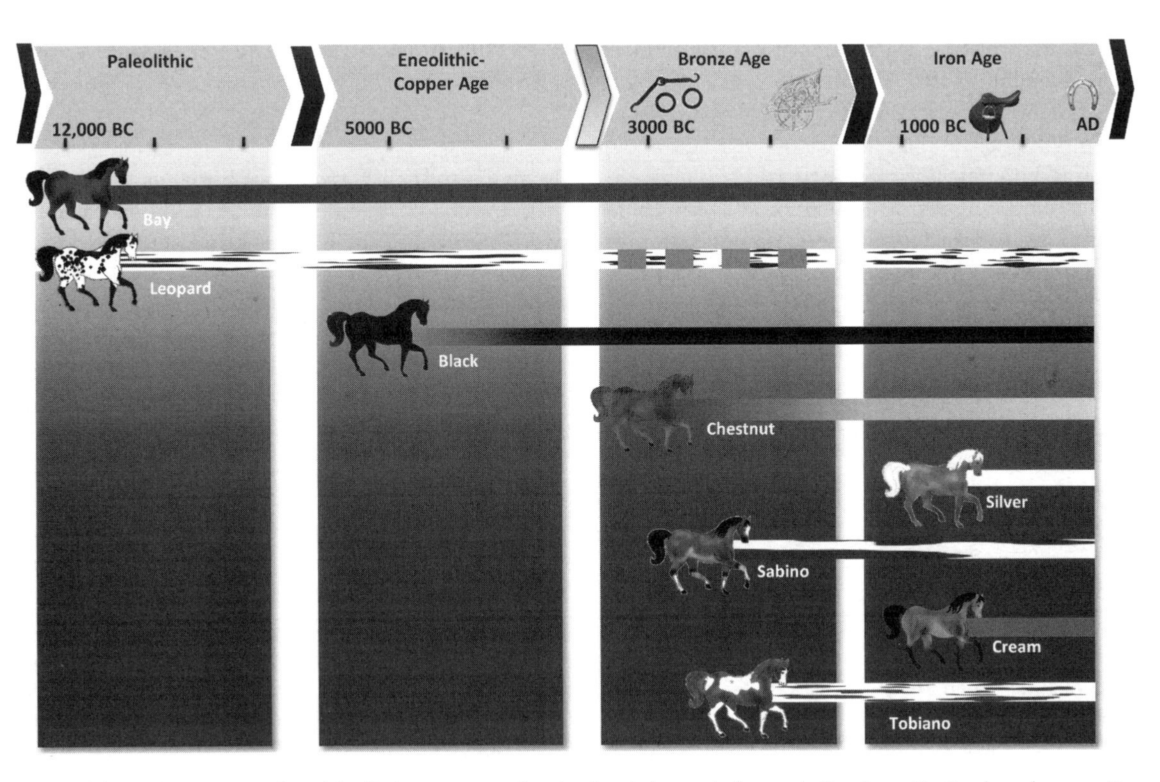

Fig. 7.2 Schematic representation of the first occurrence of coat colors in horses before and after domestication based on genetic markers in ancient remains.

pheomelanin (yellow) and eumelanin (black). These two pigments are distributed in various ways, determining the various coat colors. All coloration genes affect either the production or the distribution of these two pigments. To determine the ancient coat colors, we targeted the mutations controlling the basic coat color ("bay," "black," and "chestnut") and others responsible for the dilution of these basic colors ("silver," "cream," and "champagne"), leading to a lighter color, as well as white spotting ("tobiano," "sabino," "leopard," and "overo") patterns. For some coat colors, however, the responsible mutation is still unknown, especially for the "dun" coat color of the Przewalski's horses. Modern domestic horses show a high variability in coat coloration. No color is confined to a single breed, and the same mutation is responsible for a certain color variant across breeds, supporting the idea that mutations producing the color variants occurred prior to breed formation and may be valuable indicators for artificial selection during the domestication process.

By selecting short DNA regions including the point mutation on each gene involved in the coat colors, we analyzed the coat coloration of ancient horses using fossil remains (Pruvost *et al.* 2012). We showed that Pleistocene horses in Siberia and Germany displayed the basic bay coat color (Ludwig *et al.* 2009), which is the most common painted phenotype in Paleolithic art (fig. 7.2). At the beginning of the Holocene and until the end of the Copper Age, wild horses present both black and bay coat colors, with a slight predominance for bay (Ludwig *et al.* 2009), and so far no chestnut color has been found among predomestic horses (fig. 7.2). None of the mutations responsible for dilution (cream, champagne, silver) or the white spotting pattern was found in wild horses except for the mutation linked to the leopard spotting pattern phenotype (fig. 7.2). We estimated the frequency of the leopard allele to be higher than 0.1% in the western European horse population from the Pleistocene and suggested that this phenotype was not rare (Pruvost *et al.* 2011). The leopard complex spotting pattern is characterized by white spots that range from a few white flecks on the rump to a coat color that is almost completely white. A similar phenotype is depicted in the paintings of the "dappled horses" in the cave of Pech-Merle, France, dated to 24,700 years ago, featuring spotted horses in a frieze that includes hand outlines and abstract patterns of spots. The spotted pattern had been considered until now more symbolic than realistic because of the juxtaposition of abstract elements and the unlikelihood of finding spotted coat phenotypes in wild animals, but this interpretation might need to be revisited, as genetics indicates the existence of such a coat color in horses during the Pleistocene.

The distribution of the coat color in ancient horse populations seems to change with domestication because we could demonstrate a significant shift and an increase of coat color variants, including dilutions and spotting patterns other than leopard from Bronze Age horse bone remains. The first chestnut phenotype observed dates to the third millennium BCE in a Siberian horse, and its frequency increased rapidly, reaching a proportion of 28% in Bronze Age horse bone assemblages (Ludwig *et al.* 2009) (fig. 7.2). The three basic coat colors (bay, black, and chestnut) seem to be equally common among ancient domestic horses (Ludwig *et al.* 2009; Svensson *et al.* 2012) from Europe and Siberia (fig. 7.2).

Mutations responsible for dilution or spotting phenotypes appear later in domestic horses (fig. 7.2). Date and location of their appearance are so far known only from the study of Ludwig *et al.* (2009) and therefore are just an approximation. The sabino pattern is the first spotting phenotype found, appearing during the third millennium BCE in Siberia. It was still present in Armenia and Moldavia during the middle Bronze Age, while today it is found only in North American horses. The tobiano genotype, which is widespread today, was observed in an eastern European horse (1500–1000 BCE), demonstrating its ancient history. Both "cream dilution" (buckskin) and "silver dilution" (black silver) were not found in horse bones older than 800 BCE, which could indicate that the diluted colors were not selected at the beginning of the domestication and only appeared later. So far, only two phenotypes, champagne and overo, were not found among the ancient domestic horses tested (Ludwig *et al.* 2009; Svensson *et al.* 2012). This is not a surprising result given that overo is lethal in a homozygous state and would therefore have been under strong negative selection, and champagne is known as a recent mutation.

The analyses of genetic markers linked to coat color are a first step to the phenotypic characterization of ancient horse populations going beyond skeletal properties. Moreover, the sudden and rapid increase in coat color variation at the beginning of domestication allowed for the first time a clear-cut distinction of the onset of the domestication of horses. In the future, whole-genome analyses of ancient horse remains will allow both a much more comprehensive phenotypic characterization of ancient horses and a refined understanding of their evolution.

Genetics of Mongolian and Przewalski's Horses

With Contributions by Tsendsuren Oyunsuren

Mongolia today is home to two horse populations, the native Mongolian domestic horse (*E. caballus*) and the wild Przewalski's horse. The native Mongolian domestic horse population, estimated at >2.6 million (National Statistical Office of Mongolia 2014), consists of five main lines named Galshir, Tes, Myangad, Datkhad, and Shankh. These lines differ in body conformation, size, coat colors, speed, and endurance. The Przewalski's horse became extinct in the wild in the 1960s but was reintroduced into its native habitat in Mongolia in 1992.

The genetic diversity and evolutionary relationship of domestic and Przewalski's horses have been studied and debated for decades. A variety of genetic studies have been performed targeting the mtDNA control region (the fastest-evolving DNA region known in mammals), microsatellite loci, and single nucleotide polymorphisms (SNPs) to assess the evolutionary relationship between domestic and Przewalski's horses (George and Ryder 1986; Ishida *et al.* 1995; Oakenfull and Ryder 1998; Goto *et al.* 2011). Interestingly, these horses can breed and produce viable offspring (Short *et al.* 1974) despite their karyotype differences (chromosome number of Przewalski's horse $2n = 66$, domestic horse $2n = 64$) (Bernischke *et al.* 1965; Ryder 1978; Trommerhausen-Smith *et al.* 1979; Ryder 1994). Contradictory hypotheses were proposed regarding the question of whether the Przewalski's horse was the direct ancestor of the domestic horse (Mohr 1959; Ryder 1994; Bowling *et al.* 2003; Lau *et al.* 2009). It was shown that the estimated mtDNA nucleotide sequence divergence between domestic and Przewalski's horses is 0.05–0.5%, suggesting a relatively ancient genetic divergence of these two types of horses (Ryder 1994). However, the ancestral status of the mtDNA of the Przewalski's horse, based on studies of the mtDNA control region, was not confirmed (Ishida *et al.* 1995).

Recently, full mitochondrial genomes and a part of the nuclear genome of Przewalski's and domestic horses have been sequenced using massively parallel sequencing technology, showing that the Przewalski's horse population has a higher diversity than expected. The mean autosomal diversity was determined to be 0.195%, thus higher than that of the thoroughbred horse (0.05%) (Goto *et al.* 2011). These analyses suggested that Przewalski's and domestic horse have both retained joint ancestral genes since their divergence.

The complete mitochondrial genomes of six Przewalski's horses have been sequenced in two studies published simultaneously (Goto *et al.* 2011; Achilli *et al.* 2012). In one study (Goto *et al.* 2011), four Przewalski's mitochondrial genomes were obtained and assigned to three haplogroups named I, II, and III. Haplogroups I and II are similar and cluster with the Przewalski-specific F haplogroup, which Achilli *et al.* (2012) defined on the basis on the analysis of the mitogenomes of two Przewalski's and eighty-one domestic horses across Asia, Europe, the Middle East, and America, whereas haplogroup III is most closely related to the J and K haplogroups of domestic horses from Europe and the Middle East. Thus, when a large number of mitochondrial genomes of horses are analyzed such as in the study of Achilli *et al.* (2012), it is clear that the mitochondrial genome sequences of the Przewalski's horses are contained within the diversity of the domestic horses. This result indicates that there was gene flow between the ancestors of the present-day populations or common ancestry.

The high variability of microsatellites is useful for revealing genetic diversity and has been used to explore the genetic relationship among Przewalski's horses (Breen *et al.* 1994; Oakenfull and Ryder 1998). Twelve microsatellite loci were examined for parentage testing and were shown to discriminate two Przewalski's individuals with a probability of 99.3% (Munkhtuya 2005). Moreover, this study also showed that the heterozygosity in the Przewalski's horse population was moderate compared to most domestic horse breeds (He = 0.63; Munkhtuya 2005).

Based on the sequence of 14.3 kb of the Y chromosome of 52 horses of 15 breeds and of one Przewalski's horse, it was shown that the Przewalski's horse belongs to a distinct lineage because no sequence variability was observed among all domestic horses, whereas the Przewalski's sequence differed by six nucleotides (Lindgren *et al.* 2004). Thus it was estimated that the divergence between the Przewalski's horse and the domestic horse occurred between 120,000 and 240,000 years ago (Lindgren *et al.* 2004).

Pairwise genetic distance and multidimensional scaling of genome-wide SNP analyses of domestic and Przewalski's horses showed that they are genetically distinct from all domestic horses, including the native Mongolian horse, despite a close evolutionary relationship (McCue *et al.* 2012). Finally, a recent study where whole genomes of ancient horses (about 560,000–780,000 and 43,000 years old) were sequenced, as well as the genomes of a domestic and a Przewalski's horse, showed that the domestic and the Przewalski's horse are distinct lineages and estimated that the two populations diverged between 38,000 and 72,000 years ago (Orlando *et al.* 2013). Furthermore, in the genome of the Przewalski's horse that was sequenced in that study, no traces of gene flow from domestic horse breeds were detected. The final conclusion has to await the sequencing of more genomes, wild and domestic ones, and in particular those of the Mongolian breeds.

Even though the diversity of the Mongolian horse population has not yet been fully explored with the most advanced techniques, the combination of various studies involving the analysis of microsatellite loci, inter simple sequence repeat (ISSR) markers, and nuclear SNPs indicated that Mongolian horses have the highest heterozygosity and genetic diversity among all domestic horse breeds analyzed so far (Tozaki *et al.* 2003; Voronkova *et al.* 2011; McCue *et al.* 2012; Tsendsuren *et al.* 2013*b*). This is also seen at the level of the mitochondrial control region because the Mongolian horse population harbors 53 mitochondrial haplotypes (Tsendsuren *et al.* 2013*a*). All of these findings suggest that the Przewalski's horse is not the ancestral type of the domestic horse, but that the two belong to sister lineages within the genus *Equus*, which originated 4.0–4.5 million years ago (Orlando *et al.* 2013).

One Hundred Thousand Years of Phylogeography of the Eurasiatic Wild Ass Based on Ancient DNA

With Contributions by E. Andrew Bennett, Thierry Grange, and Eva-Maria Geigl

Paleogenetics, the analysis of traces of DNA preserved in fossilizing calcified tissue of the remains of organisms, provides a temporal perspective on the evolution of the genetic diversity of populations. These data can be invaluable for the design of conservation biology strategies. They also complement paleontological studies, which are often made difficult owing to the scarcity of bone remains, the lack of reference skeletons, and other taphonomic and morphological biases (for a review, see Geigl and Grange 2012).

The Asiatic wild ass is on the way to extinction before having been well studied (International Union for Conservation of Nature 2013). Asiatic wild asses hold a distinct biological, ecological, and cultural importance in arid and high-altitude plains and deserts. In order for a large herbivore to be able to survive in such barren environments and to develop their characteristic extreme endurance, enormous physiological adaptations are necessary; however, many of these remain to be explained.

The global population of Asiatic wild asses is estimated at ~54,000 individuals, with the largest population in the Mongolian Gobi (commonly referred to as *E. h. hemionus* or "dziggetai") and much smaller populations in India (*E. h. khur*), Turkmenistan (*E. h. kulan*), and Iran (*E. h. onager*), which have been assigned subspecies status because of some morphological differences and geographic isolation (Denzau and Denzau 1999; Groves and Grubb 2011). Kiang (*E. kiang*) are currently considered a different species and seem still numerous on the high-altitude plains of the Tibetan Plateau, with an estimated number of between 60,000 and 200,000 (St-Louis and Côté 2009).

To unravel the phylogeographic history of the Asiatic wild ass, the kiang, and their extinct European counterpart *E. hydruntinus*, we analyzed around 250 samples—archaeological bones, museum specimens, and modern animals from both zoos and the wild—ranging in time and space over a period of 100,000 years and the entire Eurasian continent, from western France to India and Mongolia. We concentrated our analysis on the mitochondrial hypervariable region, the high mutation rate of which guarantees a high-resolution phylogenetic analysis and reveals the evolution of maternal lineages. Moreover, this DNA sequence is characterized by a deletion of 28 nucleotides that is specific for the Asiatic wild ass and can be used as a bar code. We obtained 145 DNA sequences, 8 of which were older than 10,000 years and one ~100,000 years old. Fifty-eight of our samples originated from southwest Asia, including regions within the present-day borders of Syria, which contain climatic conditions that are highly unfavorable for DNA preservation (Bennett *et al.*, in review). Thus DNA from these samples is extremely degraded, and only short DNA fragments at minimal quantities are preserved, if at all. The surprisingly high success rate of our study is to be attributed to a continuous and profound effort to adapt and optimize DNA extraction and analysis methods and to eliminate contamination with modern DNA, which is a major problem of ancient DNA analyses (Pruvost and Geigl 2004; Pruvost *et al.* 2005; Champlot *et al.* 2010; Bennett *et al.* 2014).

The DNA sequences we obtained were subject to Bayesian phylogenetic reconstructions (Drummond and Rambaut 2007) and spatial principal component analysis (Jombart *et al.* 2008). The ancient DNA sequences in particular allowed us to identify nine distinct clades (haplogroups) showing a clear phylogeographic structure (Bennett *et al.*, in review). Our analyses reveal multiple aspects of the evolution of this species. First, the highest diversity over the last 8,000 years was found with three distinct clades in the area of present-day Iran, suggesting that the radiation of the hemiones took place in this area and that colonization of Eurasia started from here. One clade that seems to have gone extinct as recently as the Bronze Age comprises the population that once inhabited the Caucasus and northern Iran. The other clade that had diverged from the latter one several hundreds of thousands of years ago comprises the present-day Iranian onagers, a population that roamed the highlands of Iran in prehistoric times but still is present today in the reserve of Bahram-i-Goor. The hemiones from northern Iran and Turkmenistan belong to a third clade that diverged earlier from the two others more than half a million years ago. The fact that they belong to the same clade demonstrates that these populations, which are currently assigned to different subspecies or even species—that is, *E. h. onager* and *E. h. kulan* (Groves and Grubb 2011)—are not separated and have experienced gene flow throughout the past until recently. The individuals that were introduced in Israel in the first half of the 1990s founding the breeding core in the Hai-Bar Yotvata Reserve were from Iran and Turkmenistan, respectively, and were found to belong to the Iranian and the Iran-Turkmen clades, respectively. The Indian khurs cluster in a distinct clade that is related to the Turkmen-Iranian kulans but separated from the latter ones for hundreds of thousands of years.

The conquest of the Tibetan and Mongolian Steppes appears to have occurred in two waves. The first wave seems to represent the continuation of the migration of the ancestors of the Asiatic wild asses out of Africa that is dated, on the basis of the fossil record, to one million years ago (Forsten 1992; Eisenmann 2004; Bernor *et al.* 2010). All other clades diverged later from the clade in which the present-day Tibetan kiangs and some lineages of the Mongolian dziggetais (populations of Asiatic wild ass) cluster. The subbranches of this clade are found in both the kiang and the dziggetai populations. The fact that kiang and dziggetai share a common clade shows that gene flow occurred recently between these two populations, thus questioning the assignment of *E. kiang* as a distinct species (Groves and Grubb 2011). There is a second clade of dziggetais that is related to the Iranian haplogroups and reflects a more recent colonization event from southwest Asia. We did not detect any kiang lineage in this clade, which suggests that it is the result of a second migration wave of hemiones out of Iran that did not reach the Tibetan highlands and did not hybridize with the kiangs. Altogether these data suggest the direction of interbreeding between kiangs and dziggetais: it could be explained with female kiangs migrating back from the Tibetan Plateau to Mongolia and interbreeding with the dziggetais.

Syria was inhabited in the past by a distinct clade that separated from the Caucasian-Iranian onagers around 300,000 years ago and much earlier from the hemiones from neighboring Turkey. This clade went extinct in the twentieth century when the last Syrian onagers, the *E. h. hemippus*, died in European zoos. The Syrian hemippus had a distinctive physical appearance with short ears and was the smallest and most gracile of all hemiones. We showed that bigger animals present in Syria 4,000 years ago belonged to the same clade, indicating that size reduction was a recent phenotypic evolution of the species in an ecological pessimism due to impoverished living conditions, possibly following climatic changes and competition with domestic life stock (Bennett *et al.*, in review).

Since the Middle Pleistocene, Europe was populated by the now extinct *E. hydruntinus*, a gracile wild ass that disappeared some time during the Holocene, and for which paleontological determination is particularly difficult (Geigl and Grange 2012). We genotyped several Pleistocene bone remains from western France, Romania, and Turkey that were assigned on morphological grounds to this species. The results clearly demonstrate that *E. hydruntinus* is a hemione, as all analyzed remains showed the hemione-specific 28-bp deletion. Moreover, the individual that lived 100,000 years ago in western Europe, classified as *E. hydruntinus*, clustered together with an early twentieth-century Iranian onager to form a clade, which had diverged from the eastern European / Anatolian clade roughly half a million years ago. This latter clade is constituted by Holocene hydruntines, which lived in Romania and Anatolia roughly 7,000 years ago. This clade is sufficiently distinct from the other clades in that it could well have the characteristic morphological features that are commonly attributed to *E. hydruntinus*, suggesting that Europe was colonized by hemiones through at least two colonization waves before and after the last two glaciation maxima. We conclude from this finding that the Magdalenian cave art representations, like the famous "Panneau de l'Hémione" in the Lascaux cave in France, were indeed representations of contemporaneous animals and not inaccurate or symbolic representations of horses, as some currently believe on the basis of the assumption that hemiones were not present in western Europe at this time (Bennett *et al.*, in review).

Genetic Diversity of the Reintroduced Asiatic Wild Ass in Israel: From the Breeding Core to the Current Wild Population

With Contributions by Shirli Bar-David, Sharon Renan, Tomer Guieta, Edith Speyer, Naama Shahar, Gili Greenbaum, Idan Goodman, Amos Bouskila, and Alan Templeton

The endemic Middle East subspecies of Asiatic wild ass (*E. h. hemippus*) became extinct at the beginning of the twentieth century as a result of hunting and habitat loss (Groves 1986). In 1968, the Israel Nature and Parks Authority (INPA) established a breeding core of the Asiatic wild ass at the Hai-Bar Yotvata Reserve from six Persian wild asses (*E.h. onager*; three males, three females), and five from the Turkmenian subspecies (*E. h. kulan*; two males, three females). The animals were kept at the reserve and were allowed to interbreed (Saltz and Rubenstein 1995). Between 1982 and 1993, the INPA released the wild ass from the breeding core into the Negev Desert: 28 wild asses (14 males, 14 females) were reintroduced to Makhtesh Ramon and 10 wild asses (3 males, 7 females) to the Paran area, about 35 km south of Makhtesh Ramon. Prior to the Paran reintroduction (in 1991), a genetic study of the animals in the breeding core (17 males, 13 females; referred as the founding population) was conducted (Sinai 1994), and blood samples have been preserved since then. Since the reintroduction onset, the wild ass population has grown and expanded its range. The current population (wild population) is estimated at >250 individuals in the Negev Desert and Arava Valley.

The samples of the founding population provide a unique opportunity to explore the genetic and evolutionary processes that have occurred in the wild ass population since reintroduction. We compared the genetic diversity of the wild population with that of the founding population, four generations since reintroduction onset (generation time is 7.4 years; Saltz and Rubenstein 1995). We further examined the spatial-genetic structure of the wild population in order to understand the course of development of genetic structure and the factors affecting it.

The study was conducted by analyzing 30 blood samples of the founding population, and fecal samples of the wild population collected in the field, during 2009–2012 using noninvasive genetic techniques. DNA from the samples was extracted and amplified by a set of mtDNA (Gueta *et al.* 2014) and microsatellite primer pairs (Nielsen *et al.* 2007), following the techniques of Renan *et al.* (2012). Expected heterozygosity (H_E) and observed heterozygosity (H_O) of the founding population, calculated on the basis of nine microsatellites, were compared to measures of the *E. h. onager* from the European breeding groups documented previously (Nielsen *et al.* 2007). Both H_E and H_O of the founding population were significantly higher than the corresponding values found in the European *E. h. onager* breeding cores. The mean number of alleles per locus among the founders was lower than the European breeding cores, however. The relative high genetic diversity in the founding population in terms of heterozygosity measures is due to admixture between the subspecies/subpopulations, while the low diversity in terms of allele numbers is probably due to the recent population bottleneck in Israel (S. Bar-David *et al.*, unpublished data).

Fifty-nine samples of the wild population were analyzed using six microsatellite markers (Renan 2014). The number of alleles per locus did not differ significantly between the founding and wild populations. Most of the alleles in these loci of the founding population remained in the wild population, but there was a significant difference in the allele frequencies between the populations (Renan 2014). The wild population experienced considerable genetic drift, as indicated by substantial shifts in allele frequencies in most alleles. Model simulations suggest that the polygamous mating system of the Asiatic wild ass could induce the increased genetic drift, affecting the population's genetic diversity (Renan 2014). Decrease in the genetic diversity over time could have significant consequences for the long-term persistence of the wild ass in Israel.

We examined the spatial-genetic structure of the wild population and related it to landscape features within the wild ass range of distribution (Nezer 2011; Davidson *et al.* 2013). We collected 393 fecal samples of the wild population from 79 documented sites in the Negev Desert and Arava Valley and analyzed them using mtDNA markers. Three mtDNA haplotypes were identified in the wild population, the same haplotypes as in the founding population (Gueta *et al.* 2014). Two of these haplotypes were found mostly among the Turkmenian, and the other mostly among the Persian wild ass sequences (Bennett *et al.*, in review).

Fecal samples were delimitated to six "subpopulations" according to their geographical locations. The mtDNA haplotype frequency was calculated for each subpopulation. The "east subpopulation," at the wave front of the wild population's distribution, was reported to be significantly different than the rest of the population in terms of haplotype frequencies (AMOVA, $\Phi ST = 0.13, P = 0.04$). The east region is characterized by high-quality habitat patches (high density of *Acacia spp.* trees) and low landscape connectivity to the rest of the area. These landscape features possibly led to the relative isolation of the east subpopulation. Dispersers from the release area probably initiated the subpopulation and remained in the "new area" (Gueta *et al.* 2014).

Spatial-genetic structure, following a founder effect, can arise when a population is expanding its range, as demonstrated by theoretical models (e.g., Excoffier and Ray 2008) and a few empirical studies (e.g., Neuwald and Templeton 2013). It might diminish over time owing to gene flow and additional range expansion. Yet even a temporary genetic structure may influence the response of the population to selection and help maintain genetic diversity in the total population, thereby affecting the population's long-term persistence.

Life on the Edge: What We Know—and Don't Know—about the Genetics of the African Wild Ass

With Contributions by Sónia Rosenbom and Albano Beja-Pereira

The African wild ass was once distributed across a wide geographic range—north and west into Sudan, Egypt, and Libya—but the species is now circumscribed to the Horn of Africa (the deserts of Ethiopia, Eritrea, and Somalia and possibly Sudan) (Moehlman *et al.* 2008). Historically, three subspecies were recognized according to both geographic distribution and morphology. The Atlas wild ass (*E. a. atlanticus*) was found in the Atlas region of northwestern Algeria and adjacent parts of Morocco and Tunisia; the Nubian wild ass (*E. a. africanus*) inhabited the Nubian Desert of northeastern Sudan, from east of the Nile River to the shores of the Red Sea, south to the Atbara River and into northern Eritrea; and the Somali wild ass (*E. a. somaliensis*) lived in regions of southern Eritrea, Ethiopia, and Somalia (Moehlman *et al.* 2008). The Atlas wild ass is believed to have gone extinct in historic times (Groves 1986) as a result of intensive hunting, and the Nubian wild ass is probably now extinct in the wild or extremely rare. Populations of the Somali wild ass subsist in Ethiopia and Eritrea; however, anthropogenic pressures and possible hybridization with the domestic donkey have been driving the African wild ass to the brink of extinction. Currently, the number of African wild asses in the wild is estimated at <600 animals (see chap. 12).

The few molecular studies that included African wild ass were designed either to assess the *Equus* genus phylogeny (George and Ryder 1986; Oakenfull *et al.* 2000) or to unravel the putative ancestor of the domestic donkey (Beja-Pereira *et al.* 2004; Kimura *et al.* 2011). Up to now, however, no studies focusing exclusively on the genetics of extant African wild ass populations have ever been published. Such a fact might be mainly attributed to two reasons: the near-inaccessibility of extant populations in remote and harsh habitats in northeast Africa, and the limitations to sample collection imposed by current Convention on International Trade in Endangered Species (CITES) Appendix I status.

Nonetheless, the first study using samples belonging to free-ranging African wild ass individuals from Sudan (as well as three captive individuals) allowed the identification of this species as the likely ancestor of the domestic donkey; more specifically, it identified the Nubian wild ass as the probable ancestor of one domestic donkey lineage (termed Clade 1) and a relative of the Somali wild ass, probably already extinct, as the ancestor of the other lineage (termed Clade 2) (Beja-Pereira *et al.* 2004).

More recently, a study published by Kimura *et al.* (2011) was the first to include a relevant number of noninvasive (fecal) samples of free-ranging African wild ass belonging to populations in Ethiopia and Eritrea. This study analyzed mtDNA data from modern, historical, and ancient African wild ass and domestic donkey specimens in order to clarify questions regarding donkey ancestry and domestication. Median-joining network of obtained haplotypes revealed two distinct mitochondrial domestic donkey haplogroups previously reported (Chen *et al.* 2006; Kimura *et al.* 2011), and a third clade incorporating all 33 modern Somali wild ass specimens and previously published Somali wild ass sequences. Sequences belonging to ancient and historic Nubian wild ass (identified on the basis of morphology) fell inside the wider variation of a domestic donkey Clade 1, reinforcing the hypotheses of this subspecies as a possible ancestor of the domestic donkey. Four new haplotypes were found in the thirty-three analyzed specimens, and haplotype diversity of the Somali wild ass clade was only 0.7417 + 0.0444 (compared with 0.9309 + 0.0102 and 0.8212 + 0.0268 for domestic donkeys' Clades 1 and 2, respectively), suggesting a low genetic variability among present-day Somali wild ass. The observed haplotypes shared by both Eritrean and Ethiopian populations reveals the absence of geographical structure at the mitochondrial level. A historic Somali specimen from Berbera, Somalia, from the late nineteenth century yielded amplifiable mtDNA, and the obtained sequence was identical to that of a contemporary haplotype from Eritrea and Ethiopia, demonstrating a degree of continuity in the mitochondrial variability of Somali wild ass within this region over the last 120 years.

In a recent study by Rosenbom *et al.* (2015), three African wild ass populations in Ethiopia (Afdera and Hillu) and Eritrea (Asmera) as well as several Asiatic wild ass populations (representing both *E. hemionus* and *E. kiang* species) were sampled using a noninvasive approach. Given the current geopolitical scenario in areas where the Nubian wild ass may still occur, sampling of this subspecies was not achievable. Sequence analyses of a 410-base-pair fragment of the fast-evolving mitochondrial DNA control region revealed the Eritrean population of Asmera as presenting the highest values of both haplotype and nucleotide diversity among African wild ass populations.

Clear differentiation among available Nubian and obtained Somali haplotypes was found; however, three out of four haplotypes belonging to *E. a. somaliensis* were shared between populations in Eritrea and Ethiopia, confirming no geographical structure for this subspecies. Despite the sampling effort ($n = 85$), no new haplotypes were found, which is a clear indication that variability in the wild was most probably fully captured in this study. Demographic dynamics analyses revealed that populations in Ethiopia and Eritrea remained relatively stable over time but at lower maternal effective population sizes when compared with focal populations of *E. hemionus* in Iran and China.

Given the actual scenario of decreasing numbers and extreme endangerment, the next step for the African wild ass will necessarily be the assessment of relevant parameters for the conservation and management of this emblematic species. In this way, the same populations in Ethiopia and Eritrea are now being further investigated. Previously tested horse microsatellites (Rosenbom *et al.* 2012) are being used in order to assess levels of genetic diversity, effective population size, geographic structure, and migration. Preliminary results corroborate the level of diversity of the Eritrean population of Asmera being considerably higher than for both Ethiopian populations. This value is also in line with previously reported values for a captive population of the Somali wild ass, using a similar set of microsatellite loci (Rosenbom *et al.* 2012). Obtained results for geographic structuring corroborates the idea of weak structuring among Eritrean and Ethiopian populations of the African wild ass, as previously observed in mitochondrial DNA haplotypes, with only 14% of the captured variation being attributed to differences among populations. These results raise the possibility that up until recently gene flow still existed among studied populations.

Future perspectives on genetic studies on the African wild ass include the increase in the number of sampled individuals from extant populations and, more importantly, the use of new genomic tools that will allow the analyses of genome-wide data in order to continue answering questions regarding this critically endangered equid species.

Has Genetic Conservation Been as Successful as "the Numbers"? The Conservation of Cape Mountain Zebra Genetics

With Contributions by Halszka Hrabar and Eric H. Harley

Mountain zebra are divided into two subspecies, the Cape mountain zebra (*E. z. zebra*) of South Africa and Hartmann's mountain zebra (*E. z. hartmannae*) of Namibia (Moodley and Harley 2005). The individual population histories of these two subspecies differ significantly in scale and intensity. Hartmann's mountain zebra underwent a serious decline in the first half of the 1900s (Novellie *et al.* 2002), yet the arid conditions in Namibia prevented large-scale human settlements and hence limited habitat destruction. In the last two decades, tight control of hunting and the installation of artificial watering points resulted in an increase in Hartmann's mountain zebra numbers such that the total population size in 2002 was estimated to be over 25,000 (Novellie *et al.* 2002).

By contrast, in South Africa, where Cape mountain zebra were once widespread in the mountains of the western and eastern Cape provinces (Millar 1970*a*,*b*), excessive hunting and habitat loss to agriculture (intensive human settlement dates to the seventeenth century) left their numbers in a critical status in the 1950s, with <80 individuals remaining in only three relic populations. Since the 1950s, numbers have gradually increased through active conservation programs, with a metapopulation approach through translocations to ensure continued population growth (Novellie *et al.* 2002) (see chap. 14).

Despite the improved status, however, the security of the subspecies is still questionable, as a lack genetic biodiversity is still a key threat. A primary concern is that all reintroduced subpopulations except for one (De Hoop Nature Reserve) originate from only one of the natural relic populations (Mountain Zebra National Park [MZNP]). Two-thirds of the entire genotype is therefore located in just two populations (Moodley 2002), namely, the Kammanassie and Gamka populations (which together make up <7% of individuals in the metapopulation). It is also a problem that all Cape mountain zebra populations are isolated from one another because of fencing (i.e., there is no gene flow between populations).

The level of genetic variation in each individual relict population is lower than in the subspecies overall, which implies that higher levels of diversity still exist within the Cape mountain zebra gene pool. For example, De Hoop, the population of mixed (MZNP/Kammanassie) origin, have the highest genetic variation for any Cape mountain zebra population (Moodley and Harley 2005). Translocations from the Kammanassie and Gamka populations have not occurred in the past, as the management aim has been to first allow the populations to increase sufficiently in size (to around 90 individuals in Gamkaberg). Despite formal protection since 1923 and 1971, respectively, however, both populations still consist of <50 individuals and are at risk owing to limited suitable habitat availability (Watson *et al.* 2005). For a decade now, it has been recognized that the populations can no longer be managed in a hands-off manner, and viable management options need to be implemented urgently (e.g., purchasing or leasing neighboring land to increase the available suitable habitat) to reduce the risk of quasi-extinction of the population (Watson *et al.* 2005). But progress has been minimal, and the previously less favored option of translocating the populations to other protected areas within the region with more suitable habitat is now a necessary consideration. The situation has reached a point where the need to expand these genetically valuable populations outweighs the ideal to conserve them in their original habitat.

An additional concern is the large number of small populations (particularly on private land) because the risk of inbreeding depression and increasing loss of genetic diversity from genetic drift is higher in small populations (Frankham 1996; Bowland *et al.* 2001). The total heterozygosity in the Cape mountain zebra population, based on measurements at 15 microsatellite loci, was only 0.236 (Moodley and Harley 2005), which equates to about three-quarters of gene loci being homozygous and compares with a total heterozygosity of 0.476 for Hartmann's mountain zebra. Apart from the resulting decreased evolutionary potential from the lack of adequate genetic variation, inbreeding depression may result in decreased fecundity and especially increased susceptibility to disease. This problem appears to be already manifesting itself in the Cape mountain zebra population, where there is marked increase in the prevalence of equine sarcoids (Sasidharan 2006; Sasidharan *et al.* 2011).

Of the 52 populations, 11 are smaller than the recommended minimum founder size of 14 individuals

(Novellie *et al.* 2002), and 31 populations consist of fewer than 30 individuals. The recommendation that reinforcement of existing populations receive priority over the establishment of new populations, and that one or two animals be added to subpopulations once every five to ten years in order to avoid inbreeding depression (Novellie *et al.* 1996, 2002), has also not been adhered to. This is largely because of the cost of investing in a large number of animals (according to personal communication with private land owners).

The problem of low genetic diversity in Cape mountain zebra populations is exacerbated by their social structure, whereby the behavior of the males tends to reduce the effective population size (a fraction of the males can dominate matings for an extended period of time). Females within a breeding herd are usually unrelated, however, thereby increasing the potential genetic diversity of offspring produced by a single harem. The loss of genetic integrity through human-driven hybridization between the two subspecies has been considered to be a threat in the past (Novellie *et al.* 2002). A permit system for the transportation of wildlife, maintained by provincial conservation agencies, has reduced this threat, however, as the system ensures that Hartmann's mountain zebras cannot legally be introduced into the range of the Cape subspecies. The capacity of provincial conservation agencies to undertake the necessary measures to enforce such segregation is questionable, though, and cases of hybridization are still reported.

No restrictions exist for the co-occurrence of plains zebra and Cape mountain zebra, as historic records show the two species did occur in sympatry when the Cape mountain zebra moved onto the flats from time to time for grazing (Skead 2007, 2011). The threat from hybridization between the two species has also not been a great concern, as fertile hybrids are thought to be unlikely owing to the relatively large difference in the number of chromosomal pairs (44 vs. 32 in plains zebra vs. Cape mountain zebra, respectively) compared to plains and Grevy's zebra (44 vs. 46), which do produce fertile hybrids (Ryder 1978; Cordingley *et al.* 2009). In 2014, hybrids were recorded in Mountain Zebra National Park, however, which serves as a key source population. All plains zebra and known hybrids have subsequently been culled from this population, but the two species still coexist in at least six other populations. Hybridization is most likely to occur in poor habitats or small populations, where low mate availability and skewed sex ratios may lead to exclusion of some individuals from mating (Mace and Waller 1998; Jansson *et al.* 2007), which is the case for a number of Cape mountain zebra populations. It is therefore vital that the fertility of Cape mountain zebra–plains zebra hybrids be determined so that the potential degree of threat in coexisting populations can be assessed. Regulations regarding their coexistence also need to be reassessed, and a system needs to be in place to test for hybridization in individuals so as to ensure that no movement or sale of hybrids occurs (many private owners will already only purchase DNA-tested animals).

In conclusion, conservation efforts to decrease the threat of hybridization and increase genetic diversity within the entire Cape mountain zebra metapopulation remains an urgency, as the current distribution of two-thirds of the entire genotype in just 2 of the 52 populations leaves the population highly vulnerable to genetic loss. Much of the population is at risk of inbreeding depression owing to the small size of many populations and the isolated nature of all populations. We therefore advocate a management strategy that (1) entails the mixing of relic populations, (2) focuses on increasing the size of small populations, (3) ensures gene flow between populations, and (4) addresses the threat of hybridization with plains zebra.

Genetic Diversity in Populations of Feral Horses on Public Lands in the Western United States

With Contributions by E. Gus Cothran, Brian W. Davis, Anas Khanshour, Eleanore Conant, Reid C. Booker, and Rytis Juras

Feral horses are domestic individuals that have returned to a free-living condition, either through escaping from a controlled domestic environment or by being released into the wild by owners who can no longer care for them. Many species of domestic animals have undergone this feral transition, and in most cases the free-living animals quickly return to what are essentially native phenotypes and behaviors. We examined 12 autosomal microsatellite markers distributed across the horse genome to evaluate genetic variation in feral horse populations in the United States. These 175 populations consisted of over 7,000 total horses and ranged over 10 states, with some populations being represented two or more times over the sampling time period.

Observed heterozygosity, which is a measure of individual genetic variation, ranged from 0.49 to 0.87, which spans the range of variation for domestic breeds. The same could be said for all variability measures, however; on average, genetic variation levels tend to be lower in the feral horse populations. A variety of factors can influence genetic variability in these relatively small populations. In addition to variability measures, we obtained estimates for the percentage of individuals sampled within the population that do not fit the genetic profile of the remainder of the population and may represent recent immigrants into the herd. We also obtained the appropriate management level (AML) for each herd. This is the manager-calculated number of horses that the herd area can support. It is thus an estimate of target population size (actual census numbers are almost never available). The herd will increase in size above the AML between "gathers" (roundups of horses, which are mainly used to remove what are considered to be excess animals), but the population size is usually reduced to a number well below the AML at the time of a gather.

Variability was significantly associated with population size as measured by AML. Population sizes of the herds have fluctuated over time. When a herd was sampled on more than one occasion, levels of variation at the different time periods were compared. In most cases no change was evident, mainly because sampling periods spanned less than a generation interval. There also was a significant association between the estimated number of migrants in a population and genetic variation, with variation increasing as the number of migrants increased. Various measures of genetic distance and genetic differentiation among the populations were examined, and genetic distance was compared to geographic distance based upon GPS coordinates for each herd (supplied by the US Bureau of Land Management). In general, there is no strong correlation between genetic distance and geographic distance, but those herds that are geographically closest to each other average lower genetic distances between them. Some herds show no close relationship to any other herd or to any domestic breed. These are mostly herds with variability levels outside those near the mean herd values.

Possible ancestry of each feral herd from domestic breeds was estimated by comparison to 65 domestic horse breeds, focusing on those from the New World. Most herds appeared to be of widely mixed ancestry with no definitive breed ancestor. This result is consistent with the history of most areas where the feral herds are located. When a clear ancestral breed could be identified, it was almost always a common North American breed such as the Quarter Horse or Morgan. Only a handful of herds show evidence of Spanish ancestry, with these herds tending to be more isolated than others. This is significant because the first feral horses of the American West were of Spanish origin, but most of the Spanish influence now appears to be gone.

The general management trend for feral herds on public lands has been a reduction in population size. Based upon the analyses conducted, this practice will inevitably lead to a long-term loss of genetic variability within individual Herd Management Areas. This loss of variation can be mitigated by a low level of exchange of individuals from geographically close herds. This process will tend to homogenize the herds, but it would take many generations. Some consideration should be made to manage a few of the herds to preserve them as they are for historical conservation, or to retain breed type in the cases of the small number of herds with evident Spanish horse ancestry.

Conclusion

Genetic analysis has enabled the description of both the evolution of wild equid populations in the present and in the recent and more ancient past, as well as revealed the consequences of conservation policies. Much remains to be done. In particular, the reduction of the costs of next-generation sequencing renders possible the genomic analyses of many individuals. Including multiple markers would allow expanding on this current work to complete our understanding of the past demography of populations and the relationships between them to an extent beyond what can be achieved solely through mitochondrial or microsatellite markers. This opportunity should be taken quickly to obtain a global view of these populations, which would allow for a more enlightened approach to future conservation management.

ACKNOWLEDGMENTS

While the opening page of this chapter lists the lead author of each section, many talented contributors also coauthored this chapter. The authors also wish to thank anonymous referees for their comments and corrections.

REFERENCES

Achilli, A., A. Olivieri, P. Soares, H. Lancioni, B. Hooshiar Kashani, *et al.* 2012. Mitochondrial genomes from modern horses reveal the major haplogroups that underwent domestication. Proceedings of the National Academy of Sciences of the USA 109:2449–2454.

Allshire, R.C., and G.H. Karpen. 2008. Epigenetic regulation of centromeric chromatin: Old dogs, new tricks? Nature Reviews Genetics 9:923–937.

Beja-Pereira, A., P.R. England, N. Ferrand, S. Jordan, A.O. Bakhiet, M.A. Abdalla, M. Mashkour, J. Jordana, P. Taberlet, and G. Luikart. 2004. African origins of the domestic donkey. Science 304:1781.

Bennett, E.A., D. Massilani, G. Lizzo, J. Daligault, E.M. Geigl, and T. Grange. 2014. Library construction for ancient genomics: Single strand or double strand? Biotechniques 56:289–300.

Bennett, E.A., S. Champlot, J. Peters, B.A.B. Arbuckle, S.Bar-David, *et al.* In review. Wild asses of Asia and Europe: Phylogeography from the Late Pleistocene to the present.

Bernischke, K.N., N. Malouf, R.J. Low, and H. Heck. 1965. Chromosome complement: Differences between *Equus caballus* and *Equus przewalskii*, Poliakoff. Science 148:382–383.

Bernor, R., M. Armour-Chelu, H. Gilbert, T. Kaiser, and E. Schulz. 2010. Equidae. Pages 685–721 *in* L. Werdelin and B. Sanders (Eds.), Cenozoic mammals of Africa. University of California Press, Berkeley, USA.

Bowland, A.E., K.S. Bishop, P.J. Taylor, J. Lamb, F.H. van der Bank, E. van Wyk, and D. York. 2001. Estimation and management of genetic diversity in small populations of plains zebra (*Equus quagga*) in KwaZulu-Natal, South Africa. Biochemical Systematics and Ecology 29:563–583.

Bowling, A.T., W. Zimmermann, O. Ryder, C. Penado, S. Peto, L. Chemnick, N. Yasinetskaya, and T. Zharkikh. 2003. Genetic variation in Przewalski's horses, with special focus on the last wild caught mare, 231 Orlitza III. Cytogenetic and Genome Research 102:226–234.

Breen, M., P. Downs, Z. Irvin, and K. Bell. 1994. Intrageneric amplification of horse microsatellite markers with emphasis on the Przewalski's horse (*E. przewalskii*). Animal Genetics 25:401–405.

Carbone, L., S.G. Nergadze, E. Magnani, D. Misceo, M. Francesca Cardone, *et al.* 2006. Evolutionary movement of centromeres in horse, donkey, and zebra. Genomics 87:777–782.

Champlot, S., C. Berthelot, M. Pruvost, E.A. Bennett, T. Grange, and E.M. Geigl. 2010. An efficient multistrategy DNA decontamination procedure of PCR reagents for hypersensitive PCR applications. PLoS ONE 5:e13042.

Chen, S.Y., F. Zhou, H. Xiao, T. Sha, S.F. Wu, and Y.P. Zhang. 2006. Mitochondrial DNA diversity and population structure of four Chinese donkey breeds. Animal Genetics 37:427–429.

Cieslak, M., M. Reissmann, M. Hofreiter, and A. Ludwig. 2011. Colours of domestication. Biolical Reviews of the Cambridge Philosophical Society 86:885–899.

Cordingley, J.E., S.R. Sundaresan, I.R. Fischhoff, B. Shapiro, J. Ruskey, and D.I. Rubenstein. 2009. Is the endangered Grevy's zebra threatened by hybridization? Animal Conservation 12:505–513.

Davidson, A., Y. Carmel, and S. Bar-David. 2013. Characterizing wild ass pathways using a non-invasive approach: Applying least-cost path modelling to guide field surveys and a model selection analysis. Landscape Ecology 28:1465–1478.

Denzau, G., and H. Denzau. 1999. Wildesel. Jan Thorbecke Verlag GmbH, Stuttgart, Germany.

Drummond, A.J., and A. Rambaut. 2007. BEAST: Bayesian evolutionary analysis by sampling trees. BMC Evolutionary Biology 7:214.

Eisenmann, V. 2004. Equus: An evolution without lineages? Paper presented at the 18th International Senckenberg Conference, VI International Palaeontological Colloquium in Weimar. Late Neogene and Quaternary Biodiversity and Evolution: Regional Developments and Interregional Correlations. Weimar, Selbstverlag der Alfred-Wegner-Stiftung Berlin, Germany.

Excoffier, L., and N. Ray. 2008. Surfing during population expansions promotes genetic revolutions and structuration. Trends in Ecology and Evolution 23:347–351.

Forsten, A., 1992. Mitochondrial-DNA time-table and the evolution of *Equus*: Comparison of molecular and palaeontological evidence. Annales Zoologici Fennici 28:301–309.

Frankham, R. 1996. Relationship of genetic variation to population size in wildlife. Conservation Biology 10:1500–1508.

Geigl, E.M., and T. Grange. 2012. Eurasian wild asses in time and space: Morphological versus genetic diversity. Annals of Anatomy 194:88–102.

George, M., and O.A. Ryder. 1986. Mitochondrial DNA evolution in the genus *Equus*. Molecular Biology and Evolution 3:535–546.

Goto, H., O.A. Ryder, A.R. Fisher, B. Schultz, S.L. Kosakovsky Pond, A. Nekrutenko, and K.D. Makova. 2011. A massively parallel sequencing approach uncovers ancient origins and high genetic variability of endangered Przewalski's horses. Genome Biology and Evolution 3:1096–1106.

Groves, C.P. 1986. The taxonomy, distribution, and adaptations of recent equids. Paper presented at the Equids in the Ancient World Conference. Ludwig Reichert Verlag, Wiesbaden, Germany.

Groves, C.P., and P. Grubb. 2011. Ungulate taxonomy. Johns Hopkins University Press, Baltimore, USA.

Gueta, T., A.R. Templeton, and S. Bar-David. 2014. Development of genetic structure in a heterogeneous landscape over a short time frame: The reintroduced Asiatic wild ass. Conservation Genetics 15:1231–1242.

Guthrie, R.D. 2006. Human-horse relations using Paleolithic art: Pleistocene horses drawn from life. Pages 61–77 *in* S.L. Olsen, S. Grant, A.M. Choyke, and L. Bartosiewicz (Eds.), Horses and humans: The evolution of human-equine relationships. Archaeopress, Oxford, UK.

Hazkani-Covo, E., R.M. Zeller, and W. Martin. 2010. Molecular poltergeists: Mitochondrial DNA copies (numts) in sequenced nuclear genomes. PLoS Genetics 6:e1000834.

International Union for Conservation of Nature. 2013. Guidelines for reintroductions and other conservation translocations. Version 1.0. Species Survival Commission, Gland, Switzerland.

Ishida, N., T. Oyunsuren, S. Mashima, H. Mukoyama, and N. Saitou. 1995. Mitochondrial DNA sequences of various species of the genus *Equus* with special reference to the phylogenetic relationship between Przewalskii's wild horse and domestic horse. Journal of Molecular Evolution 41:180–188.

Jansson, G., C.G. Thulin, and A. Pehrson. 2007. Factors related to the occurrence of hybrids between brown hares Lepus europaeus and mountain hares L-timidus in Sweden. Ecography 30:709–715.

Jombart, T., S. Devillard, A.B. Dufour, and D. Pontier. 2008. Revealing cryptic spatial patterns in genetic variability by a new multivariate method. Heredity 101:92–103.

Jónsson, H., M. Schubert, A. Seguin-Orlando, A. Ginolhac, L. Petersen, *et al.* 2014. Speciation with gene flow in equids despite extensive chromosomal plasticity. Proceedings of the National Academy of Sciences of the USA 111:18,655–18,660.

Kimura, B., F.B. Marshall, S. Chen, S. Rosenbom, P.D. Moehlman, *et al.* 2011. Ancient DNA from Nubian and Somali wild ass provides insights into donkey ancestry and domestication. Proceedings of the Royal Society of London B: Biological Sciences 278:50–57.

Lau, A.N., L. Peng, H. Goto, L. Chemnick, O.A. Ryder, and K.D. Makova. 2009. Horse domestication and conservation genetics of Przewalski's horse inferred from sex chromosomal and autosomal sequences. Molecular Biology and Evolution 26:199–208.

Lindgren, G., N. Backstrom, J. Swinburne, L. Hellborg, A. Einarsson, K. Sandberg, G. Cothran, C. Vila, M. Binns, and H. Ellegren. 2004. Limited number of patrilines in horse domestication. Nature Genetics 36:335–336.

Lippold, S., M. Knapp, T. Kuznetsova, J.A. Leonard, N. Benecke, *et al.* 2011. Discovery of lost diversity of paternal horse lineages using ancient DNA. Nature Communications 2:450.

Ludwig, A., M. Pruvost, M. Reissmann, N. Benecke, G.A. Brockmann, *et al.* 2009. Coat color variation at the beginning of horse domestication. Science 324:485.

Mace, R.D., and J.S. Waller. 1998. Demography and population trend of grizzly bears in the Swan Mountains, Montana. Conservation Biology 12:1005–1016.

McCue, M.E., D.L. Bannasch, J.L. Petersen, J. Gurr, E. Bailey, *et al.* 2012. A high density SNP array for the domestic horse and extant Perissodactyla: Utility for association mapping, genetic diversity, and phylogeny studies. PLoS Genetics 8:e1002451.

Millar, J.C.G. 1970*a*. Census of Cape mountain zebra. 1. African Wildlife 24:17–25.

Millar, J.C.G. 1970*b*. Census of Cape mountain zebra. 2. African Wildlife 24:105–114.

Moehlman, P.D., H. Yohannes, R. Teclai, and F. Kebede. 2008. *Equus africanus*. IUCN Red List of Threatened Species. Version 2009.2. International Union for Conservation of Nature, Gland, Switzerland. Available at www.iucnredlist.org.

Mohr, E. 1959. Das Urwildpferd. A Ziemsen Verlag, Wittenberg, Germany.

Montefalcone, G., S. Tempesta, M. Rocchi, and N. Archidiacono. 1999. Centromere repositioning. Genome Research 9:1184–1188.

Moodley, Y. 2002. Population structuring in southern African zebras. Pages 953–968 *in* Conservation genetics. University of Cape Town, Cape Town, South Africa.

Moodley, Y., and E.H. Harley. 2005. Population structuring in mountain zebras (*Equus zebra*): The molecular consequences of divergent demographic histories. Conservation Genetics 6:953–968.

Munkhtuya, B. 2005. Behaviour and ecology implication for conservation of Equid species in Mongolia. Thesis, Charles University, Prague, Czech Republic.

National Statistical Office of Mongolia. 2014. Number of livestock. Mongolian Statisitical Information Service Database. Available at www.1212.mn/en.

Nergadze, S.G., M. Rocchi, C.M. Azzalin, C. Mondello, and E. Giulotto. 2004. Insertion of telomeric repeats at intrachromosomal break sites during primate evolution. Genome Research 14:1704–1710.

Nergadze, S.G., M.A. Santagostino, A. Salzano, C. Mondello, and E. Giulotto. 2007. Contribution of telomerase RNA retrotranscription to DNA double-strand break repair during mammalian genome evolution. Genome Biology 8:R260.

Nergadze, S.G., M. Lupotto, P. Pellanda, M. Santagostino, V. Vitelli, and E. Giulotto. 2010. Mitochondrial DNA insertions in the nuclear horse genome. Animal Genetics 41:176–185.

Neuwald, J.L., and A.R. Templeton. 2013. Genetic restoration in the eastern collared lizard under prescribed woodland burning. Molecular Ecology 22:3666–3679.

Nezer, O. 2011. The use of predicted distribution model of the Asiatic wild ass (*Equus hemionus*) for sustainable management of the Negev and the Arava. Faculty of Civil and Environmental Engineering, Technion–Institute of Technology, Haifa, Israel.

Nielsen, R.K., C. Pertoldi, and V. Loeschcke. 2007. Genetic evaluation of the captive breeding program of the Persian wild ass. Journal of Zoology 272:349–357.

Novellie, P., P.S. Millar, and P.H. Lloyd. 1996. The use of VORTEX simulation models in a long term programme of reintroduction of an endangered large mammal, the Cape mountain zebra (*Equus zebra zebra*). Acta Oecologica 17:657–671.

Novellie, P., M. Lindeque, P. Lindeque, and P.H. Lloyd. 2002. Status and action plan for the mountain zebra (*Equus zebra*). Pages 28–42 *in* P.D. Moehlman (Ed.), Equids: Zebras, asses and horses. Status survey and conservation action plan. International Union for Conservation of Nature, Gland, Switzerland.

Oakenfull, E.A., and O.A. Ryder. 1998. Mitochondrial control region and 12S rRNA variation in Przewalski's horse (*Equus przewalskii*). Animal Genetics 29:456–459.

Oakenfull, E.A., H.N. Lim, and O.A. Ryder. 2000. A survey of equid mitochondrial DNA: Implications for the evolution, genetic diversity and conservation of *Equus*. Conservation Genetics 1:345–355.

Orlando, L., V. Eisenmann, F. Reynier, P. Sondaar, and C. Hänni. 2003. Morphological convergence in Hippidion and Equus (Amerhippus) South American equids elucidated by ancient DNA analysis. Journal of Molecular Evolution 57 Supplement 1:S29–40.

Orlando, L., D. Male, M.T. Alberdi, J.L. Prado, A. Prieto, A. Cooper, and C. Hanni. 2008. Ancient DNA clarifies the evolutionary history of American Late Pleistocene equids. Journal of Molecular Evolution 66:533–538.

Orlando, L., J.L. Metcalf, M.T. Alberdi, M. Telles-Antunes, D. Bonjean, *et al.* 2009. Revising the recent evolutionary history of equids using ancient DNA. Proceedings of the National Academy of Sciences of the USA 106:21,754–21,759.

Orlando, L., A. Ginolhac, G. Zhang, D. Froese, A. Albrechtsen, *et al.* 2013. Recalibrating *Equus* evolution using the genome sequence of an early Middle Pleistocene horse. Nature 499:74–78.

Piras, F.M., S.G. Nergadze, V. Poletto, F. Cerutti, O.A. Ryder, T. Leeb, E. Raimondi, and E. Giulotto. 2009. Phylogeny of

horse chromosome 5q in the genus Equus and centromere repositioning. Cytogenetics and Genome Research 126:165–172.
Piras, F.M., S.G. Nergadze, E. Magnani, L. Bertoni, C. Attolini, L. Khoriauli, E. Raimondi, and E. Giulotto. 2010. Uncoupling of satellite DNA and centromeric function in the genus *Equus*. PLoS Genetics 6:e1000845.
Pruvost, M., and E.-M. Geigl. 2004. Real-time quantitative PCR to assess the authenticity of ancient DNA amplification. Journal of Archeological Science 31:1191–1197.
Pruvost, M., T. Grange, and E.-M. Geigl. 2005. Minimizing DNA contamination by using UNG-coupled quantitative real-time PCR on degraded DNA samples: Application to ancient DNA studies. BioTechniques 38:569–575.
Pruvost, M., R. Bellone, N. Benecke, E. Sandoval-Castellanos, M. Cieslak, *et al.* 2011. Genotypes of predomestic horses match phenotypes painted in Paleolithic works of cave art. Proceedings of the National Academy of Sciences of the USA 108:18,626–18,630.
Pruvost, M., M. Reissmann, N. Benecke, and A. Ludwig. 2012. From genes to phenotypes—Evaluation of two methods for the SNP analysis in archaeological remains: Pyrosequencing and competitive allele specific PCR (KASPar). Annales of Anatomy 194:74–81.
Purgato, S., E. Belloni, F.M. Piras, M. Zoli, C. Badiale, *et al.* 2014. Centromere sliding on a mammalian chromosome. Chromosoma 124:277–287.
Raimondi, E., F.M. Piras, S.G. Nergadze, G.P. Di Meo, A. Ruiz-Herrera, M. Ponsa, L. Ianuzzi, and E. Giulotto. 2011. Polymorphic organization of constitutive heterochromatin in *Equus asinus* (2n = 62) chromosome 1. Hereditas 148:110–113.
Renan, S. 2014. From behavioral patterns to genetic structure: The reintroduced Asiatic wild ass (*Equus hemionus*) in the Negev Desert. Thesis, Ben-Gurion University of the Negev, Beersheba, Israel.
Renan, S., E. Speyer, N. Shahar, T. Gueta, A.R. Templeton, and S. Bar-David. 2012. A factorial design experiment as a pilot study for noninvasive genetic sampling. Molecular Ecology Resources 12:1040–1047.
Rieder, S. 2009. Molecular tests for coat colours in horses. Journal of Animal Breeding and Genetics 126:415–424.
Rocchi, M., N. Archidiacono, W. Schempp, O. Capozzi, and R. Stanyon. 2012. Centromere repositioning in mammals. Heredity 108:59–67.
Rosenbom, S., V. Costa, B. Steck, P. Moehlman, and A. Beja-Pereira. 2012. Cross-species genetic markers: A useful tool to study the world's most threatened wild equid—*Equus africanus*. European Journal of Wildlife Research 58:609–613.
Rosenbom, S., V. Costa, N. Al-Araimi, E. Kefena, A.S. Abdel-Moneim, M.A. Abdalla, A. Bakhiet, and A. Beja-Pereira. 2015. Genetic diversity of donkey populations from the putative centers of domestication. Animal Genetics 46:30–36.
Ryder, O.A. 1978. Chromosomal polymorphism in *Equus hemionus*. Cytogenetics and Cell Genetics 21:177–183.
Ryder, O.A. 1994. Genetic studies of Przewalski's horses and their impact on conservation. Pages 75–92 *in* L. Boyd,and K.A. Houpt (Eds.), Przewalski's horse. State University of New York Press, Albany, USA.
Salem, A.H., J.S. Myers, A.C. Otieno, W.S. Watkins, L.B. Jorde, and M.A. Batzer. 2003. LINE-1 preTa elements in the human genome. Journal of Molecular Biology 326:1127–1146.
Saltz, D., and D.I. Rubenstein. 1995. Population-dynamics of a reintroduced Asiatic wild ass (*Equus hemionus*) herd. Ecological Applications 5:327–335.
Sasidharan, S.P. 2006. Sarcoid tumours in Cape mountain zebra (*Equus zebra zebra*) populations in South Africa: A review of associated epidemiology, virology and genetics. Transactions of the Royal Society of South Africa 61:11–18.
Sasidharan, S.P., A. Ludwig, C. Harper, Y. Moodley, H.J. Bertschinger, and A.J. Guthrie. 2011. Comparative genetics of sarcoid tumour-affected and non-affected mountain zebra (*Equus zebra*) populations. South African Journal of Wildlife Research 41:36–49.
Short, R.V., A.C. Chandley, R.C. Jones, and W.R. Allen. 1974. Meiosis in interspecific equine hybrids II: The Przewalski horse / domestic horse hybrid (*Equus przewalskii* X *E. caballus*). Cytogenetics and Cell Genetics 13:465–478.
Sinai, J. 1994. A program for assembling a herd from a large breeding core, for the purpose of reintroduction—The Asiatic wild-ass *Equus hemionus* in Hai-Bar Yotvata. Thesis, Hebrew University of Jerusalem, Jerusalem, Israel.
Skead, C.J. 2007. Historical incidence of the larger land mammals in the broader eastern Cape. Centre for African Conservation Ecology, Nelson Mandela Metropolitan University, Port Elizabeth, South Africa.
Skead, C.J., A. Boshoff, G.I.H. Kerley, and P. Lloyd. 2011. Historical incidence of the larger land mammals in the broader western and northern Cape. Centre for African Conservation Ecology, Nelson Mandela Metropolitan University, Port Elizabeth, South Africa.
St-Louis, A., and S.D. Côté. 2009. *Equus kiang* (Perissodactyla: Equidae). Mammalian Species 835:1–11.
Svensson, E.M., Y. Telldahl, E. Sjoling, A. Sundkvist, H. Hulth, T. Sjovold, and A. Gotherstrom. 2012. Coat colour and sex identification in horses from Iron Age Sweden. Annales of Anatomy 194:82–87.
Tozaki, T., N. Takezaki, T. Hasegawa, N. Ishida, M. Kurosawa, M. Tomita, N. Saitou, and H. Mukoyama. 2003. Microsatellite variation in Japanese and Asian horses and their phylogenetic relationship using a European horse outgroup. Journal of Heredity 94:374–380.
Trifonov, V.A., R. Stanyon, A.I. Nesterenko, B. Fu, P.L. Perelman, *et al.* 2008. Multidirectional cross-species painting illuminates the history of karyotypic evolution in Perissodactyla. Chromosome Research 16:89–107.
Trommerhausen-Smith, A., J.P. Hughes, and D.P. Neeley. 1979. Cytogenetic and clinical findings in mares with gonadal dysgenesis. Journal of Reproduction and Fertility Supplement 27:271–276.
Tsendsuren, T., V.N. Voronkov, T. Ariuntuul, G.E. Sulimova, and T. Janchiv. 2013*a*. Genetic structure of Mongolian horse. Proceedings of the Institute of Biology MAS 29:4–7.
Tsendsuren, T., V.N. Voronkov, T. Ariuntuul, G.E. Sulimova, and T. Janchiv. 2013*b*. Genetic diversity of the mitochondrial DNA control region in Mongolian horse populations. Proceedings of the Institute of Biology MAS 29:14–18.

Vilstrup, J.T., A. Seguin-Orlando, M. Stiller, A. Ginolhac, M. Raghavan, *et al.* 2013. Mitochondrial phylogenomics of modern and ancient equids. PLoS ONE 8:e55950.

Voronkova, V.N., T. Tsedev, and G.E. Sylimova. 2011. Comparative analysis of the informativeness of ISSR markers for estimating genetic diversity of horse breeds. [In Russian.] Genetika 47:1131–1134.

Wade, C.M., E. Giulotto, S. Sigurdsson, M. Zoli, S. Gnerre, *et al.* 2009. Genome sequence, comparative analysis, and population genetics of the domestic horse. Science 326:865–867.

Watson, L.H., H.E. Odendaal, T.J. Barry, and J. Pietersen. 2005. Population viability of Cape mountain zebra in Gamka Mountain Nature Reserve, South Africa: The influence of habitat and fire. Biological Conservation 122:173–180.

Yang, F., B. Fu, P.C. O'Brien, T.J. Robinson, O.A. Ryder, and M.A. Ferguson-Smith. 2003. Karyotypic relationships of horses and zebras: Results of cross-species chromosome painting. Cytogenetics and Genome Research 102:235–243.

Yang, F., B. Fu, P.C. O'Brien, W. Nie, O.A. Ryder, and M.A. Ferguson-Smith. 2004. Refined genome-wide comparative map of the domestic horse, donkey and human based on cross-species chromosome painting: Insight into the occasional fertility of mules. Chromosome Research 12:65–76.

The Roles of Humans in Horse Distribution through Time

SANDRA L. OLSEN

The human–horse relationship has been a key factor in alterations of the horse's range, its introduction to new regions, and its redistribution in areas previously occupied before the Holocene (12,000 BP to present). This chapter provides a synopsis of the ways in which the changing roles of the horse in human societies and the movements of people have combined to produce the current geographic coverage of this amazingly adaptable species. Today, horses are found on every continent in the world with the exception of Antarctica, but this has not always been the case. Before human intervention, wild equid species dispersed from one landmass to the next via land bridges when sea levels were low. Mountains, deserts, and rainforests impeded the expansion of wild equids, as grasslands are typically their preferred habitat, but large bodies of water presented the most challenging barriers. After domestication, humans became the dominant agent of dispersal, introducing them into radically different environments.

Horses played a variety of vital roles in societies, including those related to diet, transportation, work, religious, property / commodity, military, and sports. Because horses conveyed people, either on their backs or in vehicles, the dispersal of the two species was often synchronized. The movement of horses from one region to the next was accomplished through migration, exploration, colonization, trade, marriage exchanges such as dowry and bride price, theft and counting coup, military campaigns, war booty, gifts or tribute, and accidents such as shipwrecks. Numerous innovations facilitated horse transport, with the most significant being improved roads and shipping methods. Domestication of grain provided a supply of fodder when needed, especially in marginal environments, but horses can survive on relatively poor-quality pastures. Documenting the presence of horses in a region is accomplished through identification of their remains or trappings in archaeological contexts, images in art, and written accounts. Tracking the arrival of horses in new lands beyond their original Holocene native range is not always easy or reliable, however, because it is often dependent on inconsistent or incomplete records. The spread of horses is reported here with the goal of shedding light on the timing of their initial occurrences around the world and their lasting impact.

The first equid to be associated with hominins in the archaeological record was *Hipparion*. *Hipparion* was a relatively large (about 13 and a half hands) grazing horse that retained four toes, although three were vestigial and did not touch the ground. Its footprints were found juxtaposed with those of the early hominin *Australopithecus* at the Pliocene trackway of Laetoli, Tanzania, which dates to 3.6 million years ago (Leakey and Hay 1979). It is impossible to develop any scenarios of interaction between the two genera, however, as the volcanic ash preserved tracks from a wide array of animals that lived in the area at the time and happened to be passing through.

North America was the homeland of the lineage that gave rise to the genus *Equus*, which appeared as early as 4 to 4.5 million years ago (Orlando *et al.* 2013). Fossils of *Equus* appear on every continent except Australia and Antarctica. *Equus* fossils of the stenonid type, probably the ancestors of zebras, dispersed from North America into Eurasia by the Late Pliocene, 2.6 million years ago (Forstén 1988; Webb and Hemmings 2006). These zebra predecessors first appear in the African record in the Omo beds, Member G, Shungura Formation, Ethiopia, about 2.3 million years ago (Eisenmann 1976; Hooijer 1976), where they were contemporaneous with robust Australopithecine hominins.

The oldest record for the association between the genus *Homo* and the modern African wild ass, *E. africanus*, extends to the Early Pleistocene (1.5–1.7 million years ago) in Bed II, Olduvai Gorge, Tanzania (Churcher 1982). Although Olduvai Bed II also contains large quantities of stone tools made by *Homo erectus,* the taphonomic conditions make it difficult to identify which bone accumulations represent natural deposits as opposed to those created by the hominins. There is speculation as to whether Olduvai Bed I hominins were hunting or merely scavenging large herbivores (Potts and Shipman 1981), but by the time of Bed II it is assumed hunting was performed regularly. The modern Asiatic wild ass, *E. hemionus*, appeared in Asia in the Middle to Late Pleistocene.

Caballine equids (true horses) most likely originated in North America. The oldest clearly identified one, *E. scotti*, from Rock Creek, Texas, dates to 1.2 million years ago (Webb and Hemmings 2006). The earliest documented caballine horse in Europe, *E. mosbachensis,* was present by one million years ago (Eisenmann 1992). *Equus* died out at the end of the Pleistocene in North America, as did camelids, mastodons, and mammoths. This continent remained without horses until Columbus introduced domestic ones to the New World on his second voyage in 1493.

Paleolithic Equid Hunting in Europe

For hundreds of millennia, there were few landmark events in the human–horse relationship. The connection was restricted to one of predator versus prey as hunters dispatched wild equids with stone and wooden weapons. This early link was confined to the Pleistocene ranges of both groups, primarily in Eurasia, but also briefly in the New World, when paleoindians and paleohorses overlapped for a few millennia before the close of the Ice Age.

During the Pleistocene, European equids were actively hunted by various species of *Homo.* Cave deposits and faunal assemblages in open-air sites indicate that the horse was regularly on their list of game animals, which included reindeer, red deer, bison, aurochs, ibex, saiga antelope, and mammoths, as well as more exotic extinct species.

The site of Dmanisi, Georgia, provides documentation that hominins and *Equus* coexisted in Eurasia for as much as 1.77 million years. In its faunal assemblage, remains of *E. stenonis*—along with a short-necked giraffe, giant ostrich, probocidian, rhinoceros, and giant cheetah—indicate a steppe or savanna environment, but the frequent occurrence of deer also demonstrates the presence of forests (Gabunia *et al.* 2000; Lordkipanidze *et al.* 2005; Vekua and Lordkipanidze 2010). The much-debated hominin fossils found at Dmanisi are now referred to as *H. erectus* (Lordkipanidze *et al.* 2013). Although stone tools similar to those from Olduvai Gorge, Bed I, occur in the deposits at Dmanisi, the formation of the site is complex. The high frequencies of carnivores—including large felids, canids, bears, and hyenas in addition to carnivore-gnawed bone, coprolites, and burrows—make it likely that nonhuman predators played a significant role in the accumulation of the faunal assemblage (Tappen *et al.* 2002). Thus it is currently unclear whether the Dmanisi hominins consumed horse meat.

Dating to about 400,000 years ago, a site in a lignite mine near Schöningen, Germany, yielded an amazing discovery in 1995 (Thieme 2005, 2007) that provides much more compelling evidence for predation by humans. Through unique conditions of preservation, eight Lower Paleolithic wooden spears, grooved wooden tools, and large quantities of stone artifacts were found lying in close proximity to the remains of at least twenty wild horses (*E. mosbachensis*). One spear was still embedded in a horse pelvis, and the horse bones exhibited numerous cut marks from butchery. The site produced the oldest known wooden artifacts, shedding light on how early humans were already able to dispatch such large game by this time. Horses dominated the faunal assemblage (90% of the identifiable bone fragments), and the site provides convincing evidence that they were hunted, butchered, and presumably eaten by *Homo heidelbergensis.*

The upper Danube Valley and surrounding area report high frequencies of *Equus* remains in archaeological sites dating to the Middle Paleolithic, when Neanderthals occupied Europe (Olsen 2003*a*). In the European Upper Paleolithic (40,000–12,000 BP), after *Homo sapiens* replaced Neanderthals, horses were represented in art as well as in archaeological contexts by their bones. Images of horses were plentiful in the cave art of France, at sites like Niaux, Pech Merle, Chauvet (Chauvet *et al.* 1996), Cosquer (Clotte and Courtin 1996), and Lascaux; and in Spain at Tito Bustillo, Ekain, Altamira, and others (Leroi-Gourhan 1967; Sieveking 1979; Olsen 2003*a*; Guthrie 2006). Dun-coated horses with erect manes were often depicted running or being shot with spears. Small pieces of mobile art included numerous examples of horse heads and whole figurines carved in antler or ivory (Bahn and Vertut 1997; Olsen 2003*a*).

Open-air Upper Paleolithic sites in central Germany, on the southern edge of the North European Plateau, have yielded high concentrations of horse bones. Horses clearly preferred this steppe environment, at 200–500 m elevation, to the high alpine tundra of the Thuringian Forest or the damp Elbe Valley lowlands (Olsen 2003*a*). In central Russia, Soffer (1985) reported that 48% of the 21 Upper Paleolithic sites surveyed contained horse remains.

One of the most famous examples of Middle to Upper Paleolithic horse hunting sites is Solutré, in east-central France (Olsen 1989; Combier and Montet-White 2002). The site is a 9-m-deep deposit, consisting mostly of densely packed horse bones with some stone tools spread over a hectare in area. Here, from 50,000 to 12,500 BP, hunters drove bands of wild horses into a cul-de-sac in the limestone ridge known as Roche de Solutré to be taken with spears in this natural corral (Olsen 1989). The horses were likely deflected up against the ridge as they passed through the valley during their biannual migrations between their winter home on the Saône River and their summer pastures in the Massif Central. Stone tools found in the strata indicate that different Middle and Upper Paleolithic cultures—including the Mousterian, Aurignacian, Gravettian, Solutrean, and Magdalenian—all utilized this place to slaughter wild horses.

Across much of Europe, the numbers of European wild horses (*E. ferus*) declined greatly at the end of the Pleistocene, but tiny relict populations continued to be widely dispersed throughout the Neolithic (Benecke 2006; Olsen 2006*a*,*b*). Arbogast *et al.* (2002) conducted a survey of the faunal remains from 133 Early Neolithic Linearbandkeramik (LBK) period sites (5500–4900 BCE) and found a small presence in France, Germany, Austria, Czech Republic, Hungary, Slovak Republic, Ukraine, Moldova, and Romania. The work of Boyle (2006) demonstrates the continuation of this low-frequency and broad distribution for Middle to Late Neolithic sites in western Europe between 4900 and 3800 BCE. She found that horses constituted on average only about 0.2% of the bones from sites in this time span. Unless there is evidence of domestication at these early dates, it is likely these remains represent wild horses.

The Appearance of the Horse in the British Isles

Regarding the United Kingdom, the oldest horse remains, as well as one of the earliest hominin occupations, were found at the site of Pakefield, East Anglia. There researchers retrieved flint artifacts and a rich fauna that included two species of horse considered to extend to 700,000 BCE (Parfitt *et al.* 2005). The 500,000-year-old Lower Paleolithic Eartham Pit at Boxgrove, near Chichester, West Sussex, produced a horse scapula with possible damage from a spear point, suggesting that hominins were hunting horses in the region by this time. A horse butchery site, also at Boxgrove, contained at least seven separate flint tool knapping episodes that took place while the carcass was being processed (Roberts and Parfitt 1999). Remains of *Homo heidelbergensis* and hundreds of stone tools have also been found during the many years of excavation at Boxgrove.

Sporadic skeletal finds indicate the continued presence of the horse in Britain through the Late Pleistocene to the first half of the Holocene. For example, at Pin Hole Cave, in the Creswell Crags Gorge, North Midlands, and at Kent's Cavern near Torquay, Devonshire, horse skeletal elements dating to 55,000–45,000 BCE have been recovered. Robin Hood Cave, also in Creswell Crags, is especially interesting in terms of early horse–human relationships. In addition to the recovery of horse bones dating to 32,000–24,000 BCE, an engraving of a horse head and neck delicately carved on a rib fragment was discovered there in 1876 (Sieveking 1987) (fig. 8.1). Based on style, the engraved bone was likely made by a Magdalenian artist between 11,000 and 13,000 years ago. It is the only depiction of an animal dating to the Upper Paleolithic in the United Kingdom. Horses continued to exist in Britain, albeit in small numbers, through the early Holocene, at sites like Kendrick's Cave, Wales (10,437–8928 calibrated years [cal] BCE), Seamer Carr, Yorkshire (10,007–8723 cal BCE) (Clutton-Brock and Burleigh 1991*a*), and Cavall's Cave, in the Wye Valley (6445–6106 cal BCE) (Bronk Ramsey *et al.* 2002). A horse mandible found

Fig. 8.1 Incised image of a horse head from the Upper Paleolithic at Creswell Crags, England. Photo by Jill Cook, British Museum

at Kendrick's Cave in 1880 was incised with numerous chevrons from a sharp flint implement (Sieveking 1987). Only the anterior portion survived, and its function and how it was displayed are unknown.

During extremely cold intervals in the Pleistocene, sea levels were sufficiently low for a broad land bridge known as Doggerland to connect Great Britain to the European mainland at the Netherlands and west coasts of Germany and Denmark (Cunliffe 2001). Doggerland persisted as a marshy tundra with rivers and lakes through the Mesolithic period until around 6200 BCE. During its existence, animals could migrate between the continent and Great Britain. As the North Sea gradually rose, it left individual islands on the high ground, but then finally inundated Doggerland. An enormous tsunami occurring between 6225 and 6170 BCE, created by the last of three submarine landslides along the coast of Norway known collectively as the Storegga Slides (Bondevik *et al.* 2003), probably finished off any traces of animals or people still living on the terrestrial remnants of Doggerland.

Evidence for the horse in the United Kingdom after the end of the seventh millennium BCE does not appear until much later, with a rash of dates between 1500 and 800 BCE (Bendrey 2012). A date for horse remains from Grimes Graves (2859–1630 cal BCE) (Clutton-Brock and Burleigh 1991*b*) has recently been shown to be too early (Higham *et al.* 2007; Bendrey 2012). Newly calibrated radiocarbon dates for this material place it at 55–384 CE. Because there is a long temporal gap, it is likely that the European wild horse (*E. ferus*) died out completely in the United Kingdom and was only reintroduced as a domesticate much later.

Ireland differs from the United Kingdom because it has been an island from about 14,000 BCE, when rising sea levels inundated the land bridge that connected it to Great Britain. Horses have been reported at Shandon Cave dating to 30,977–29,275 cal BCE (Hedges *et al.* 1997; Woodman *et al.* 1997), but not again until domestic ones finally appear. For decades, the first domestic horse was thought to be that found at Newgrange, dating to around 2000 BCE (Wijngaarden-Baker 1974). Recent radiocarbon dating of skeletal remains from the site resulted in a much younger age range of ca. 67–220 CE, meaning they date to the Iron Age rather than Bronze Age (Bendrey *et al.* 2013).

Summary for Old World Equids of the Pleistocene

During the Pleistocene, human hunters appear to have had little impact on the distribution of horses. Shrinkage of their range in the Old World and extinction in the New World coincide with major climatic change at the close of the Ice Age that altered vegetation zones globally. Shifts in ranges of other herbivores, like the reindeer—as well as the extinction of mammoths, mastodons, and North American camelid—are relatively contemporaneous with changes in the distribution of the horse. The argument that small numbers of paleoindians caused megafaunal extinction in North America at the end of the Pleistocene (Martin

1973) downplays the comparable herbivore extinction (mammoth, woolly rhinoceros, giant deer, musk oxen) and range changes (horse, reindeer, chamois) in Eurasia that occurred simultaneously. That humans and *Equus* coexisted in Eurasia for nearly 1.8 million years is demonstrated by the *E. stenonis* remains in the faunal assemblage at the site of Dmanisi (Gabunia *et al.* 2000). Despite the long duration of shared habitats for horses and humans in Eurasia, the horse's range only shrank significantly when climate and vegetation zones markedly changed globally at the close of the Pleistocene.

Wild Horses and Humans Collide in the New World

There is convincing evidence that paleoindians hunted Pleistocene horses in North America (Grayson and Meltzer 2003; Webb and Hemmings 2006), even if humans were probably not the major cause of their New World extinction. Hemmings (2004) has identified nearly 50 Late Pleistocene archaeological sites containing the remains of *Equus*. Webb and Hemmings (2006), however, caution against the mere co-occurrence of stone tools and *Equus* skeletal material as definitive proof of an interaction between humans and horses, as opposed to merely coincidental deposition in the same location. Clovis people in the terminal Pleistocene created most of the archaeological sites containing horse bones, and those with later deposits did not yield horse bones in situ in a post-Clovis archaeological stratum. This accords well with paleontological sites in North America that indicate extinction had occurred by 11,000 BP. Webb and Hemmings (2006) rank the level of confidence for indication of human involvement with the horse at individual sites in increasing order: (1) blood residues on stone tools, (2) butchery and burning of bones, and (3) artifacts made from horse bones. Guthrie (2003, 2006) has provided a strong argument against humans being the main cause for the extinction of the North American horse at the close of the Pleistocene by demonstrating a reduction in equid body size, a response likely triggered by the changing climate and vegetation.

The early equid called *Hippidion* migrated from North to South America in the Late Pliocene (ca. 2.6 million years ago) and apparently survived into the early Holocene, until at least 8,000 years ago (Alberdi and Prado 1993). Its large distribution incorporated modern-day Venezuela, Bolivia, Peru, Uruguay, Brazil, Chile, and Argentina. It is fairly clear that *Hippidion* and paleoindians occupied South America at the same time, as the archaeological site of Monte Verde, Chile, dates to ca. 14,000 years ago (Dillehay *et al.* 2008). *Hippidion* remains have been found in caves that were occupied by humans, but contemporaneity and direct association can be difficult to demonstrate. If so, then Pleistocene and even early Holocene hunters could have taken them as prey. Examples include Cueva del Milodon (*Mylodon* was a giant ground sloth that went extinct around 10,000 years ago), in Chilean Patagonia, where *Hippidion* bones were recovered (Martinic 1996). At Piedra Museo, in Santa Cruz Province, Argentina, organic residues on spear points have been identified as belonging to *Hippidion* (Ramirez Rozzi *et al.* 2000).

Domestication of the Horse

Domestication can be viewed as the initial catalyst that eventually enabled humans to greatly expand the geographic distribution of horses well beyond their Pleistocene range and to reintroduce them into areas where they had vanished in the early Holocene (Olsen 2006*a*,*b*). When compared to many other common domesticates—including dogs, sheep, goats, cattle, and pigs—horses were late in being brought fully under human control. Because the Eurasian steppe cultures were familiar with horses through hunting them, it is no wonder they were most likely the first people to tame and eventually domesticate them. In the Dneister Valley, Ukraine, domestic cattle and pigs had been introduced in the sixth millennium BCE (Anthony 2007), and some have argued that the idea of domesticating the horse may have originated among steppe cattle herders. Between 5200 and 4600 BCE, people living in the lower Don and north Caspian region were burying their deceased with domestic cattle and sheep, as well as with horses of indeterminate status. The percentage of horse meat in the diet rose to over 50%, and horse figurines began to appear. At the Eneolithic site of Dereivka, Ukraine, dating to 4470–3530 BCE (Rassamakin 1999), 63% of the faunal remains were from horse (Bibikova 1986). Even though a horse skull from Dereivka exhibiting metal bit wear has been shown to be an Iron Age intrusion (Anthony 2007), Dereivka still reflects an increased dependence on the horse that could reflect an early attempt to bring this species under control.

Verification for horse domestication dates only to about 5,500 years ago, when the Copper Age Botai people in north-central Kazakhstan were keeping, breeding, and even milking horses (Olsen 2003*a*,*b*, 2006*a*,*b*; Olsen *et al.* 2006; Outram *et al.* 2009). Evidence includes large, permanent settlements sustained by an economy heavily focused on horse pastoralism. Over

90% of the enormous faunal assemblages of the Botai sites consist of horse remains that exhibit cut marks indicative of heavy butchering. Soil micromorphological analysis (French and Kousoulakou 2003) revealed that roof construction materials in houses incorporated horse manure (Olsen *et al.* 2006). Corrals made of upright posts contained high concentrations of nitrogen and sodium from manure and urine. A soft thong bridle may have caused light wear on a lower premolar (Outram *et al.* 2009). One of the most common types of tool in the Botai culture was a "thong-smoother" for straightening and stretching rawhide thongs, manufactured from horse mandibles (Olsen 2001). Horse keepers would have found thongs essential for even the most basic equipment, such as bridles, hobbles, lassoes, and pole snares. That the Botai revered their horses is clear from the number of horse head and neck burials in ritual pits (Olsen 2000), as well as a mass grave containing 4 humans and at least 14 horses (Zaibert 1993). Lastly, and importantly, deuterium and lipids extracted from Botai pottery sherds indicate that these Copper Age people were milking their mares, just as modern Kazakhs do in the region (Outram *et al.* 2009).

The Initial Expansion of the Domestic Horse's Range

The domestic horse probably spread fairly rapidly across that enormous highway known as the Eurasian Steppe because there were few geographic obstructions. The steppes had already proven to be an ideal habitat for its progenitor, the European wild horse (*E. ferus*). Recent research on mitochondrial DNA (mtDNA) indicates that there are large numbers of matrilines for domestic horses (Jansen *et al.* 2002; Achilli *et al.* 2012), probably because along that vast Eurasian Steppe corridor, and perhaps elsewhere, wild mares were being brought into domestic herds and bred. Studies of the horse Y-chromosome, on the other hand, show extremely low nucleotide diversity (Wallner *et al.* 2013). This could be due to the small number of male founders, or later selective breeding strongly favoring particular sires (Bendrey 2014).

Other parts of the Old World were much slower to adopt horse pastoralism and riding in particular. The fact that there were still small, relict populations of European wild horse sprinkled across Europe in the mid-Holocene makes it difficult to distinguish new domestic arrivals from wild individuals, especially in the fourth to third millennium BCE. A more or less contiguous range of the wild horse probably extended west from the Eurasian Steppe across northern Europe, but western Europe, with the exception of Germany, has revealed few horse remains in the early to middle Holocene. Bendrey (2012) has done a thorough study outlining the distribution of the horse in the archaeological record of Europe. One relatively clear boundary seems to be the Balkan Mountains (Benecke 2006; Greenfield 2006), south of which no horses, wild or domestic, have been found after the Pleistocene and before the Bronze Age.

The Near East

In the Near East, the distribution of horses appears sparse until the first millennium BCE (Olsen 2010*a*). At the site of Abu Hureyra, Syria, a single horse humerus from a context dating to 9000–6000 BCE was found (Legge and Rowley-Conwy 2000). Tappeh Zageh, in north-central Iran, produced horse bones at around 5000 BCE (Mashkour *et al.* 1999). Otherwise, skeletal remains that date to the Copper and Bronze Ages have been found in the southern Levant. For Mesopotamia, the earliest evidence dates to about 2500 BCE. Horses finally appear in southern Iran around 2000 BCE. On the Konya Plain of Turkey, an interesting situation occurs at the famous site of Çatal Höyük, where the onager (*Equus hemionus*), extinct ass (*E. hydruntinus*), and the wild horse (*E. ferus*) all coexisted during the Neolithic, in the seventh to eighth millennium cal BCE (Martin and Russell 2006). *E. ferus* was also positively identified at the Early Neolithic site of Aşikh Höyük (Vigne *et al.* 1999), and many Chalcolithic (Copper Age) sites in Anatolia have yielded horse remains, but their wild/domestic status is uncertain.

The oldest domestic horse figurines from the Near East come from Tell es-Sweyhat, Syria (2300–2200 BCE) (Holland 2006). Three or possibly four partial horse figurines with flowing manes were found at the site. The best example is a stallion that is nearly complete. In addition to the long mane, a perforation for a harness in its muzzle demonstrates that this delicate clay model clearly represents a domestic horse. Unfortunately, the pieces do not definitively prove whether living horses were present; the figurines could have been imported or made by a traveler who had seen horses elsewhere. That there are several examples suggests familiarity with horses. The faunal material from the site produced the remains of onagers and possibly asses, but not *E. ferus* (Weber 1997). Cuneiform tablets from the Third Dynasty Sumerian city of Ur mention the horse specifically by ca. 2150–2000 BCE (Olsen 2010*a*). A clay tablet, also from Ur and dating to the same time, shows a man riding a horse (Owen 1991).

The Far East

In Asia, the genus *Equus* was present in the Pleistocene and Holocene. *E. sanmeniensis* overlapped with hominins in the Middle Pleistocene (Cai *et al.* 2013), including at the famous Beijing Man site of Zhoukoudian (Gaboardi *et al.* 2005). The Asiatic wild horse, now known as Przewalski's horse (*E. ferus przewalskii*), existed before and after domesticated horses were present in Asia, although the maximum extent of its range is not documented. In the last few centuries, before it became endangered, it probably occupied at least the western half of Mongolia, Xinjiang Uyghur Autonomous Region of China, an area northwest of the great bend of the Yellow River, and possibly part of Tibet (Bökönyi 1974) (see chap. 12). Based on horse remains in Neolithic sites, it is possible that at one time the Przewalski's horse occupied a much larger region. By the time Western scientists reported the existence of the Przewalski's horse, its range had shrunk tremendously and it was approaching extinction. According to Bannikov (1958) (Groves 1994), in the late nineteenth to early twentieth century, the species had only been spotted in Djungaria, or, in other words, northern Xinjiang and southwestern Mongolia.

When horse remains appear in the Holocene archaeological record in China, it is not always clear whether they were derived from the Przewalski's horse or domestic animals unless they are associated with trappings, a chariot, or carriage fittings. Remains of horses, which are equivocal in terms of their status as wild or domestic, have been found in Middle Neolithic sites in Shaanxi, Shanxi, Inner Mongolia, and Gansu from 5000–2000 BCE (Liu and Chen 2012). By the Shang Dynasty, 1550–1050 BCE, domestic horses were depicted in art, referenced in texts, and sacrificed and placed in tombs with trappings and chariots (Linduff 2006). The use of cavalry began in the fourth century BCE. The Chinese were not well known for their horse breeding, so they had to supply their cavalries from the hostile nomads to the north or, in the Han Dynasty, from Dayuan, in the fertile Ferghana Valley of today's Uzbekistan (Creel 1965). In 100 BCE, the Han Emperor Wudi's troops forced the people of Dayuan to give them 3,000 horses, but only about 1,000 of them survived the long journey across Asia. These were the so-called Heavenly Horses (Creel 1965).

Domestic horses reached the Korean Peninsula via China, probably by war booty as well as through trade and migration. Korea, in turn, supplied Japan with horses. The Han Dynasty expanded its territory in all directions, including the Korean Peninsula. In 108 BCE, they invaded the Gojoseon Kingdom, whose capital was at Pyongyang. Both sides employed cavalries in battle. Gojoseon pit burials in the Taedong River basin contained both horse trappings and carriage fittings at this time (Barnes 1999:209).

Until recently, most scholars believed the horse to be a relative newcomer to Japan, arriving only in the Kofun period, around the fifth century CE, because earlier Chinese texts reported that horses were not observed. The classical Chinese account in the *Nihon Shoki* that Empress Jingu was given a horse by the defeated king of Silla was compiled very late, in 720 CE. In fact, wild horses have occupied Japan well into the distant past. Lower sea levels at various points in ancient times connected Japan to the mainland, enabling the free flow of wild animals during the Holocene, across what is today underwater. *Hipparion* was present in the Pliocene, from five million years BP. Even as recently as the Middle Pleistocene (0.78–0.127 million years BP) small caballine equids could be found there. Localities yielding Pleistocene horse remains include Tuki-noki, in Akita prefecture; Hitati-mati, in Ibaraki prefecture; Keisei, Tyosen, and Kuroi, in Hyogo prefecture; Kotari, Shioda, and Tskinoki, all on the island of Honshu; as well as in Kyushu.

The earliest evidence for domestic horses on Japanese islands dates to the Late Jōmon period (2000–1000 BCE) during a peak of maritime activity. In the faunal remains from 532 Jōmon shell mounds (Kidder 2007), horses ranked fourth most common, after deer, wild boars, and dogs. The horses were mostly older individuals, buried complete after death and lacking indications of butchering, suggesting they were used for transportation rather than food. The early archaeological horse remains associated with the Jōmon culture were from small individuals, but by the Yayoi period (250 BCE–300 CE), they were somewhat larger, signaling either another influx from the outside or the autochthonous development of a new type. Because the Yayoi people probably immigrated by ships from either the coast of Jiangsu, where the Han were living at the time, or the Korean Peninsula (Hammer and Horai 1995), they likely brought the larger horses with them. The initial onset and source of horseback riding in Japan have been hotly disputed (Ledyard 1975; Kidder 1985). During the Middle Kofun period (400–475 CE), burial chambers began to contain horse trappings and iron armor (Barnes 2007), probably reflecting a professional cavalry.

In written records, sources for obtaining ancient Japanese horse populations were multiple and included China, the Korean Peninsula, Mongolia, and possibly southeastern Russia, via Sakhalin. By the second century CE, even Arabian horses were reported to be arriving via China (Friday 2004:96–97). Recent mtDNA and microsatellites studies of seven modern native Japanese horse populations contradict the ancient texts, however, indicating that they were all derived from Mongolian sources (Kakoi *et al.* 2007).

Egypt as the Entry Point for Africa

It is considered well established that domestic horses first entered the African continent through Egypt in the 1600s BCE. The fortress of Buhen, in northern Sudan, yielded a relatively complete horse skeleton that radiocarbon analysis dated indirectly to 1675 BCE (Raulwing and Clutton-Brock 2009). Found in 1958, there has always been some hesitation in acceptance of the date because it is earlier than horse remains anywhere else in Egypt and is located in the distant south. Slightly later are the multiple skeletons and skulls that have been uncovered in Manfred Bietak's excavations of the Hyksos capital of Avaris, today known as Tell el Dab'a, in the Delta, dating to ca. 1660–1600 BCE (Bietak 1991). In addition to the horse, these Asian foreigners are credited with introducing the chariot and composite bow to Egypt. The Hyksos were mostly Western Semitic speakers, likely from Canaan, but may have included some Indo-Europeans, like the Hurrians, who were noted horsemen. War booty in Asian campaigns appears to have been the most significant source for new horses as well as for chariots once the Hyksos were defeated. The records of Thutmoses III's 14 campaigns in the Hall of the Annals at Karnak reveal that, with his victory at his first battle at Megiddo, ca. 1457 BCE, the Egyptian army returned with 2,041 mares, 191 foals, 6 stallions, and 924 chariots (Olsen 2010*b*:46–47). It was immediately after this time that a modern-looking Arabian horse begins to be depicted in New Kingdom tomb art, strongly suggesting that Asia was the origin for this breed.

Early images of horses and chariots can be found in rock art at sites all across the Sahara thought to date to no earlier than 1000 BCE. Some of the best examples are located in Tassili-n-Ajjer, Algeria, and the Acacus Mountains, Fezzan District, southwest Libya, but they can also be found at the Aïr of Niger, northern Mali, and west all the way to the Atlantic Coast. The Garamantes, a Berber people in the Fezzan between 1500 BCE and 0 CE (Lutz and Lutz 1995), were credited with using chariots and creating the rock art along their trade routes, but other Saharan cultures probably adopted the horse and wheeled vehicles as well. Donkeys and oxen would have been used more for haulage, as the horse was a relatively rare status symbol and was probably not that suited to the deep desert.

The Horse Appears in the Arabian Peninsula

There is currently much debate about when the horse entered the Arabian Peninsula. A recent discovery of a large stone animal effigy by Bedouins at Al Magar, south-central Saudi Arabia, has led to a claim that the figure represents a horse rather than an onager, African wild ass, domestic donkey, or even a nonequid animal (Harrigan 2012). Although the top of the head is missing, traces of a slight ridge or crest on the neck just behind the fracture line could possibly represent the base of an erect mane, like that found on wild equids. Onagers were probably plentiful during the Holocene Wet Phase, ca. 8000–2500 BCE, based on their prevalence in Neolithic rock art (Olsen 2013). They seem to have disappeared or become quite rare when the climate reverted to its current arid condition, by 2500 BCE, because they are not depicted in later petroglyphs. An indentation on the Al Magar animal's muzzle that has been called a bridle or harness may just be the indication of a coat color change, such as a "mealy muzzle" often found on wild equids and some domesticated ones with a dun coat. European Upper Paleolithic cave artists often indicate this color demarcation on the wild horse muzzles with an incised line (Olsen 2003*a*). A raised ridge over the shoulders could depict a withers stripe on a domestic donkey, rather than a horse collar, but it could also be a cuff or architectural feature added to support the figure were it tenoned into a building's masonry wall. The body is attenuated with few anatomical details behind the ridge. A date as early as 7000 BCE was derived from a carbon sample taken in the vicinity of where local pastoralists purportedly found the effigy, but because the sample was not in situ in a sealed context with the sculpture, it is unclear how closely the date correlates with the statue. Middle Paleolithic stone tools and recent Islamic graves also occur in the area.

The best evidence so far for the horse in Arabia comes from rock art, which unfortunately cannot be directly dated either. Several examples of petroglyphs of domestic equids (probably horses) pulling chariots in northern Saudi Arabia appear to predate written script. With their characteristic splayed four-spoked wheels and recumbent, back-to-back beasts, they are reminiscent of horse and chariot images found in central Asia attributed to the Bronze Age (Olsen 2010*c*, 2013) (fig. 8.2). A rough date estimate for these petroglyphs is the

Fig. 8.2 Bronze Age petroglyph of a chariot pulled by two horses at Jubbah, Saudi Arabia. Note the recumbent, back-to-back position of the horses and the splayed four-spoked chariot wheels. Photo by Richard T. Bryant

second millennium BCE. Examples from North Africa are similar in style (MacDonald 2009). Later horse and chariot depictions are shown in profile and could be influenced by Egyptian or Assyrian styles. The ancient Arabic script known as Thamudic, which is unknown before the eighth century BCE, accompany these depictions. The horses shown from this period resemble the Arabian breed in several characteristics, like ones found in New Kingdom tombs in Luxor, Egypt, from about 1400 BCE on (Olsen 2010*b*).

Roads and Ships Facilitate Travel of Horses and Humans

One of the advantages of riding, which may have begun in the Eurasian steppes soon after domestication, in the fourth millennium BCE, is that the horse and the rider can cross long distances in a short time compared to pedestrian travel. Terrain is only a deterrent in the most extreme cases of high mountains, large marshes, or significant bodies of water. Speed and access are greatly increased with improvements like roads and bridges where the route becomes more level and direct, and with fewer obstacles. The famous Persian "royal road" began in Sardis, western Turkey, and terminated some 1,600 miles later at their capital, Susa, near the head of the Persian Gulf (Casson 1994). It was first developed by the Assyrians, but refined by the Persians mainly for their government couriers. The road was open to all travelers, including both horsemen and various types of wheeled vehicles. Regular travelers averaged eighteen miles a day and made the trip in three months, but the elite Persian dispatch, which operated much like the Pony Express, covered the distance in about one-fifth the time. The vast network of Roman roads, stretching from Syria to the far reaches of Scotland and across North Africa, was even more impressive in terms of the movement of troops and horses. Consisting of an estimated 50,000 miles, their designs included straight, graded, paved, and well-drained highways, as well as tunnels and bridges. A mounted courier could cover 360 miles in 2.5 days, a speed not matched until the nineteenth century. Only the contemporaneous Chinese Han Dynasty (200 BCE to 200 CE) could compete with the Romans in road building, especially in terms of challenging mountainous terrains. It was recorded that some of their gravel roads were 50 feet wide and could manage 9 chariots abreast (Casson 1994).

Boats also transported horses up and down rivers and across large bodies of water. The ancient Egyptians carried horses and chariots up the Nile in wooden sailing ships, as depicted in the relief in the tomb of Paheri, at El Kab, for example (Wilkinson 1996). In the scene, the horses and their chariot are on the deck rather than in the cabin, where the rest of the livestock would have been kept. Shipbuilding also involved several important innovations that allowed horses to finally make marine crossings that heretofore had been an obstacle to their dispersal. The Viking knarr was a short, sturdily built ship designed for hauling livestock and supplies across the Atlantic. Capable of carrying 120 tons and covering 75 miles a day, these were the vessels that would have brought the horse to Iceland and Greenland. The Spanish explorers and colonists brought large quantities of livestock to the New World on their ships. Atlantic voyages were particularly treacherous for horses, and it was typical for half of them to die before landing on the opposite shore. The horses were kept in stalls in the dark, dirty hold. Sometimes they were placed in slings that allowed them to sway with the rolling tides to reduce seasickness and get them off their feet to avoid leg injuries. When they arrived at their destination, the surviving horses had to be hoisted up to the deck and led off on a plank or lowered into the water to swim ashore.

Repopulating the New World

The reintroduction of horses into the New World by European explorers and colonists was made possible by advancements in seafaring. The transport of horses on ships overcame natural barriers, but shipwrecks sometimes led to the creation of feral populations like that on Sable Island, near Nova Scotia. Although the horse populations on other North American islands in the Atlantic—like Chincoteague, Assateague (Keiper 1985), and Shackleford Banks Islands—could also have originated with shipwreck survivors, the records sufficient to prove this are lacking.

The earliest westward expansion of horses into the Atlantic was by Scandinavians, particularly Norwegians, who sailed far out in search of new land. They initially settled in Iceland in 874 CE, then moved on to Greenland, and even to Canada. During the Medieval Warm Period, between 900 and 1200 CE, the climate in the North Atlantic was much warmer than today. The Norse began bringing horses to Iceland sometime between 874 and 935 CE from their Norwegian homeland and colonies in Ireland, the Isle of Man, and the western isles of Scotland. The Viking ships that transported the horses, sheep, goats, pigs, and cattle to Iceland would have been the deep-keeled knarr, rather than the dragon ships most often associated with the Norse (Brown 2001).

Recent mtDNA studies show that Icelandic horses exhibit affinities with today's Exmoor, Shetland, Scottish Highland, and Connemara ponies, as well as those breeds descended from Viking horses, such as the Faroe, Norwegian Fjord, and Nordlandshest ponies (Jansen *et al.* 2002; Björnsson and Sveinsson 2006; McCue *et al.* 2012). They also share a relationship with Mongolian horses, probably because of close trade ties between the Norse in Scandinavia and Russians, who had access to horses from the East (Nolf 2012).

Icelandic horses are the most isolated breed in the world. The Icelandic parliament, or Althing, voted in 982 CE to prohibit additional importation of horses, partly because of the degradation of the environment due to overgrazing. Over the following 1,000 years, the breed has undergone modification from natural selection as it adapted to the extreme climate and terrain, becoming more purebred. Fortunately, they originated from northern European types that were already well suited to endure severe winters. For the most part, Icelandic horses were left to fend for themselves with only modest monitoring. Large numbers of females were kept in a valley with a choice stallion as a means of controlled breeding (Sundkvist 2002). The sagas report long overland rides, and a complex network of riding trails linked remote settlements. The people of Iceland depended primarily on horses for transportation until the 1900s. Given the lack of success in raising cattle in many areas, horses provided meat as well. In 1783, 75% of the Icelandic horse population was destroyed because of a volcanic explosion, but it has since replenished itself.

After founding settlements on Iceland, the Scandinavians ventured farther west to establish colonies on Greenland that endured for about 450 years. The archaeological remains of over 400 Norse farms have been discovered in southern Greenland. Cattle, sheep, goats, and pigs were imported to the eastern and western settlements for food, horses for transportation, and dogs for herding and hunting (Arneborg *et al.* 2012; Nelson *et al.* 2012). Recent isotopic research on the diet of 27 Norse Greenlanders (Arneborg *et al.* 2012) supports the theory that overgrazing and deterioration of the climate made farming and pastoralism less feasible and necessitated their resorting to more fishing and seal hunting. The last known Norse occupation of Greenland was in 1430. Perhaps they simply abandoned these hostile conditions and returned to their homeland. Whether or for how long their abandoned horses survived there is unknown. Based on current evidence, the Norse only transported horses as far west as Greenland and failed to bring them to North America (Ingstad 1985; Nydal 1989; Ingstad and Stine Ingstad 2000).

The first introduction of domesticated horses into the Americas is then attributed to Spanish explorers and conquistadors. Columbus apparently did not bring horses on his first voyage, not knowing whether he would succeed in finding land. The next year, however, in 1493, he returned to Hispaniola with 17 ships. One of his ships carried 20 fighting horses (stallions) and 5 mares from Granada. Another ship transported 24 horses, 14 of which were male and 10 female. The total number of domestic horses to reach the New World for the first time is unknown because additional, privately owned horses were on other ships in his fleet, and no count is provided for them. Beginning with Columbus, efforts were made to establish horse-breeding colonies on the Caribbean Islands in order to produce a ready supply in the New World for the Spanish conquistadors and cavalries. The Colonial Spanish type of horse was small, around 13–14 hands, and highly variable in color.

On February 18, 1519, Hernán Cortés led a fleet of 11 Spanish ships from Cuba to the east coast of the Yucatan Peninsula of Mexico. Four of the ships were large galleons, and the remainder consisted of smaller brigantines. Along with 530 Europeans, Cortés brought 16 horses and numerous fighting dogs (Wood 2000). The horses and dogs, as well as the Spanish munitions, allowed a small European army to defeat the mighty Aztecs. The Spanish colonized Mexico City, which was surrounded by land suitable for grazing livestock. By about 1550, it was reported that there were as many as 10,000 horses in the vicinity (Crosby 1972; Clutton-Brock 2003:99).

History repeated itself when, on 16 November 1532, Francisco Pizarro and his 106 infantrymen and 62 mounted cavalrymen quickly defeated 80,000 Inca soldiers and took their emperor, Atahualpa, prisoner. It was the Incas' first encounter with these enormous beasts and the Spanish weapons, so their soldiers lacked experience and fled quickly in horror. Local subjugated tribes probably provided significant support in defeating the Inca.

In 1535, shortly after the Inca fell to the Spaniards, Pedro de Mendoza founded Buenos Aires, Argentina. Starvation led the colonists to flee to Paraguay, however, leaving behind small numbers of horses that immediately became feral. By 1580, when Don Juan de Garay brought new settlers to Buenos Aires, he found horses to be commonplace on the landscape (Crosby 1972; Clutton-Brock 2003). The indigenous people of the Pampas and Patagonia took advantage of the situation, readily abandoning horticulture and developing an equestrian lifestyle. *Baguales,* or feral horses, were so plentiful that they were basically free for anyone willing to capture them, and horse meat became an important source of nutrition.

Francisco Vazquez Coronado led an expedition northward from Mexico City to the Grand Canyon, New Mexico, Texas, Kansas, and Nebraska in 1540–1542 in search of the Seven Cities of Cibola and then Quivera, with the hope of finding a civilization rich in gold. With him were 250 Spanish horsemen, 70 infantry, and 300 or so friendly Indians. There were also perhaps 1,000 African and Indian servants to monitor the extra horses; drive the mules, cattle, sheep, and swine; and carry provisions (Winship 1904). He may have had as many as 558 horses, but only 2 were mares, so his impact in terms of establishing herds in the New World was minimal.

In a similar quest for Quivera, in 1601, Don Juan de Oñate y Salazar also penetrated the Great Plains as far as Arkansas City, Kansas, bringing with him 350 horses and mules. Unlike Coronado, his herd did include several mares, so his contribution to the distribution of horses in North America could have been more significant.

Hernando de Soto was an important figure in terms of the spread of horses in the New World. He is best known for his expedition to the southeastern United States between 1539 and 1542, in which he penetrated as far as what is now western North Carolina and Arkansas. De Soto had already gained considerable experience in the New World, having arrived in Panama at the age of 14. Trained to ride in Spain, he served as a young cavalryman under Alviar Nunez de Balboa in Panama and Nicaragua. De Soto later became wealthy in his partnership with Hernan Ponce de Leon and Francisco Companion by horse ranching and trading Indian slaves. It was de Soto who provisioned Francisco Pizarro's expedition to South America with horses, and de Soto accompanied him as second in command in the conquest of the Inca (Wood 2000).

Incredibly rich from his booty of Inca gold, de Soto returned to Spain and in 1537 convinced Emperor Charles V to make him governor of Cuba. With the goal of exploring the land called "La Florida," he set sail from Havana in 1539, purchasing some 200 horses for his expedition in Cuba. After his 9 ships landed at Tampa Bay, his men journeyed through the southeast for 4 years, covering some 4,000 miles. In a bloody battle at Mabila, a fortress controlled by Chief Tuskaloosa in southern Alabama, de Soto lost 220 men, most of his supplies and equipment, and the majority of his horses.

For over 100 years after their initial arrival in 1493, the numbers of horses in the New World remained small, but as Spanish missions and colonies began to breed them, their populations gradually expanded, especially around Santa Fe, New Mexico. The Spanish allowed their Indian slaves and workers to manage the horses, and even ride them as needed, but the indigenous people were forbidden by law to possess firearms or horses. The Pueblo Revolt of 1680 caused the Spanish to flee the Santa Fe area, leaving many horses behind until their return in 1694. During that interval, Pueblo tribes began breeding horses for their own use and for trade. They sold them to the Kiowa and Comanche, who quickly adapted them to their nomadic lifestyle and bison hunting. Horses eventually started drifting from the east as well. The Cheyenne moved from Minnesota to South Dakota and Montana in the early eighteenth century and introduced the horse to the Lakota in 1730.

The Shoshone had obtained horses from the Ute, Comanche, and Spanish settlers in the early 1700s, before any other northern plains tribe. Their reputation as some of the best Indian horsemen led Lewis and Clark in 1805 to seek them out in order to barter for horses. Lewis and Clark traded guns for 29 Shoshone horses, but when they inspected the animals, it was revealed that they were all weak and young, with sore backs (Ronda 1984:138–154). Giving up only the worst members of their herd, the Shoshone proved to be savvy traders.

Within a few decades, tribes throughout the plains and as far north as Alberta, Canada, adopted the horse. Because of their tremendous advantage in bison hunting, horses transformed life on the prairies for Native Americans. Still lacking wheels, the travois was transferred from the dog to the horse for moving large objects such as household goods, tipis, and bison meat and hides.

In addition to trade, the Indians acquired horses from each other through theft, which was considered a measure of ability, courage, and cleverness. Counting coup was a way warriors kept track of their brave acts, including touching an enemy warrior in battle without being harmed or stealing an enemy's horse or weapon (Linderman 2002). Notches in a coup stick, feathers in a headdress, or stripes on a shirt were methods by which an individual's coups were recorded. In the plains especially, where horses were important for transportation, hunting bison, and warfare, acquiring the swiftest horses was essential, so a warrior gained considerable status if he captured particularly fine individuals for his own tribe.

Many accounts of how the Spanish horse changed the course of history in the New World could be given. Horses flowed up from the Caribbean to the southeastern part of what is today the United States, but most entered from Mexico, into New Mexico, Texas, and the Great Plains. Taller European horses, such as Thoroughbreds, entered with British colonists along the East Coast of North America.

The Horse's Recent Arrival to Australia

Australia was late on the scene in terms of receiving horses. History records the initial arrival at Botany Bay in 1788, with the First Fleet. On 13 May 1787, eleven British ships, commanded by Captain Arthur Phillip, headed for Australia to establish a penal colony. The voyage, via Rio de Janeiro and Cape Town, took about 250 days. Because Cape Town was the last port, they purchased livestock for the colony, including one stallion and three mares. The horse population remained small in and around Sydney for about 20 years, where the animals were used primarily for work and transportation. In 1810, however, horse racing became a popular sport, leading to thousands of imported Thoroughbreds. Additional horses were shipped to Australia from Britain, South Africa, and Indonesia. Droughts often led settlers to abandon their farms and neglect their fences, which resulted in the rise in numbers of feral horses known as brumbies. Today, there are over 400,000 brumbies across Australia, causing serious environmental degradation (Dobbie *et al.* 1993; Dawson *et al.* 2006; see also chap. 10).

Conclusion

The global spread of horses by human agency has a long, rich history involving a wide variety of means, including riding, driving, and transporting on ships, by rail, and, today, even by air. The goals and purposes for moving horses have been diverse. Horses were transferred from one place to another to provide transportation and work, to serve military functions, to be commodities, to serve as religious and status symbols, for sports, and more. Although it is impossible to account for every case of expansion and exchange of horse populations, this chapter has attempted to summarize and elucidate the process. Today, horses are less important in developed countries for work and transportation, but they maintain substantial popularity in sports and leisure activities around the world. Horses of certain breeds, like the Thoroughbred and Arabian, can be tremendously valuable, resulting in shipping prized individuals around the world. In the future, movements of horses will most likely be stimulated by equestrian sports and goals centered on refining breeds, rather than for exploration, warfare, and other more traditional goals. Humans have curtailed the natural long-distance migrations of many wild equids, but the reintroduction of the Przewalski's horse into Mongolia and China brings a new facet to the process (Bouman and Boyd 1994; see also chaps. 13 and 14).

ACKNOWLEDGMENTS

The author would like to extend her gratitude to Jason Ransom and Petra Kaczensky for their tremendous efforts in organizing the International Wild Equid Conference and in editing this volume. The editors and anonymous reviewers contributed valuable recommendations on the original manuscript, for which the author is very appreciative. She also wishes to thank the University of Veterinary Medicine, Vienna, Austria, and the courteous and hard-working conference staff who diligently ensured the success of the conference.

REFERENCES

Achilli, A., A. Olivieri, P. Soares, H. Lancioni, B.H. Kashani, *et al.* 2012. Mitochondrial genomes from modern horses reveal the major haplo-groups that underwent domestication. Proceedings of the National Academy of Sciences of the USA 109:2449–2454.

Alberdi, M.T., and J.L. Prado. 1993. Review of the genus *Hippidion* Owen, 1869 (Mammalia: Perissodactyla) from the Pleistocene of South America. Zoological Journal of the Linnean Society 108:1–22.

Anthony, D. 2007. The horse, the wheel, and language: How Bronze Age rides from the Eurasian steppes shaped the modern world. Princeton University Press, Princeton, NJ, USA.

Arbogast, R.-M., B. Clavel, S. Lepetz, P. Méniel, and J.-H. Yvinec. 2002. Archéologie du cheval. Editions Errance, Paris, France.

Arneborg, J., N. Lunnerup, J. Heinemeier, J. Mohl, N. Rud, and A.E. Sveinbjornsdottir. 2012. Norse Greenland dietary economy ca. AD 980–ca. AD 1450: Introduction. Greenland isotope project: Diet in Norse Greenland AD 1000–AD 1450. Journal of the North Atlantic Special Volume 3:1–39.

Bahn, P.G., and J. Vertut. 1997. Journey through the Ice Age. University of California, Berkeley, USA.

Bannikov, A.G. 1958. Distribution géographique et biologie du cheval sauvage et du chameau de Mongolie (*Equus przewalskii* et *Camelus bactrianus*). Mammalia 22:152–160.

Barnes, G.L. 1999. The rise of civilization in East Asia: The archaeology of China, Korea, and Japan. Thames and Hudson, London, UK.

Barnes, G.L. 2007. State formation in Japan: Emergence of a fourth century ruling elite. Routledge, New York, USA.

Bendrey, R. 2012. From wild horses to domestic horses: A European perspective. World Archaeology 44:135–157.

Bendrey, R. 2014. Population genetics, biogeography, and domestic horse origins and diffusions. Journal of Biogeography 41:1441–1442.

Bendrey, R., N. Thorpe, A. Outram, and L.H. van Wijngaarden-Bakker. 2013. The origins of domestic horses in north-west Europe: New direct dates on the horses of Newgrange, Ireland. Proceedings of the Prehistoric Society 79:91–103.

Benecke, N. 2006. Late prehistoric exploitation of horses in central Germany and neighboring areas—The archaeozoological record. Pages 195–208 *in* S.L. Olsen, S. Grant, A.M. Choyke, and L. Bartosiewicz (Eds.), Horses and humans: The evolution of human-equine relationships. International Series 1560. British Archaeological Reports, Oxford, UK.

Bibikova, V.I. 1986. Appendix 3: On the history of horse domestication in southeast Europe. Pages 163–182 *in* D. Telegin (Ed.), Dereivka: A settlement and cemetery of Copper Age horse keepers on the middle Dnieper. International Series 287. British Archaeological Reports, Oxford, UK.

Bietak, M. 1991. Egypt and Canaan during the middle Bronze Age. Bulletin of the American Schools of Oriental Research 281:27–72.

Björnsson, G., and H. Sveinsson. 2006. The Icelandic horse. Edda, Reykjavik, Iceland.

Bökönyi, S. 1974. The Przevalsky horse. Souvenir Press, London, UK.

Bondevik, S., S. Dawson, A. Dawson, and Ø. Lohne. 2003. Record-breaking height for 8,000-year-old tsunami in the North Atlantic. Eos: Transactions of the American Geophysical Union 84:289, 293.

Bouman, I.J., and L. Boyd. 1994. Reintroduction. Pages 255–263 *in* L. Boyd and K. Houpt (Eds.), Przewalski's horse: The history and biology of an endangered species. State University of New York Press, Albany, USA.

Boyle, K.V. 2006. Neolithic wild game animals in western Europe: The question of hunting. Pages 10–23 *in* D. Serjeantson and D. Field (Eds.), Animals in the Neolithic of Britain and Europe. Oxbow, Oxford, UK.

Bronk Ramsey, C., T.H.F. Higham, D.C. Owen, A.W.G. Pike, and R.E.M. Hedges. 2002. Radiocarbon dates from the Oxford AMS system: Archaeometry datelist 31. Archaeometry 44:1–149.

Brown, N.M. 2001. A good horse has no color: Searching Iceland for the perfect horse. Stackpole, Mechanicsburg, PA, USA.

Cai, B.-Q., S.-H. Zheng, J.C. Liddicoat, and Q. Li. 2013. Review of the litho-, bio- and chronostratigraphy in the Nihewan Basin, Hebei, China. Pages 218–242 *in* X. Wang, L.J. Flynn, and M. Fortelius (Eds.), Fossil mammals of Asia: Neogene biostratigraphy and chronology. Columbia University Press, New York, USA.

Casson, L. 1994. Travel in the Ancient World. Johns Hopkins University Press, Baltimore, USA.

Chauvet, J.-M., E.B. Deshamps, and C. Hillaire. 1996. Dawn of art: The Chauvet cave. The oldest known paintings in the world. Harry N. Abrams, New York, USA.

Churcher, C.S. 1982. Oldest ass recovered from Olduvai Gorge, Tanzania, and the origin of asses. Journal of Paleontology 56:1124–1132.

Clotte, J., and J. Courtin. 1996. The cave beneath the sea: Paleolithic images at Cosquer. Harry N. Abrams, New York, USA.

Clutton-Brock, J. 2003. Horses in history. Pages 83–102 *in* S. L. Olsen (Ed.), Horses through time. Roberts Rinehart, Boulder, CO, USA.

Clutton-Brock, J., and R. Burleigh. 1991*a*. The mandible of a Mesolithic horse from Seame Carr, Yorkshire, England. Pages 238–241 *in* R.H. Meadows and H.-P. Uerpmann (Eds.), Equids in the Ancient World, vol. 2. Ludwig Reichert Verlag, Wiesbaden, Germany.

Clutton-Brock, J., and R. Burleigh. 1991*b*. The skull of a Neolithic horse from Grime's Graves, Norfolk, England. Pages 242–249 *in* R.H. Meadows and H.-P. Uerpmann (Eds.), Equids in the Ancient World, vol. 2. Ludwig Reichert Verlag, Wiesbaden, Germany.

Combier, J., and A. Montet-White. 2002. Solutré 1968–1998. Mémoire de la Société Préhistorique Française No. 30. Société Préhistorique Française, Paris, France.

Creel, H.G. 1965. The role of the horse in Chinese history. American Historical Review 70:647–672.

Crosby, A.W. 1972. The Columbian exchange: Biological and cultural consequences of 1492. Greenwood Press, Westport, CT, USA.

Cunliffe, B. 2001. Facing the ocean: The Atlantic and its peoples. Oxford University, Oxford, UK.

Dawson, M.J., C. Lane, and G. Saunders. 2006. Proceedings of the National Feral Horse Management Workshop. Invasive Animals Cooperative Research Center, University of Canberra, Canberra, ACT, Australia.

Dillehay, T.D., C. Ramírez, M. Pino, M.B. Collins, J. Rossen, and J.D. Pino-Navarro. 2008. Monte Verde: Seaweed, food, medicine, and the peopling of South America. Science 320:784–786.

Dobbie, W.R., D.M. Berman, and M.L. Braysher (Eds.). 1993. Managing vertebrate pests: Feral horses. Australia Government Publishing Service, Canberra, ACT, Australia.

Eisenmann, V. 1976. Equidae from the Shugur Formation. Pages 225–233 *in* Y. Coppens, T.C. Howell, G.L. Isaac, and R.E.E. Leakey (Eds.), Earliest man and environments in the Lake Rudolf Basin. University of Chicago Press, Chicago, USA.

Eisenmann, V. 1992. Origins, dispersals, and migrations of *Equus* (Mammalia, Perissodactyla). Pages 161–170 *in* W. von Koenigswald and L. Werdelin (Eds.), Mammalian migration and dispersal events in the European Quaternary. Courier Forschungsinstitut Senckenberg, Frankfurt, Germany.

Forstén, A. 1988. Middle Pleistocene replacement of stenonid horses by caballoid horses—Ecological implications. Palaeogeography, Palaeoclimatology, Palaeoecology 65:23–33.

French, C., and M. Kousoulakou. 2003. Geomorphological and micromorphological investigations of palaeosols, valley sediments and a sunken floored dwelling at Botai, Kazakhstan. Pages 105–114 *in* M. Levine, A.C. Renfrew, and K. Boyle (Eds.), Prehistoric steppe adaptation and the horse. McDonald Institute for Archaeological Research, Cambridge, UK.

Friday, K.F. 2004. Samurai, warfare and the state in early medieval Japan. Routledge, New York, USA.

Gaboardi, M., T.T. Deng, and Y. Wang. 2005. Middle Pleistocene climate and habitat change at Zhoukoudian, China, from the carbon and oxygen isotopic record from herbivore tooth enamel. Quaternary Research 63:329–338.

Gabunia, L., A. Vekua, D. Lordkipanidze, R. Ferring, A. Justus, *et al.* 2000. Current research on the hominid site of Dmanisi. Études et Recherches Archéologiques de l'Université de Liège 92:13–27.

Grayson, D.K., and D.J. Meltzer. 2003. A requiem for North American overkill. Journal of Archaeological Science 30:585–593.

Greenfield, H. 2006. The social and economic context for domestic horse origins in southeastern Europe: A view from Ljuljaci in the central Balkans. Pages 221–244 *in* S.L. Olsen, S. Grant, A.M. Choyke, and L. Bartosiewicz (Eds.), Horses and humans: The evolution of human-equine relationships. International Series 1560. British Archaeological Reports, Oxford, UK.

Groves, C.P. 1994. Morphology, habitat, and taxonomy. Pages 39–59 *in* L. Boyd and K.A. Houpt (Eds.), Przewalski's horse: The history and biology of an endangered species. State University of New York Press, Albany, USA.

Guthrie, R.D. 2003. Rapid body size decline in Alaskan Pleistocene horses before extinction. Nature 426:169–171.

Guthrie, R.D. 2006. New carbon dates link climatic change with human colonization and Pleistocene extinctions. Nature 441:207–209.

Hammer, M.F., and S. Horai. 1995. Y chromosomal DNA variation and the peopling of Japan. American Journal of Human Genetics 56:951–962.

Harrigan, P. 2012. Discovery at al-Magar. Saudi Aramco World 63:2–9.

Hedges, R.E.M., P.B. Pettitt, C. Bronk Ramsey, and G.J. van Klinken. 1997. Radiocarbon dates from the Oxford AMS system: Archaeometry datelist 24. Archaeometry 39:445–471.

Hemmings, A. 2004. The organic clovis: A single continent-wide cultural adaptation. Thesis, University of Florida, Gainesville, USA.

Higham, T.F.G., C. Bronk Ramsey, F. Brock, D. Baker, and P. Ditchfield. 2007. Radiocarbon dates from the Oxford AMS system: Archaeometry datelist 32. Archaeometry 49:S1–S60.

Holland, T.A. 2006. Excavations at Tell es-Sweyhat, Syria, vol. 2, pt. 1: Text. Archaeology of the Bronze Age, Hellenistic, and Roman remains at an ancient town of the Euphrates River. Oriental Institute Publications 125. University of Chicago Press, Chicago, USA.

Hooijer, D.A. 1976. Evolution of the Perissodactyla of the Omo Group deposits. Pages 209–213 *in* Y. Coppens, F.C. Howell, G.L. Isaac, and R.E.E. Leakey (Eds.), Earliest man and environments in the Lake Rudolf Basin. University of Chicago Press, Chicago, USA.

Ingstad, H. 1985. Historic background and the evidence of the site at L'Anse aux Meadows. Norwegian University Press, Oslo, Norway.

Ingstad, H., and A. Stine Ingstad. 2000. The Viking discovery of America: The excavation of a Norse settlement in L'Anse aux Meadows, Newfoundland. Breakwater, St. Johns, Newfoundland.

Jansen, T., P. Forster, M.A. Levine, H. Oelke, M. Hurles, C. Renfrew, J. Weber, and K. Olek. 2002. Mitochondrial DNA and the origins of the domestic horse. Proceedings of the National Academy of Sciences of the USA 99:10,905–10,910.

Kakoi, H., T. Tozaki, and H. Gawahara. 2007. Molecular analysis using mitochondrial DNA and microsatellites to infer the formation process of Japanese native horse populations. Biochemical Genetics 45:375–395.

Keiper, R.R. 1985. The Assateague ponies. Tidewater, Centreville, MD, USA.

Kidder, J.E., Jr. 1985. The archaeology of the early horse-riders in Japan. Transactions of the Asiatic Society of Japan (Third Series) 20:89–123.

Kidder, J.E., Jr. 2007. Himiko and Japan's elusive chiefdom of Yamatai: Archaeology, history, and mythology. University of Hawaii Press, Honolulu, USA.

Leakey, M.D., and R.L. Hay. 1979. Pliocene footprints in the Laetolil Beds at Laetoli, northern Tanzania. Nature 278:317–323.

Ledyard, G. 1975. Galloping along with the horseriders: Looking for the founders of Japan. Journal of Japanese Studies 1:217–254.

Legge, A.J., and P.A. Rowley-Conwy. 2000. The exploitation of animals. Pages 423–471 *in* A.M.T. Moore, G.C. Hillman, and A.J. Legge (Eds.), Village on the Euphrates. Oxford University Press, Oxford, UK.

Leroi-Gourhan, A. 1967. Treasures of prehistoric art. Harry N. Abrams, New York, USA.

Linderman, F.B. 2002. Plenty-Coups, Chief of the Crows. University of Nebraska Press, Lincoln, USA.

Linduff, K.M. 2006. Imaging the horse in early China: From the table to the stable. Pages 303–322 *in* S. L. Olsen, S. Grant, A.M. Choyke, and L. Bartosiewicz (Eds.), Horses and humans: The evolution of human-equine relationship. International Series 1560. British Archaeological Reports, Oxford, UK.

Liu, L., and X. Chen. 2012. The archaeology of China: From the late Paleolithic to the early Bronze Age. Cambridge University Press, Cambridge, UK.

Lordkipanidze, D., A. Vekua, R. Ferring, G.P. Rightmire, J. Agusti, *et al.* 2005. The earliest toothless hominin skull. Nature 434:717–718.

Lordkipanidze, D., M.S. Ponce de Leon, A. Margvelashvili, Y. Rak, G.P. Rightmire, *et al.* 2013. A complete skull from Dmanisi, Georgia, and the evolutionary biology of early *Homo*. Science 342:326–331.

Lutz, R., and G. Lutz. 1995. The secret of the desert: The rock art of Messak Sattafet and Messak Mellet, Libya. Universitätsbuchhandlung Golf Verlag, Innsbruck, Germany.

MacDonald, M.C.A. 2009. Wheels in a land of camels: Another look at the chariot in Arabia. Arabian Archaeology and Epigraphy 20:156–184.

Martin, L., and N. Russell. 2006. The equid remains from Neolithic Catalhöyük, central Anatolia: A preliminary report. Pages 115–126 *in* S.L. Olsen, S. Grant, A.M. Choyke, and L. Bartosiewicz (Eds.), Horses and humans: The evolution of human-equine relationships. International Series 1560. British Archaeological Reports, Oxford, UK.

Martin, P.S. 1973. The discovery of America. Science 179:969–974.

Martinic, M. 1996. La Cueva del Milodon (Ultima Esperanza, Patagonia Chilena): Un siglo de descubrimientos y estudios referidos a la vida primitiva en el sur de America. Journal de la Société des Américanistes 82:311–323.

Mashkour, M., M. Fontugue, and C. Hatte. 1999. Investigation on the evolution of subsistence economy in the Qazvin Plain (Iran) from the Neolithic to the Iron Age. Antiquity 73:65–76.

McCue, M.E., D.L. Bannasch, J.L. Petersen, J. Gurr, E. Bailey, *et al.* 2012. A high density SNP array for the domestic horse and extant Perissodactyla: Utility for association mapping, genetic diversity, and phylogeny studies. PLoS Genetics 8:e1002451.

Nelson, D.E., J. Heinemeier, N. Lynnerup, Á.E. Sveinbjörnsdóttir, and J. Arneborg. 2012. An isotopic analysis of the diet of the Greenland Norse. Journal of the North Atlantic Special Volume 3:93–118.

Nolf, P.S. 2012. Detecting Icelandic horse origins. Icelandic Horse Quarterly 4:18–23.

Nydal, R. 1989. A critical review of radiocarbon dating of a Norse settlement at L'Anse aux Meadows, Newfoundland, Canada. Radiocarbon 31:976–985.

Olsen, S.L. 1989. A theoretical approach to Upper Palaeolithic horse hunting strategies at Solutre, France. Presented at the Society for American Archaeology Conference, Atlanta, GA, USA.

Olsen, S.L. 2000. Reflections of ritual behavior at Botai, Kazakhstan. Pages 183–207 *in* K. Jones-Bley, M. Huld, and A. Della Volpe (Eds.), Proceedings of the eleventh annual UCLA Indo-European conference. Journal of Indo-European Studies Monograph 35. Institute for the Study of Man, Washington, DC, USA.

Olsen, S.L. 2001. The importance of thong-smoothers at Botai, Kazakhstan. Pages 197–206 *in* A. Choyke and L. Bartosiewicz (Eds.), Crafting bone: Skeletal technologies through time and space. International Series 937. British Archaeological Reports, Oxford, UK.

Olsen, S.L. 2003*a*. Horse hunters of the Ice Age. Pages 35–56 *in* S.L. Olsen (Ed.), Horses through time. Rowman and Littlefield, Lanham, MD, USA.

Olsen, S.L. 2003*b*. The exploitation of horses at Botai, Kazakhstan. Pages 83–104 *in* M. Levine, C. Renfrew, and K. Boyle (Eds.), Prehistoric steppe adaptation and the horse. McDonald Institute for Archaeological Research, Cambridge, UK.

Olsen, S.L. 2006*a*. Early horse domestication on the Eurasian steppe. Pages 245–269 *in* M.A. Zeder, D.G. Bradley, E. Emshwiller, and B.D. Smith (Eds.), Documenting domestication: New genetic and archaeological paradigms. University of California Press, Berkeley, USA.

Olsen, S.L. 2006*b*. Early horse domestication: Weighing the evidence. Pages 81–113 *in* S.L. Olsen, S. Grant, A.M. Choyke, and L. Bartosiewicz (Eds.), Horses and humans: The evolution of human-equine relationships. International Series 1560. British Archaeological Reports, Oxford, UK.

Olsen, S.L. 2010*a*. The horse's debut in the Near East. Pages 24–39 *in* S.L. Olsen and C. Culbertson (Eds.), A gift from the desert. International Museum of the Horse, Kentucky Horse Park, Lexington, USA.

Olsen, S.L. 2010*b*. The proto-Arabian horse appears in ancient Egypt. Pages 40–49 *in* S.L. Olsen and C. Culbertson (Eds.), A gift from the desert. International Museum of the Horse, Kentucky Horse Park, Lexington, USA.

Olsen, S.L. 2010*c*. The horse in Arabia. Pages 50–61 *in* S.L. Olsen and C. Culbertson (Eds.), A gift from the desert. International Museum of the Horse, Kentucky Horse Park, Lexington, USA.

Olsen, S.L. 2013. Stories in the rocks: Exploring Saudi Arabian rock art. Carnegie Museum of Natural History, Pittsburgh, PA, USA.

Olsen, S.L., B. Bradley, D. Maki, and A. Outram. 2006. Community organisation among Copper Age sedentary horse pastoralists of Kazakhstan. Pages 89–111 *in* D. Peterson, L.M. Popova, and A.T. Smith (Eds.), Beyond the steppe and sown: Proceedings of the 2002 University of Chicago conference on Eurasian archaeology. Colloquia Pontica 13. Brill, Leiden, Netherlands.

Orlando, L., A. Ginolhac, G. Zhang, D. Froese, A. Albrechtsen, *et al.* 2013. Recalibrating *Equus* evolution using the genome sequence of an early Middle Pleistocene horse. Nature 499:74–78.

Outram, A.K., N.A. Stear, R. Bendrey, S. Olsen, A. Kasparov, V. Zaibert, N. Thorpe, and R.P. Evershed. 2009. The earliest horse harnessing and milking. Science 323:1332–1335.

Owen, D.I. 1991. The "first" equestrian: An Ur III glyptic scene. Acta Sumerologica 13:259–273.

Parfitt, S.A., R.W. Barendregt, M. Breda, I. Candy, M.J. Collins, *et al.* 2005. The earliest record of human activity in northern Europe. Nature 438:1008–1012.

Potts, R., and P. Shipman. 1981. Cutmarks made by stone tools on bones from Olduvai Gorge, Tanzania. Nature 291:577–580.

Ramirez Rozzi, F.V., F. d'Errico, and M. Zarate. 2000. Le site paléo-indien de Piedra Museo (Patagonie): Sa contribution au débat sur le premier peuplement du continent américan. Comptes Rendus de l'Académie des Sciences, Series IIA Earth and Planetary Science 331:311–318.

Rassamakin, Y. 1999. The Eneolithic of the Black Sea steppe: Dynamics of cultural and economic development 4500–2300 BC. Pages 59–182 *in* M. Levine, Y. Rassamakin, A. Kislenko, and N. Tatarintseva (Eds.), Late prehistoric exploitation of the Eurasian steppe. McDonald Institute for Archaeological Research, University of Cambridge, Cambridge, UK.

Raulwing, P., and J. Clutton-Brock. 2009. The Buhen horse: Fifty years after its discovery (1958–2008). Journal of Egyptian History 2:1–106.

Roberts, M., and S. Parfitt. 1999. Boxgrove: A Middle Pleistocene hominid site at Eartham Quarry, Boxgrove, West Sussex. English Heritage Archaeological Report. English Heritage, London, UK.

Ronda, J.P. 1984. Lewis and Clark among the Indians. University of Nebraska Press, Lincoln, USA.

Sieveking, A. 1979. The cave artists. Thames and Hudson, London, UK.

Sieveking, A. 1987. A catalogue of palaeolithic art in the British Museum. British Museum Press, London, UK.

Soffer, O. 1985. The Upper Paleolithic of the central Russian plain. Academic Press, Orlando, FL, USA.

Sundkvist, A. 2002. Herding horses: A model of prehistoric horsemanship in Scandinavia—and elsewhere? Presented at the PECUS: Man and Animal in Antiquity Conference. Swedish Institute, Rome, 9–12 September.

Tappen, M., D.S. Adler, C.R. Ferring, M. Gabunia, A. Vekua, and C.C. Swisher III. 2002. Akhalkalaki: The taphonomy of an Early Pleistocene locality in the Republic of Georgia. Journal of Archaeological Science 29:1367–1391.

Thieme, H. 2005. Die ältesten Speere der Welt-Fundplätze der frühen Altsteinzeit im Tagebau Schöningen. Archäologisches Nachrichtenblatt 10:409–417.

Thieme, H. 2007. Warum ließen die Jäger die Speere zurück? Pages 188–190 *in* H. Thieme (Ed.), Die Schöninger Speere—Mensch und Jagd vor 400 000 Jahren. Konrad Theiss Verlag, Stuttgart, Germany.

Vekua, A., and D. Lordkipanidze. 2010. Dmanisi (Georgia)—Site of discovery of the oldest hominid in Eurasia. Bulletin of the Georgian National Academy of Sciences 4:158–164.

Vigne, J.D., H. Buitenhuis, and S.J.M. Davis. 1999. Les premiers pas de la domestication animale à l'ouest de l'Euphrate: Chypre et l'Anatolie central. Paléorient 25:49–62.

Wallner, B., C. Vogl, P. Shukla, J.P. Burgstaller, T. Druml, and G. Brem. 2013. Identification of genetic variation on the horse Y chromosome and the tracing of male founder lineages in modern breeds. PLoS ONE 8:e60015.

Webb, S.D., and C.A. Hemmings. 2006. Last horses and first humans in North America. Pages 11–23 *in* S.L. Olsen, S. Grant, A.M. Choyke, and L. Bartosiewicz (Eds.), Horses and humans: The evolution of human-equine relationships. International Series 1560. British Archaeological Reports, Oxford, UK.

Weber, J.A. 1997. Faunal remains from Tell es-Sweyhat and Tell Jajji Ibrahim. Pages 133–142 *in* R. Zettler *et al.* (Eds.), Subsistence and settlement in a marginal environment: Tell es-Sweyhat, 1989–1995 preliminary report. MASCA Research Papers in Science and Archaeology 14. University of Pennsylvania, Philadelphia, USA.

Wijngaarden-Bakker, L.H. 1974. The animal remains from the Beaker settlement at Newgrange, Co. Meath: First report. Proceedings of the Royal Irish Academy 74C:313–383.

Wilkinson, J.G. 1996. The ancient Egyptians—Their life and customs, vols. 1 and 2. Reprinted from original 1853 edition. Random House, London, UK.

Winship, G.P. 1904. The journey of Coronado, 1540–1542, from the city of Mexico to the Grand Canon of the Colorado and the buffalo plains of Texas, Kansas and Nebraska, as told by himself and his followers. Translated and edited with an introduction by George Parker Winslip. Allerton, New York, USA.

Wood, M. 2000. Conquistadors. University of California Press, Berkeley, USA.

Woodman, P.C., M. McCarthy, and N. Monaghan. 1997. The Irish quaternary fauna project. Quaternary Science Reviews 16:129–159.

Zaibert, V. 1993. Eneolit Uralo-Irtyshskovo. Akademiya Nauk Respubliki Kazakhstan, Almaty, Kazakhstan.

Human Dimensions of Wild Equid Management: Exploring the Meanings of "Wild"

JOHN D.C. LINNELL, PETRA KACZENSKY, AND NICOLAS LESCUREUX

Most everyone who has ever ridden a domestic horse or seen any of the wild equid species in their natural habitat has felt some strong emotional response to the encounter. The prehistoric cave paintings of wild horses in southern France, the historic horse cultures of central Asia, the role of modern-day horses as companion animals, or the efforts being used to conserve and reintroduce wild species all testify to the existence of a strong and enduring connection between people and equids that goes far beyond basic utilitarian needs. Much has been written about the complex relationships between humans and domestic horses both in past (Meyer 2013) (see chap. 8) and present times (e.g., Robinson 1999; Keaveney 2008), and reviewing this work is outside the scope of this chapter. Rather, we explore the human relationship with wild equids, especially those aspects relevant for present-day wild equid management. This has proven to be a challenging task, as (1) defining "wild" in the case of equid management is far from trivial and (2) there was an extreme paucity of studies, requiring us to construct the following review from diverse and highly fragmented sources from many disciplines. In the resulting chapter, we present both a review of what is known about the topic as well as a wider discussion about the conceptual issues related to defining the borders of the "wild" in the context of conserving large mobile animals in the Anthropocene.

Wild, Feral, and Domestic: A Confusion of Terms

Equid taxonomy is far from being fully resolved, and the literature contains a confusing diversity of Latin names. Based on the most recent taxonomic assessment, there are seven extant species that are widely considered "wild"; however, there are considerable controversies surrounding the "wildness" of free-ranging populations of two species: domestic horses and donkeys. All free-ranging horses and donkeys have a domestic origin despite their various backgrounds as primitive breeds (e.g., Exmoor ponies), animals introduced by early settlers to the New World (e.g., North American mustangs and burros, and Australian brumbies), recent escapees, or animals (e.g., Konik and Heck horses) from conservation initiatives that seek to re-create the ecological role of the extinct wild horse (*Equus ferus*). Terminology is further complicated by the ongoing scientific efforts to fully elucidate the relationship

Fig. 9.1 When is a wild equid really wild? A. Strictly speaking, all mustangs in the United States are feral domestic horses, but they are frequently referred to as "wild horses" because they are free ranging and receive no direct support from humans. B. Khulan, or Asiatic wild asses, are a wild species that have never been domesticated and range over vast areas, such as in southern Mongolia. C. Przewalski's horse is a wild species that went extinct in the wild, was bred in captivity for a century, and has now been returned to the wild in Mongolia. D. Konik horses are a domestic horse species that have been bred to resemble the extinct tarpan, which is often claimed to be a descendent of the extinct *E. ferus* that roamed prehistoric Europe. Konik horses range on many fenced nature reserves in Europe but are increasingly being referred to as "wild horses" within rewilding projects. Photos by Jason I. Ransom (A), Petra Kaczensky (B), and John D.C. Linnell (C and D)

between domestic forms and their wild ancestors (fig. 9.1). Many questions need to be resolved, including: When did the wild horse become extinct in Eurasia? Was the twentieth-century tarpan the wild ancestor of the domestic horse, a domesticated horse turned feral, or a hybrid between the wild and domestic form (Olsen 2006; Kavar and Dovč 2008; Sommer *et al.* 2011; Warmuth *et al.* 2011; Bendrey 2012)? What are the functional and taxonomic differences between mustangs in North America and the equid species that went extinct at the end of the Pleistocene? Any discussion of wild equid management and conservation cannot be made by only focusing on the first set of seven wild species, as they are now outnumbered by massive populations of free-ranging horses and donkeys.

People variously refer to free-roaming horses and donkeys as "feral" or "wild." From a traditional natural science perspective, wild (of nondomestic origin) and feral (free-living individuals from a species with domestic origin but without owners) refer to origin and past human influence. The wider public tends to use these two terms interchangeably, however, or focuses more on their way of living than their origins. The media and public also frequently apply the term "wild" to free-ranging herds of domestic horses (subject to management and with clearly identifiable owners) that are subject to extensive forms of husbandry. To add to the confusion, "wild horse" is often used to refer to Przewalski's horse, a distinct species. Furthermore, the public often uses "domesticated" and "tame" as synonyms, whereas in the natural science context, domestication means a change of the gene pool as a result of selective breeding, and taming merely means a change in behavior, such as habituation. Consequently, terms are used rather differently in lay versus scientific literature, causing confusion and misperceptions. Although the use of terms can be controversial, the need for clarity is also important. For this chapter we thus chose to use the term "feral" (without prejudice) for free-roaming populations of *E. caballus* and *E. asinus* that do not have clear individual owners (e.g., mustangs and burros in North America, brumbies in Australia, the horses of the Danube delta, etc.). This distinction differentiates them from the free-ranging horse populations where there are clear owners that graze their horses in an extensive manner (e.g., horses in northern Iberia and the ponies of Exmoor and the New Forest in the southern United Kingdom). We reserve the term "wild equid" for the seven equid species without any history of domestication, although we will use the term "wild" (in quotes) when referring to how some sectors of society refer to feral or domestic populations.

Clarity from Confusion: How the Use of Terms Reveals Deeper Values

The use of certain terms clearly reflects more than different people's understandings of technical terms. It also reflects a certain set of values (Bhattacharyya *et al.* 2011). People referring to mustangs or brumbies as feral horses often view them as a recent and artificial addition to the North American or Australian faunas with little intrinsic value (e.g., as "pests"), threatening to affect native rangeland or biodiversity (Dobbie *et al.* 1993; Bies *et al.* 2011; The Wildlife Society 2011) and representing significant opportunity costs for livestock production through grazing competition (Bastian *et al.* 1999). In contrast, those referring to them as "wild" see an iconic species (symbols of a glorious past with high "cultural value" or as engineers of a new future with "rewilding value") and a valuable addition to the environment (Schwartz 2005; Rikoon 2006; Dalke 2010; Bhattacharyya *et al.* 2011; Notzke 2013).

This diversity in the way that terms are used provides key insights into the human dimensions of many of the controversies surrounding the management of wild and free-ranging equids. The last decade has seen a small number of in-depth ethnographic studies, mainly conducted from the point of view of political ecology or social geography, that have used the case study of free-ranging horse / donkey management in various contexts as a lens to explore issues related to how rural people construct their relationship with nature and with wider society. Detailed studies have been conducted in western Canada (Bhattacharyya *et al.* 2011; Notzke 2013), the eastern United States (Rikoon 2006), the US Virgin Islands (Fortwangler 2009), and Australia (Peace 2009). In all these cases there have been many controversies and conflicts between rural communities and external nature management agencies about the removal or use of lethal methods to control small herds of horses and donkeys. Although feral from a technical point of view, these herds were all termed "wild" locally. The local objections to their removal appear to have been complex, but they center on two issues: power and values. On the one hand, the conflicts all contained elements of a local community fighting what was portrayed as an uneven battle against outside and powerful government agencies, underlining how power struggles and issues of governance lie at the heart of many conservation conflicts (Redpath *et al.* 2013). On the other hand, there were clear differences in the way that local people and the government agencies valued the animals and the landscapes they inhabit. To the agencies, the animals were undesirable pests that did not belong in the "natural" environment and were associated with negative impacts on other elements of the local biodiversity. To locals, the animals were clearly regarded as valued elements of their local environment. Despite being given the label "wild," the herds were explicitly regarded as being key elements of the cultural heritage of the area. The controversy surrounding the way the horses and donkeys should be managed revealed key differences in the most basic ways that the landscapes and the place of people and animals were perceived.

Mustangs, Burros, and Brumbies

The case studies mentioned above may have been controversial, but they played out at a local level and involved small herds of feral domestic equids. They contrast with the much larger conflicts centered on the management of the larger populations of mustangs and burros in the western US states (and some Atlantic islands and coastal areas) and brumbies in Australia's outback and New Zealand. In total, feral populations number between 1.5 and 2 million animals (see chap. 10), well outnumbering all the other wild equid populations combined. In North America and Australasia, the conflicts surrounding the management of these populations are massive and have been ongoing for decades. The conflicts are played out in the public arena, involving the media and the courts, and engage some active and high-profile individuals and nongovernment organizations (Symanski 1996; Smith 2010; Bies *et al.* 2011).

Considering the size and visibility of such conflicts, there is a shocking lack of scientific study of the social and cultural aspects. Nimmo and Miller (2007) reviewed human dimensions studies on feral equid management and found no peer-reviewed papers on the subject. The situation has not changed in the intervening years, and apart from some papers on peripheral issues like the economics aspects of mustang adoption (Garrott and Oli 2013) and the experience of adopters (Bastian *et al.* 1999), there are still no published quantitative studies on attitudes of the various publics engaged in these conflicts. A few publications have explored issues related to the politics at work, and there has been one broad ethnographic study (Dalke 2010, 2011). A National Research Council (2013) review of social considerations of US mustang and burro management confirmed this lack of relevant social science research despite it having been requested two decades earlier.

In this knowledge vacuum, it is only possible to sketch the most general details of such conflict. There is no doubt that mustangs mean different things to different people. Ranchers often view them as competition for limited grazing (Bastian *et al.* 1999). Some conservationists view them as alien species that damage native habitats and endanger other species (Beever 2003), although other ecologists view them as surrogates for North American equids that went extinct in the Late Pleistocene (Donlan *et al.* 2006). They are also powerful heritage symbols to both Native Americans and those of European origin (Dalke 2010, 2011). US federal law mandates the status of mustangs as symbols of cultural heritage (US Public Law 92-195). In addition to competing ideas of what the horses represent, their management (especially concerning the use of veterinary euthanasia and lethal control) clearly attracts attention from many interest groups concerned with animal rights and animal welfare. Finally, there appears to be a substantial conflict over knowledge concerning the numbers of horses, the actions of the agency in charge of managing feral horses, horses' impact on habitats, and even their origins (Symanski 1996). In contrast to the well-studied smaller conflicts mentioned in the previous section, the controversy surrounding mustang management is a national-level conflict, with publics across the United States believing that they have a stake in how mustangs should be managed on the public lands of the western states. In sum, mustang management appears to involve conflicts over economic interests and governance as well as fundamental values, beliefs, and ethics. Broadly speaking, the same conflict elements also appear to be present in Australia and New Zealand (Symanski 1994; Nimmo and Miller 2007; Nimmo *et al.* 2007).

Free-Ranging Domestic Horses in Europe

At the end of the Pleistocene, two species of wild equid—a wild horse (*E. ferus*) and a wild ass (*E. hydruntinus*)—occupied Europe. Both species unquestionably survived through the early and middle Holocene, but their later fate is unclear because of gaps in the fossil/bone record and the challenge of discriminating between remains of the original wild species and the introduced domestic forms (Burke *et al.* 2003; Orlando *et al.* 2006; Sommer *et al.* 2011; Bendry 2012). There has been some speculation that a population of free-living asses termed "encebro" that occupied central Iberia until the sixteenth century could have been descended from *E. hydruntinus*, although the existing evidence does not support this idea, making it more likely that the encebro constituted a feral population of domestic donkeys (Orlando *et al.* 2006; Crees and Turvey 2014).

Likewise, there has been a great deal of speculation about to what extent, if any, genes from European *E. ferus* were incorporated into the diversity of domestic breeds that emerged in Europe. Although this diversity of livestock breeds is being increasingly viewed as a vital target for conservation from both the points of view of heritage and of genetic resources (Hall and Bradley 1995; Food and Agriculture Organization of the United Nations 2007; Rosenthal 2010), there is an emerging discourse that places value on the putative links of domestic breeds to European populations of *E. ferus* (Vega-Pla *et al.* 2006; Georgescu *et al.* 2011; Warmuth *et al.* 2011; Hovens and Rijkers 2013). The situation is further complicated by the historical existence of a free-living equid in the forests of central Europe termed "tarpan," which persisted in the wild until the nineteenth century. There has been much speculation that the tarpan represented some form of descendent (either in pure or introgressed form) of the original *E. ferus*, but the alternative hypothesis, that the tarpan was simply a feral domestic horse, cannot be rejected. Irrespective of the link between tarpans and *E. ferus*, there have been attempts to breed a horse that retains some of the characteristics of the tarpan (derived from bone material and historical descriptions) based on Polish domestic horses that reputedly mixed with some of the last remaining tarpans. The resulting Konik horse is widely used in nature conservation projects and is frequently called a "wild" horse (fig. 9.1D). Another initiative tried to use selective back-breeding of various breeds to produce a horse, the Heck horse, with phenotypic characteristics similar to the tarpan.

European conservation has traditionally placed a high value on open grasslands and heaths, which are associated with high species diversity and have a strong aesthetic appeal (García *et al.* 2013). Various schools of thought have placed different emphasis on whether the origins of these habitats are purely anthropogenic (from millennia of forest clearance and livestock grazing) or if they actually represent the modern version of what was an open forest: grassland mosaic maintained by large wild herbivores such as bison (*Bison bonasus*), which nearly became extinct, and the extinct wild horse and auroch (*Bos primigenius*) (Birks 2005; Kirby 2009; Kerley *et al.* 2012). Although these different schools justify their goals on different ideological platforms (conserving heritage landscapes and species diversity vs. rewilding), there has been broad support for the desirability of maintaining grazing pressure in certain landscapes. Horses have been frequently used to help achieve these goals. In many cases, some of the more hardy and primitive domestic horse breeds have been used along with other livestock to achieve site-specific goals (Van Wieren and Bakker 1998). But in other cases there has been a focus on using either Konik or Heck horses to rewild landscapes or establish "naturalistic grazing" regimes (Kirby 2009). Such projects have frequently referred to these horses as "wild" horses. The paradox is that every single one of these projects fences its horses into relatively limited areas and uses supplementary feeding in winter. Most of these populations occur in areas with no or very limited predation. Ironically, there are several large populations of domestic horses in Iberia that are extensively farmed for meat production and live under far wilder conditions, but without the epithet of "wild." These horses are routinely exposed to natural mortality, including wolf (*Canis lupus*) predation (Vos 2000; García *et al.* 2013; López-Bao *et al.* 2013). In addition, there are several other populations of free-ranging domestic horses that have been living under conditions ranging from semiferal to extensively managed, such as the ponies of Exmoor and the New Forest in the United Kingdom or the Camargue horses of southern France that are frequently labeled "wild."

In total contrast to North America and Australia, where most conservationists and nature management agencies are trying to reduce the number of free-ranging horses, there is a widespread aim by these bodies in Europe to increase their numbers. The little data that exist indicate that local attitudes toward these "wild" horse populations vary. The long-established populations of free-ranging horses in the United Kingdom, France, and Iberia are clearly objects of local pride and are incorporated into aspects of local identity, sense of place, and cultural heritage. Studies have indicated that the recently established populations of Konik horses in Lake Pape in Latvia have been associated with at least a certain degree of local skepticism (O'Rourke 2000; Schwartz 2005). But the horse issue cannot be easily separated from wider political issues (Galbreath and Auers 2009). It appears that the local population has not developed the same strong attachment to these recently introduced "wild" horses. The most high profile of these projects in the Dutch Oostvaardersplassen reserve has been highly controversial because of the legal and ethical implications associated with attempts to "de-domesticate" Konik ponies. Central to the issue is whether these animals that are free ranging, within a large fenced area, should be regarded as domestic species or as wildlife, with all the attendant consequences of obligations for veterinary care and animal welfare. Both geographers and ethicists have explored the issue in great detail (Gamborg *et al.* 2010; Lorimer and Driessen 2013*a*,*b*), with the conclusion that these de-domesticated animals provide a great challenge for our present legal and value systems.

Defining the "Wild"

So far in this chapter we have explored the thin but diverse literature on the complex relationships between humans and "wild" domestic horses. When combined, they represent a fascinating insight into how people view nature and their relationships with it. A primary theme that emerges is the blurry definition that many people have in their definition of "wild." It appears that people are willing to attach the epithet to a given population of horses even when it is managed intensively. It is enough for horses to be living in any reasonably sized natural area (fenced or unfenced) under conditions with a minimum of husbandry. The geographical and genetic origins of the horses are not barriers for them to be called "wild," although in some discourses considerable efforts are made to legitimize this designation, with reference to genetic studies that are interpreted as demonstrating some connection to the predomestication species (e.g., Hovens and Rijkers 2013). The studies also reveal how many people have integrated their relationship with their "wild" horses into their cultural heritage. Yet there are many contrasting views of horses that illustrate the diversity of ways in which people can view the same situation depending on their underlying value orientations and economic interests (Dalke 2010, 2011).

While the classical modern dualist worldview generally separates nature from culture and wild from domestic, it also generates numerous hybrids falling outside existing categories (Latour 1993; Descola 2013). Various authors have pointed that often-contradictory views of "wild" horses challenge dualism (Gamborg *et al.* 2010; Bhattacharyya *et al.* 2011; Notzke 2013). These studies on "wild" horses clearly show that many people view their interaction with the wild (as symbolized by the horse) as a part of their cultural heritage. Equally, they show that free-ranging horses do not neatly fit into either the category of domestic animal or wildlife (Gamborg *et al.* 2010; Lorimer and Driessen 2013*b*; Notzke 2013). Instead, free-ranging horses appear to be a conceptual "hybrid" in Latour's sense, as their relationship with people is too complex, diverse, and dynamic to clearly qualify them as domestic or wild animals. Among other factors, the fact that horses can become companion animals with which people can form complex emotional relationships on a one-on-one basis (Keaveney 2008) probably contributes to this special status. A similar complex situation exists for the relationships between humans, domestic dogs, and wolves (Lescureux and Linnell 2014).

These studies of horse–human relationships provide insights into basic aspects of human–nature relationships that are important for the development of theories and conceptual frameworks. They also provide some general messages relevant for the management of these populations. First, the management of horses must be seen within a wider ecological, social, economic, and cultural context. One size will not fit all, so focus should be placed on finding good pragmatic solutions to local situations rather than universal blueprints. Second, horses are special, and they cannot be managed simply by transferring practices and values from either domestic animal or wildlife management traditions. They will require a special, hybrid framework. Associated with this is a need to recognize that the public does not use words in the same ways as wildlife management professionals, and although we may need to focus on semantics within our professional arenas, there is a need to adjust language use to local contexts to avoid unnecessary provocation (Bhattacharyya *et al.* 2011). Third, horses mean different things to different people, and any solution will require open processes that build legitimate compromises (Linnell 2015). It is here that the weakness of the existing research base is most visible. Although we now have a good qualitative idea of the underlying issues and discourses, there is absolutely no idea of how widespread the different positions are, of who thinks what, and of how the different issues (of values, attitudes, economic interests, knowledge, experience, location, culture) are bundled together into coherent and contrasting world views. There is an urgent need for highly focused quantitative studies that seek to disentangle the various strands of the competing discourses as a compliment to in-depth qualitative studies. Without this research base, it is difficult to see how much progress can be made in this highly controversial field.

Conflicts with Wild Equids

There have been remarkably few formal and direct human dimension studies on wild equid–human relationships. As a result, the following sections have been pieced together from a diverse range of sources and at best represent a qualitative overview of the many ways in which people and wild equids interact and of how they are perceived. These interactions can be grouped into three categories: conflict, consumption, and viewing. Conflicts between humans and wildlife are widespread (Redpath *et al.* 2013), and equids are no exception. Plains zebra (*E. quagga*) are considered problem wildlife in parts of Africa due to the occurrence of crop raiding (Gadd 2005; Okello 2005; Anthony *et al.* 2010; Chomba *et al.* 2012). Local herders in Kenya perceive Grevy's zebra (*E. grevyi*) as competitors for pasture and water, and a majority of herders apparently sees few benefits from their presence (Sundaresan *et al.* 2012).

The same is true for Asiatic wild asses (*E. hemionus*) in the Bahram-e-Goor protected area in Iran (S. Esmaili, personal communication) and the Little Rann of Kutch in India (Dave 2010). In the Gobi regions of Mongolia or the high pastures in Ladakh, wild asses are also primarily perceived as pasture competitors by local herders (Bhatnagar *et al.* 2006; Kaczensky *et al.* 2006, 2007). Although heavily outnumbered by livestock, Asiatic wild ass and kiang (*E. kiang*) occasionally form large temporary groups (>70 in Ladakh and >1,000 in the Gobi), which locally can remove significant amounts of forage. Such groups can negatively affect local herders when depleting critical key areas like moist sedge meadows in Ladakh (Bhatnagar *et al.* 2006) or pastures set aside as winter grazing reserves in Mongolia (Kaczensky *et al.* 2007). The perceived conflict level depends on observed wild ass presence and seems to be highest in regions where wild asses are recovering in numbers or only temporarily aggregating (i.e., where people are not so accustomed to their presence). Similar conflicts are also reported for kiang in Xinjiang, China, and Asiatic wild ass in India (Dave 2010; Turghan *et al.* 2013). Conflicts between people and free-ranging Przewalski's horses (*E. ferus przewalskii*) are currently limited, as reintroduced populations are still small (Kaczensky *et al.* 2007). But some incidents of stealing of domestic mares by Przewalski's horses, hybridization, and aggression toward domestic stock and their owners have occurred (see chap. 14) and are said to have been a source of conflict prior to their extinction in the wild (Heptner *et al.* 1988).

Although multiple studies have identified conflicts between wild equids and humans in just about all areas where they overlap, these conflicts are usually relatively minor compared to the conflicts that local people have with other species in the same areas. In Africa, for example, the conflicts with zebra are minor compared to the conflicts with other species such as elephants. Despite this type of conflict, however, local people in Iran generally value wild asses and are in favor of their conservation (S. Esmaili, personal communication). The same is true for the Mongolian Gobi, where wild asses, and especially Przewalski's horses, are considered beautiful animals and are valued as a component of their natural heritage (Kaczensky *et al.* 2006; Kaczensky 2007; P. Kaczensky, unpublished data) (figs. 9.1B and 9.1C).

Consumption of Wild Equids

Wild equids were an important prey species for paleolithic and neolithic humans (Velichko *et al.* 2009). In historic times, wild equids were killed for their meat, fat, hides, or specific body parts for medical purposes. They were also considered challenging or noble game. In ancient Persia, for example, hunting wild Asiatic asses is said to have been particularly enjoyed by Bahram V, the fourteenth Sassanid King of Persia (421–438 CE), who was famous for the passion and skill with which he hunted onagers. As a result he was called Bahram-e-Goor, with "Goor" meaning onager. Hunting wild equids was also considered an exciting and manly sport in central Asia, as the following quotes illustrate. "The speed and lightness of a running wild ass should be admired. He leaves the pursuing hunter behind as if he is playing or joking" (Bannikov 1981). "The hunting of tarpan [used by the author interchangeably with Przewalski's horse] was considered a difficult and dangerous sport in which man could exhibit not only his bravery, but also test the quality of his horses" (Heptner *et al.* 1988).

Nowadays, most wild equid populations are fully protected. Only plains and mountain zebra (*E. zebra*) are still subject to regular hunting, mainly in the form of trophy hunting by foreign hunters in countries like Botswana, Namibia, South Africa, Tanzania, Zambia, and Zimbabwe (Lindsey *et al.* 2007*a*). The cost of a trophy hunt provides some indication of the extent to which hunters value the species. Zebra trophies are in the middle tier of costs, much lower than the big five—leopard, lion, elephant, rhino, and African buffalo—but higher than the smaller antelope species (Johnson *et al.* 2010).

Although African trophy hunting is far from perfectly managed, it is not regarded as a major threat to any wild equid population. In contrast, illegal hunting for bushmeat and medicinal purposes is a far greater threat. Studies have shown that zebra are killed for bushmeat in many countries, including Tanzania (Martin *et al.* 2012; Mwakatobe *et al.* 2012) and Zimbabwe (Lindsey *et al.* 2011). In Tanzania, there have been several studies of the social and economic factors influencing illegal killing of zebra and other species for bushmeat. Ceppi and Nielsen (2014) have documented that species preferences vary among groups of indigenous people as a result of both availability and cultural factors. Studies of bushmeat prices indicate that zebra are not particularly valued as food, with prices for all bushmeat being lower than for beef (Ndibalema and Songorwa 2007; Mwakatobe *et al.* 2012; Rentsch and Damon 2013). While this might indicate a low value attached to zebra meat, it is nevertheless frequently consumed (Ndibalema and Songorwa 2007). Furthermore, there is clearly some social status attached to the role of the hunter, at least among some ethnic groups, indicating that it is not only economics and the need for protein that drive zebra hunting (Lowassa *et al.* 2012). But this study did not reveal any special value attached to the procurement of zebra meat over other meats. There are relatively few records of zebra being used for medicinal purposes, although records of zebra products exist from markets in South Africa and Benin, the latter being outside the distribution range of all zebras (Djagoun *et al.* 2013; Whiting *et al.* 2013).

Both African (*E. africanus*) and Asian wild asses are reported as being heavily poached for meat and medicinal purposes (Wingard and Zahler 2006; Kebede *et al.* 2014). Body parts of African wild asses are used for treating tuberculosis, constipation, rheumatism, backache, and bone ache (Moehlman *et al.* 2015). In Iran, local people believe onager (*E. h. onager*) meat and bone oil to be a cure for osteal diseases (Tatin *et al.* 2003; S. Esmaili, personal communication). Wild ass body parts are also present in the traditional Indian pharmacy, used for arthritis among other aliments (Gupta *et al.* 2003; Mahawar and Jaroli 2008). In Mongolia, wild ass meat is regarded as inferior to other ungulate meat because it is very lean, thus only consumed when no alternative is available, and it is sold as "cheap" horse meat (Kuehn *et al.* 2006). Even so, this practice does not prevent poaching (Wingard and Zahler 2006). Kiang are also reported as being subject to poaching in Xinjiang (Turghan *et al.* 2013).

Viewing of Wild Equids

Ecotourism is an expanding industry in many areas, providing a way to see how people interact with and value different wildlife species. At present, only the three zebra species form part of regular tourist packages, as the Asiatic and African asses live in areas little visited by tourists. Reintroduced Przewalski's horses have become a major tourist attraction in Mongolia (see chap. 14). Studies of tourists' preferences have been conducted in South Africa and Kenya (Lindsey *et al.* 2007*b*; Okello *et al.* 2008; Di Minim *et al.* 2012; Maciejewski and Kerley 2014). The results are highly consistent in showing that tourists are primarily interested in seeing the large predators like lion, leopard, and cheetah, and the megaherbivores like elephant, giraffe, and African buffalo. Zebra tend to rank at the top of the second tier of preferred species along with the other medium-sized herbivores, but there is a substantial gap between these two tiers of species groups. Because zebra tend to be locally abundant and thus almost guaranteed to be frequently seen, however, it is possible that their relatively low preference scores underestimate their value to tourists, who regard them as intrinsic and iconic parts of the African landscape.

Are Wild Equids Always "Wild"?

Our narrative concerning horses has been very much dominated about a discussion of the nature of the "wild." Although the identity of wild equid species is uncontested, there are some emerging circumstances in which it is possible to question where the borders between wild and domestic living are, even for these species. The most obvious example concerns game management practices in South Africa, where many species are kept in private reserves that are used for everything from conservation, to game ranching, to trophy hunting, to wildlife viewing, or various combinations of these activities (Cousins *et al.* 2008). In some of these reserves, the wildlife is more intensively managed than in many herds of free-ranging domestic horses. The extreme example is of the Cape mountain zebra (*E. z. zebra*), whose subpopulations are all fenced. Many reserves are small and depend on human-assisted exchange of individuals for genetic integrity (see chap. 14). Another complex example concerns the Przewalski's horse. Extinct in the wild, the species was maintained in captivity for several decades before being released at a number of locations in Mongolia and China. Reintroductions have used a range of methods, including erecting fencing, providing veterinary care, and feeding released horses during the winter. An interesting question arises: When should a reintroduced population be said to have returned to the wild (Xia *et al.* 2014)?

A contrasting set of questions concerns issues of genetic purity, which can also influence how people perceive and value a species. An assumed wild ancestry or introgression of wild genes into domestic horse breeds is often used as an argument for the "higher value" of these breeds, whereas introgression of domestic genes into wild Przewalski's horse populations is a major conservation concern that may "lower the value" of these populations (Boyd 2007). But the situation is even more complicated, as all living Przewalski's horses are descended from only 13 founders, one of which was a known Przewalski's horse–domestic horse hybrid. Some of the wild-caught Przewalski's horses were also suspected not to have been entirely pure blooded, and some additional hybridization has occurred in captivity. Up to the present day, the degree of domestic introgression in the captive breeding stock of Przewalski's horses has fueled a highly emotional discussion over what constitutes a "true" Przewalski's horse among some breeders in Western countries (see chap. 11). But the discussion has been of little relevance for the recovery of the Przewalski's horse in the wild, as currently only the breeding line representing all 13 founders is large enough to provide animals for reintroductions. The Przewalski's horses released in Mongolia and China were all selected on the bases of health, age, and a high diversity of founder representation; in the receiving countries, the returned horses have never been considered anything other than true Przewalski's horses and are now valued highly. But even on the subspecies level, genetic purity may influence how a species is valued. The population of Asiatic wild asses in Israel, for example, is composed of a mixture of two apparent subspecies (*E. h. kulan* and *E. h. onager*) that now replace the original subspecies, the extinct Syrian wild ass (*E. h. hemippus*). Because of this mixed ancestry, questions have been raised about the value of this population, even though morphologically and ecologically there are no obvious differences. So what level of introgression of domestic genes or nonlocal genes is acceptable in a species or population? There is no scientific answer to that question. Rather, the key point is that by raising these questions, people are providing insights into how they view these equids.

Summarizing Interactions between People and Wild Equids

The relationship between local people and the wild equids contrasts greatly with the relationships that North

Americans, Europeans, and Australians have with free-living horses. While interactions with the horse can be intense, emotional, and polarized, the relationship with wild equids is low key. People interact with them in various ways—through conflicts, consumption, and viewing—but in all cases the wild equids occupy relatively discrete places compared to other species. None of the wild equids appear to occupy especially visible positions in local cultures, folklore, or cosmologies. Even when Westerners visit Africa, the zebras appear to attract much less attention than other species, and the Asian species of wild equids are relatively unknown. In other words, the Western appreciation for the horse has not had a spillover effect onto other equid species, with the possible exception of the Przewalski's horse—often referred to as the only truly wild horse—whose return to the wild has captured the imagination of Westerners as well as of Mongolians.

Conclusion

In this chapter, we have tried to build a coherent narrative from scarce and scattered sources. The lack of large, robust, and quantitative studies of human relationships with wild equids is shocking, although a number of in-depth qualitative studies provide valuable insights into the diversity of discourses and narratives associated with the management of equids. The human relationship with the seven wild species can be best described as low key, with a somewhat higher status given to the Przewalski's horse. Wild equids do not appear to attract any extremes of either positive or negative attitudes or responses from any societies. The human relationship with the free-ranging populations of the domestic species is far more emotional and polarized, and provides fascinating insights into the diversity of ways people can perceive wildness, nature, and the relationship between people and nature. There is much more work to be done on the human dimensions of these species, which will be useful for the development of the field and the conservation and management of the species. Equids—wild, domestic, feral, and free ranging—provide a uniquely fascinating lens to explore the diversity of relationships that people have with wildlife.

ACKNOWLEDGMENTS

J.L. and N.L. received funding from the Research Council of Norway (project 212919) and the Norwegian Institute for Nature Research. P.K. received funding from the Austrian Science Foundation (Projects P14992, P18624, and P24231).

REFERENCES

Anthony, B.P., P. Scott, and A. Antypas. 2010. Sitting on the fence? Policies and practices in managing human-wildlife conflicts in Limpopo Province, South Africa. Conservation and Society 8:225–240.

Bannikov, A.G. 1981. The Asian wild ass. [In Russian.] Lesnaya Promyshlennost, Moscow, Russia. [English translation by M. Proutkina, Zoological Society of San Diego, San Diego, CA, USA.]

Bastian, C.T., L.W. Van Tassell, A.C. Cotton, and M.A. Smith. 1999. Opportunity costs related to feral horses: A Wyoming case study. Journal of Range Management 52:104–112.

Beever, E. 2003. Management implications of the ecology of free-roaming horses in semi-arid ecosystems of the western United States. Wildlife Society Bulletin 31:887–895.

Bendrey, R. 2012. From wild horses to domestic horses: A European perspective. World Archaeology 44:135–157.

Bhatnagar, Y.V., R. Wangchuk, H.H.T. Prins, S.E. Van Wieren, and C. Mishra. 2006. Perceived conflicts between pastoralism and conservation of the kiang *Equus kiang* in the Ladakh Trans-Himalaya, India. Environmental Management 38:934–941.

Bhattacharyya, J., D.S. Slocombe, and S.D. Murphy. 2011. The "wild" or "feral" distraction: Effects of cultural understandings on management controversy over free-ranging horses (*Equus ferus caballus*). Human Ecology 39:613–625.

Bies, L., M. Hutchins, and T.J. Ryder. 2011. The Wildlife Society responds to CNN report on feral horses. Human-Wildlife Interactions 5:171–172.

Birks, H.J.B. 2005. Mind the gap: How open were European primeval forests? Trends in Ecology and Evolution 20:154–156.

Boyd, L. 2007. Workshop to discuss the future of the takhi (*Equus ferus przewalskii*) in Mongolia: A report. Mongolian Journal of Biological Sciences 5:53–54

Burke, A., V. Eisenmann, and G.K. Ambler. 2003. The systematic position of *Equus hydruntinus*, an extinct species of Pleistocene equid. Quaternary Research 59:459–469.

Ceppi, S.L., and M.R. Nielsen. 2014. A comparative study on bushmeat consumption patterns in ten tribes in Tanzania. Tropical Conservation Science 7:272–287.

Chomba, C., R. Senzota, H. Chabwela, J. Mwitwa, and V. Nyirenda. 2012. Patterns of human–wildlife conflicts in Zambia, causes, consequences and management responses. Journal of Ecology and the Natural Environment 4:303–312.

Cousins, J.A., J.P. Sadler, and J. Evans. 2008. Exploring the role of private wildlife ranching as a conservation tool in South Africa: Stakeholder perspectives. Ecology and Society 13:43.

Crees, J.J., and S.T. Turvey. 2014. Holocene extinction dynamics of *Equus hydruntinus*, a late-surviving European megafaunal mammal. Quaternary Science Reviews 91:16–29.

Dalke, K. 2010. Mustang: The paradox of imagery. Humanimalia 1:97–117.

Dalke, K. 2011. A translocal perspective: Mustang images in the cultural, economic and political landscape. Animals 1:27–39.

Dave, C.V. 2010. Understanding conflicts and conservation of Indian wild ass around Little Rann of Kachchh, Gujarat, India. Final technical report submitted to Rufford Small Grant Program, London, UK.

Descola, P. 2013. Beyond nature and culture. University of Chicago Press, Chicago, USA.

Di Minim, E., I. Fraser, R. Slotow, and D.C. MacMillan. 2012. Understanding heterogeneous preference of tourists for big game species: Implications for conservation and management. Animal Conservation 16:249–258.

Djagoun, C.A.M.S., H.A. Akpona, G.A. Mensah, C. Nuttman, and B. Sinsin. 2013. Wild mammals trade for zootherapeutic and mythic purposes in Benin (West Africa): Capitalizing species involved, provision sources, and implications for conservation. Pages 367-382 *in* R.R.N. Alves and I.L. Rosa (Eds.), Animals in traditional folk medicine: Implications for conservation. Springer, Berlin, Germany.

Dobbie, W.R., D. Berman, and M.L. Braysher. 1993. Managing vertebrate pests: Feral horses. Department of Primary Industries and Energy, Bureau of Resource Sciences, Australian Government Publishing Service, Canberra, ACT, Australia.

Donlan, C.J., J. Berger, C.E. Bock, J.H. Bock, D.A. Burney, *et al.* 2006. Pleistocene rewilding: An optimistic agenda for twenty-first century conservation. American Naturalist 168:660–681.

Food and Agriculture Organization of the United Nations. 2007. Global plan of action for animal genetic resources and the Interlaken Declaration. Presented at the International Technical Conference on Animal Genetic Resources for Food and Agriculture in Interlaken, Switzerland, 3–7 September. Commission on Genetic Resources for Food and Agriculture, Food and Agriculture Organization of the United Nations, Rome, Italy.

Fortwangler, C.L. 2009. A place for the donkey: Natives and aliens in the US Virgin Islands. Landscape Research 34:205–222.

Gadd, M. 2005. Conservation outside of parks: Attitudes of local people in Laikipia, Kenya. Environmental Conservation 32:50–63.

Galbreath, D.J., and D. Auers. 2009. Green, black and brown: Uncovering Latvia's environmental politics. Journal of Baltic Studies 40:333–348.

Gamborg, C., B. Gremmen, S.B. Christiansen, and P. Sandøe. 2010. De-domestication: Ethics at the intersection of landscape restoration and animal welfare. Environmental Ethics 19:57–78.

García, R.R., M.D. Fraser, R. Celaya, L.M.M. Ferreira, U. García, and K. Osoro. 2013. Grazing land management and biodiversity in the Atlantic European heathlands: A review. Agroforestry Systems 87:19–43.

Garrott, R.A., and M.K. Oli. 2013. A critical crossroad for BLM's wild horse program. Science 341:847 848.

Georgescu, S.E., M.A. Manea, A. Dudu, and M. Costache. 2011. Phylogenetic relationships of the Hucul horse from Romania inferred from mitochondrial D-loop variation. Genetics and Molecular Research 10:4104–4113.

Gupta, L., C.S. Silori, N. Mistry, and A.M. Dixit. 2003. Use of animals and animal products in traditional health care systems in District Kachchh, Gujarat. Indian Journal of Traditional Knowledge 2:346–356.

Hall, S.J.G., and D.G. Bradley. 1995. Conserving livestock breed biodiversity. Trends in Ecology and Evolution 10:267–270.

Heptner, V.G., A.A. Nasimovich, and A.G. Bannikov. 1988. Mammals of the Soviet Union, vol. 1: Artiodactyla and Perissodactyla. Smithsonian Institution Libraries and the National Science Foundation, Washington, DC, USA. [English translation of the original book published in 1961 by Vysshaya Shkola, Moscow, Russia].

Hovens, J.P.M., and A.J.M. Rijkers. 2013. On the origins of the Exmoor pony: Did the wild horse survive in Britain? Lutra 56:129–136.

Johnson, P.J., R. Kansky, A.J. Loveridge, and D.W. Macdonald. 2010. Size, rarity and charisma: Valuing African wildlife trophies. PLoS ONE 5:e12866.

Kaczensky, P. 2007. Wildlife value orientation of rural Mongolians. Human Dimensions of Wildlife 12:317–329.

Kaczensky, P., D.P. Sheehy, D.E. Johnson, C. Walzer, D. Lhkagvasuren, and C.M. Sheehy. 2006. Room to roam? The threat to khulan (wild ass) from human intrusion. Mongolia Discussion Papers: East Asia and Pacific Environment and Social Development Department. World Bank, Washington, DC, USA.

Kaczensky, P., N. Enkhsaikhan, O. Ganbaatar, and C. Walzer. 2007. Identification of herder-wild equid conflicts in the Great Gobi B Strictly Protected Area in SW Mongolia. Exploration into the Biological Resources of Mongolia 10:99–116.

Kavar, T., and P. Dovč. 2008. Domestication of the horse: Genetic relationships between domestic and wild horses. Livestock Science 116:1–14.

Keaveney, S.M. 2008. Equines and their human companions. Journal of Business Research 61:444–454.

Kebede, F., P.D. Moehlman, A. Bekele, and P.H. Evangelista. 2014. Predicting seasonal habitat suitability for the critically endangered African wild ass in the Danakil, Ethiopia. African Journal of Ecology 54:533–542.

Kerley, G.I.H., R. Kowlczyk, and J.P.G.M. Cromsigt. 2012. Conservation implications of the refugee species concept and the European bison: King of the forest or refugee in a marginal habitat? Ecography 35:519–529.

Kirby, K.J. 2009. Policy in or for the wilderness. British Wildlife 20:59–62.

Kuehn, R., P. Kaczensky, D. Lkhagvasuren, S. Pietsch, and C. Walzer. 2006. Differentiation of meat samples from domestic horses (*Equus caballus*) and Asiatic wild asses (*Equus hemionus*) using a species-specific restriction site in the mitochondrial cytochrome b region. Mongolian Journal of Biological Sciences 4:57–62.

Latour, B. 1993. We have never been modern. Harvard University Press, Cambridge, MA, USA.

Lescureux, N., and J.D.C. Linnell. 2014. Warring brothers: The complex interactions between wolves (*Canis lupus*) and dogs (*Canis familiaris*) in a conservation context. Biological Conservation 171:232–245.

Lindsey, P.A., P.A. Roulet, and S.S. Romanach. 2007*a*. Economic and conservation significance of the trophy hunting industry in sub-Saharan Africa. Biological Conservation 134:455–469.

Lindsey, P.A., R. Alexander, M.G.L. Mills, S. Romanach, and R. Woodroffe. 2007*b*. Wildlife viewing preferences of visitors to protected areas in South Africa: Implications for the role of ecotourism in conservation. Journal of Ecotourism 6:19–33.

Lindsey, P.A., S.S. Romanach, C.J. Tambling, K. Chartier, and R. Groom. 2011. Ecological and financial impacts of illegal bushmeat trade in Zimbabwe. Oryx 45:96–111.

Linnell, J.D.C. 2015. Defining scales for managing biodiversity and natural resources in the face of conflicts. Pages 208–218 *in* S.M. Redpath, R.J. Guitiérrez, K.A. Wood, and J.C. Young (Eds.), Conflicts in conservation: Navigating towards solutions. Cambridge University Press, Cambridge, UK.

López-Bao, J.V., V. Sazatornil, L. Llaneza, and A. Rodríguez. 2013. Indirect effects on heathland conservation and wolf persistence of contradictory policies that threaten traditional free-ranging horse husbandry. Conservation Letters 6:448–455.

Lorimer, J., and C. Driessen. 2013*a*. Bovine biopolitics and the promise of monsters in the rewilding of Heck cattle. Geoforum 48:249–259.

Lorimer, J., and C. Driessen. 2013*b*. Wild experiments at the Oostvaardersplassen: Rethinking environmentalism in the Anthropocene. Transactions of the Institute of British Geographers 39:169–181.

Lowassa, A., D. Tadie, and A. Fischer. 2012. On the role of women in bushmeat hunting—Insights from Tanzania and Ethiopia. Journal of Rural Studies 28:622–630.

Maciejewski, K., and G.I.H. Kerley. 2014. Understanding tourists' preference for mammal species in private protected areas: Is there a case for extralimital species for ecotourism. PLoS ONE 9:e88192.

Mahawar, M.M., and D.P. Jaroli. 2008. Traditional zootherapeutic studies in India: A review. Journal of Ethnobiology and Ethnomedicine 4:17.

Martin, A., T. Caro, and M. Borgerhoff Mulder. 2012. Bushmeat consumption in western Tanzania: A comparative analysis from the same ecosystem. Tropical Conservation Science 5:352–364.

Meyer, H. 2013. Zur biozönose von Mensch und Pferd—Ein historischer und kultursoziolgischer Überblick. Pferdeheilkunde 29:363–388.

Moehlman, P.D., F. Kebede, and H. Yohannes. 2015. *Equus africanus*. IUCN Red List of Threatened Species. Version 3.1. International Union for Conservation of Nature, Gland, Switzerland. Available at www.iucnredlist.org.

Mwakatobe, A., E. Røskaft, and J. Nyahongo. 2012. Bushmeat and food security: Species preference of sundried bushmeat in communities in the Serengeti-Mara ecosystem, Tanzania. International Journal of Biodiversity and Conservation 4:548–559.

National Research Council. 2013. Using science to improve the BLM Wild Horse and Burro Program: A way forward. National Academies Press, Washington, DC, USA.

Ndibalema, V.G., and A.N. Songorwa. 2007. Illegal meat hunting in Serengeti: Dynamics in consumption and preferences. African Journal of Ecology 46:311–319.

Nimmo, D.G., and K.K. Miller. 2007. Ecological and human dimensions of management of feral horses in Australia: A review. Wildlife Research 34:408–417.

Nimmo, D.G., K.K. Miller, and R. Adams. 2007. Managing feral horses in Victoria: A study of community attitudes and perceptions. Ecological Management and Restoration 8:237–241.

Notzke, C. 2013. An exploration into political ecology and nonhuman agency: The case of the wild horse in western Canada. Canadian Geographer 57:389–412.

Okello, M.M. 2005. Land use changes and human-wildlife conflicts in the Amboseli Area, Kenya. Human Dimensions of Wildlife 10:19–28.

Okello, M.M., S.G. Manka, and D.E. D'Amour. 2008. The relative importance of large mammal species for tourism in Amboseli National Park, Kenya. Tourism Management 29:751–760.

Olsen, S.L., 2006. Early horse domestication on the Eurasian Steppe. Pages 245–272 *in* M.A. Zeder, D.G. Bradley, E. Emshwiller, and B.D. Smith (Eds.), Documenting domestication: New genetic and archaeological paradigms. University of California Press, Berkeley, USA.

Orlando, L., M. Mashkour, A. Burke, C. J. Douady, V. Eisenmann, and C. Hanni. 2006. Geographic distribution of an extinct equid (*Equus hydruntinus*: Mammalia, Equidae) revealed by morphological and genetical analyses of fossils. Molecular Ecology 15:2083–2093.

O'Rourke, E. 2000. The reintroduction and reinterpretation of the wild. Journal of Agriculture and Environmental Ethics 13:145–165.

Peace, A. 2009. Ponies out of place? Wild animals, wilderness and environmental governance. Anthropological Forum: A Journal of Social Anthropology and Comparative Sociology 19:53–72.

Redpath, S.M., J. Young, A. Evely, W.M. Adams, W.J. Sutherland, *et al*. 2013. Understanding and managing conservation conflicts. Trends in Ecology and Evolution 28:100–109.

Rentsch, D., and A. Damon. 2013. Prices, poaching and protein alternatives: An analysis of bushmeat consumption around Serengeti National Park, Tanzania. Ecological Economics 91:1–9.

Rikoon, J.S. 2006. Wild horses and the political ecology of nature restoration in the Missouri Ozarks. Geoforum 37:200–211.

Robinson, I.H. 1999. The human-horse relationship: How much do we know? Equine Veterinary Journal Supplement 28:42–45.

Rosenthal, J.S. 2010. A review of the role of protected areas in conserving global domestic animal diversity. Animal Genetic Resources 47:101–113.

Schwartz, K.Z.S. 2005. Wild horses in a "European wilderness": Imagining sustainable development in the post-Communist countryside. Cultural Geographics 12:292–320.

Smith, J.T. 2010. Wild horses and BLM management issues: What to do with 30,000 symbols of the American west. Human–Wildlife Interactions 4:7–11.

Sommer, R.S., N. Benecke, L. Lougas, O. Nelle, and U. Schmolcke. 2011. Holocene survival of the wild horse in Europe: A matter of open landscape? Journal of Quaternary Science 26:805–812.

Sundaresan, S., B. Bruyere, G. Parker, B. Low, N. Stafford, and S. Davis. 2012. Pastoralists' perceptions of the endangered Grevy's zebra in Kenya. Human Dimensions of Wildlife 17:270–281.

Symanski, R. 1994. Contested realities: Feral horses in outback Australia. Annals of the Association of American Geographers 84:251–269.

Symanski, R. 1996. Dances with horses: Lessons from the environmental fringe. Conservation Biology 10:708–712.

Tatin, L., B.F. Darreh-Shoori, C. Tourenq, D. Tatin, and B. Azmayesh. 2003. The last population of the critically endangered Onager *Equus hemionus onager* in Iran: Urgent requirements for protection and study. Oryx 37:488–491.

The Wildlife Society. 2011. Final position statement: Feral horses and burros in North America. Bethesda, MD, USA. Available at http://joomla.wildlife.org/documents/positionstatements/Feral.Horses.July.2011.pdf.

Turghan, M., M. Ma, X. Zhang, T. Zhang, and Y. Chen. 2013. Current population and conservation status of the Tibetan wild ass (*Equus kiang*) in the Arjin mountain nature reserve, China. Pakistan Journal of Zoology 45:1249–1255.

Van Wieren, S.E., and J.P. Bakker. 1998. Grazing for conservation in the twenty-first century. Pages 349–363 *in* M.F. Wallis de Vries, S.E. Van Wieren, and J.P. Bakker (Eds.), Grazing and conservation management. Springer, Berlin, Germany.

Vega-Pla, J.L., J. Calderón, P.P. Rodriíquez-Gallardo, A.M. Martinez, and C. Rico. 2006. Saving feral horse populations: Does it really matter? A case study of wild horses from Donana National Park in southern Spain. Animal Genetics 37:571–578.

Velichko A.A., E.I. Kurenkova, and P.M. Dolukhanov. 2009. Human socio-economic adaptation to environment in Late Palaeolithic, Mesolithic and Neolithic eastern Europe. Quaternary International 203:1–9.

Vos, J. 2000. Food habits and livestock depredation of two Iberian wolf packs (*Canis lupus signatus*) in the north of Portugal. Journal of Zoology London 251:457–462.

Warmuth, V., A. Eriksson, M.A. Bower, J. Canon, G. Cothran, *et al.* 2011. European domestic horses originated in two Holocene refugia. PLoS ONE 6:e18194.

Whiting, M.J., V.L. Williams, and T.J. Hibbitts. 2013. Animals traded for traditional medicine at the Faraday Market in South Africa: Species diversity and conservation implications. Pages 421–474 *in* R.R.N. Alves and I.L. Rosa (Eds.), Animals in traditional folk medicine: Implications for conservation. Springer, Berlin, Germany.

Wingard, J.R., and P. Zahler. 2006. Silent steppe: The illegal wildlife trade crisis. Mongolia Discussion Papers: East Asia and Pacific Environment and Social Development Department. World Bank, Washington, DC, USA.

Xia, C., J. Cao, H. Zhang, X. Gao, W. Yang, and D. Blank. 2014. Reintroduction of Przewalski's horse (*Equus ferus przewalskii*) in Xinjiang, China: The status and experience. Biological Conservation 177:142–147.

10

Management of Free-Roaming Horses

CASSANDRA M.V. NUÑEZ, ALBERTO SCOROLLI, LAURA LAGOS, DAVID BERMAN, AND ALBERT J. KANE

The modern horse (*Equus caballus*) evolved in North America about four million years ago, dispersing into Eurasia approximately two to three million years ago. Following this emigration, several extinctions occurred in North America, as did additional migrations to Asia and return migrations to North America (see chap. 8). The final North American extinction occurred between 13,000 and 11,000 years ago (Hunt 1992). Eurasian populations persisted, and humans began domestication ~6,000–5,500 years ago (Outram *et al.* 2009) on the western Eurasian Steppe (Warmuth *et al.* 2012) and perhaps on the Iberian Peninsula (Warmuth *et al.* 2011; Achilli *et al.* 2012). Today, European free-roaming horse populations can be grouped into three classes: (1) traditional populations, (2) true feral populations, and (3) introduced populations (for the purposes of conservation management). We consider traditional populations to be free-roaming, long-established horses that are harvested by local people (United Kingdom, Spain, Portugal). Atlantic ponies from the North Iberian Peninsula and the North Atlantic Islands make up a large proportion of these animals; some consider them remnants of wild horses living in the region since the Pleistocene (Bárcena 2012). Feral populations are domestic animals that were abandoned by local farmers and today live largely unmanaged (Romania, Russia), and introduced populations are used as components of various habitat restoration projects across Europe (Latvia, the Netherlands) (Rewilding Europe 2012). In North America, horses reintroduced in 1493 spread across the plains after escape or release, forming feral herds throughout the United States and Canada (see chap. 8). In South America, horses introduced by the Spanish and Portuguese during the sixteenth century spread north into the Pampas, Patagonia, and the mountainous Andean regions. And in Australia, horses introduced by European settlers in 1787 spread into the hills around Sydney and into the north, west, and south as pastoral settlement spread across the continent (see chap. 8). In more recent history, the extirpation of their natural predators has led to even further feral horse expansion, resulting in increased human conflict as feral horses more heavily affected livestock, industry, and native wildlife.

In this chapter, we focus on the management of free-roaming horse populations occurring in different parts of the United States, South America (in Argentina), Europe (in the United Kingdom, Portugal, Spain, the Netherlands, Romania, and Russia), and Australia (in the Northern

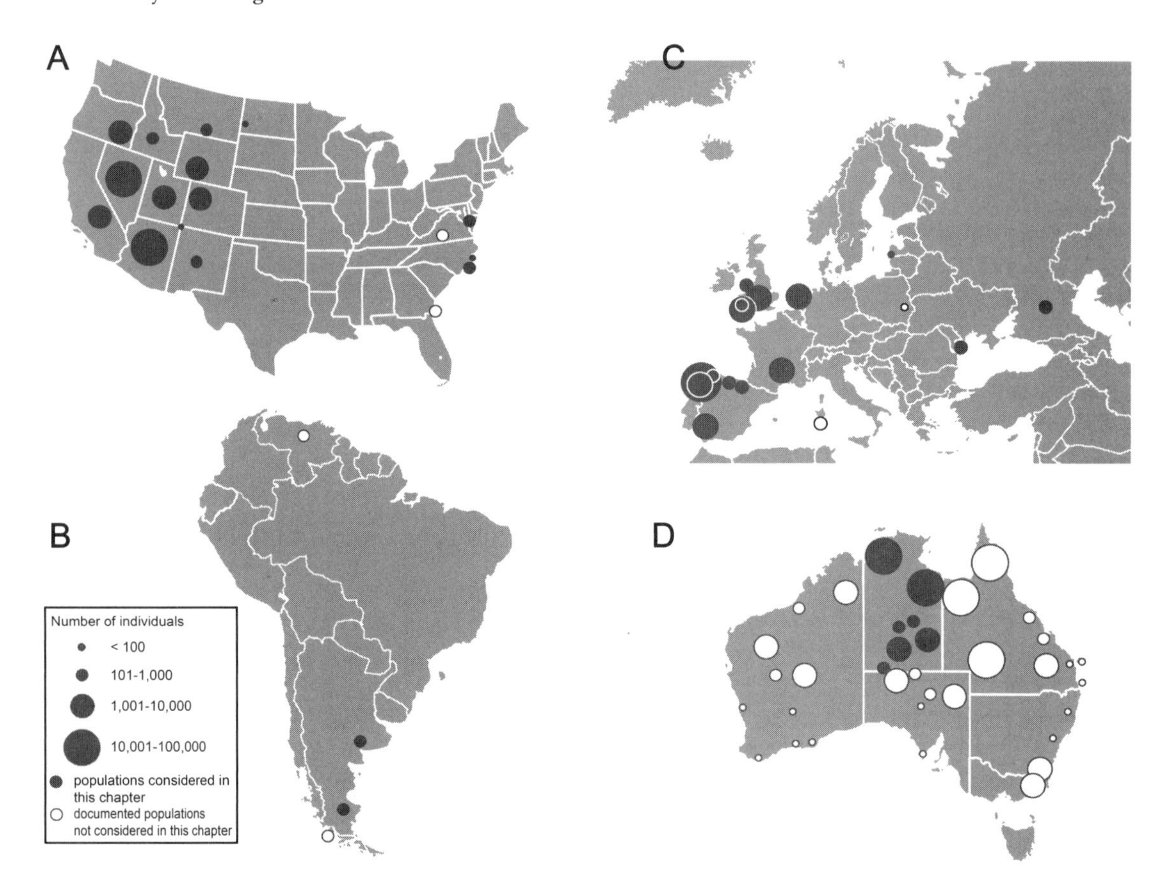

Fig. 10.1 The distribution of feral horse populations in the United States (A), South America (B), Europe (C), and Australia (D). Solid circles represent the populations discussed in this chapter; open circles represent documented populations that are not discussed. Circle size denotes the estimated number of feral horses in the different populations. *Sources*: (A) Rubenstein and Nuñez 2009, BLM 2013*b*, North Carolina Coastal Reserve and National Estuarine Research Reserve 2013, National Park Service 2013, Southern States 2013, Ransom *et al*. 2014*a*; (B) Pacheco and Herrera 1997, Scorolli and Zoratti 2007, Walton 2008, Scorolli and Lopez Cazorla 2010; (C) Iglesia 1973, Murphy and González Faraco 2002, Prieditis 2002, Dartmoor Park Authority 2006, Leblond *et al*. 2007, Tupac-Yupanqui *et al*. 2011, Matos Vieira Leite 2012, Verderers of the New Forest 2013, Winton *et al*. 2013, M. Brewer, unpublished data, M. Inza, unpublished data, L. Lorenzo, unpublished data, O. Rosu, unpublished data, Roztocze National Park, unpublished data, N. Spasskaya, unpublished data, van Dierendonck, unpublished data; (D) Dobbie *et al*. 1993

Territory) (fig. 10.1). The management strategies employed in different countries are largely dependent upon local laws, logistics, public opinion, the specific challenges horses pose in an area, and on the social and cultural importance of free-roaming horses. Significant variation in management policy exists, and depending on where they occur, horse populations are treated differently. Here we discuss human intervention with free-roaming horses in different regions as case studies of the various management options available. Several strategies are explored; however, what is feasible or acceptable often varies depending on the region, and our ability to effectively control populations with certain methods has not been sufficiently evaluated in all areas, making direct comparisons between methods difficult.

Ecological and Human Conflict

Around the world, free-roaming horses can adversely affect the local ecology. In Australia, for example, where there are no natural predators, feral horses can have particularly detrimental impacts, including the alteration of bird community composition; declines in reptile and amphibian abundance; soil loss, compaction, and erosion; vegetation trampling; reduction in plant biodiversity; tree death; damage to bog habitats and waterholes; and the expansion of invasive weeds (Nimmo and Miller 2007).

Feral horses can have similar impacts in other areas where their natural predators have been extirpated. In the western United States, for example, the effects of feral horse populations on wildlife, fish, plants, soils,

Fig. 10.2 Feral horse mares grazing in the intertidal zone of Shackleford Banks, North Carolina, USA. Photo by Cassandra M.V. Nuñez

water quality—as well as other rangeland resources such as recreation, mining, hunting, and domestic grazing—are important considerations, according to the Bureau of Land Management (BLM) (2013*b*). In the eastern United States, concern over feral horse occurrence is primarily focused on the animals' ecological impacts. On Assateague Island, off the coast of Maryland and Virginia, for example, an increase in the population from 21 horses in 1965 to 167 horses in 1975 led to concern over the animals' potential impact on the island's ecology. Subsequent studies showed that horses seemed to have little impact on some species (Rodgers 1985) but serious impacts on others (Furbish and Albano 1994). A similar population increase occurred on Shackleford Banks, a barrier island off the coast of North Carolina. Over 10 years, the horse population more than doubled, from 100 individuals in 1985 to ~220 in 1995 (Rubenstein and Nuñez 2009). A survey showed that horse-grazed marshes surrounding Shackleford Banks (fig. 10.2) had less vegetation, higher diversities of foraging birds and crabs, and a lower density and species richness of fishes than did marshes not grazed by horses.

In Argentina, high grazing pressure from feral horses can alter both taxonomic (Loydi and Zalba 2009; Loydi and Distel 2010) and functional composition of grassland plant communities (de Villalobos and Zalba 2010), facilitate invasion by alien species (Loydi *et al.* 2010; de Villalobos *et al.* 2011), and diminish reproductive success in grassland bird communities (Zalba and Cozzani 2004).

In Europe, the concern over feral horses' ecological impacts varies according to the population classes. For example, true feral horses are thought to be invasive, and so there is concern about their environmental impact. In the Biosphere Reserve in the Letea Forest, Romania, feral horses limit forest regeneration (Administraţi Rezervaţiei Biosferei Delta Dunări 2012), and on Vodnyi Island, Russia, increases in the horse population (to 419 animals in 2007) have resulted in damage to the steppe community and vegetation degradation (Prishutova 2010; Minoransky and Uzdenov 2011). Damage to crops and pastures (Russell 1976; Bárcena and Hermida 2003; Bento Gonçalves *et al.* 2011), horse–vehicle collisions (Bento Gonçalves *et al.* 2011; Valero *et al.* 2013; Verderers of the New Forest 2013), decreases to forest regeneration, and the subsequent competition with the forestry industry (Magdalena and Vidal 1984; Bárcena and Hermida 2003) have also resulted in considerable conflict over the occurrence of traditional horse populations in the United Kingdom, Spain, and Portugal.

Despite conflicts with their environment and with humans, free-roaming horses are often considered to be an integral component of an area's heritage, culture, and natural history; regardless of where they occur, they often symbolize freedom, wildness, and open space. In the United States, for example, Congress mandated that feral horses living on public lands be legally designated as "wild" horses that are "living symbols of the historic and pioneer spirit of the West" (US Congress 1971). As such, it is illegal to hunt, harass, or kill feral horses in the United States, grossly affecting management options. In Argentina, there is considerable debate about how to best manage feral horse populations. In rural areas, horses are sometimes shot to reduce competition with livestock; these methods are strongly opposed by those who consider horses to be pets or cultural icons (C. Zoratti, personal communication). Similarly, some advocates in Australia do not support the culling of feral horses; in New South Wales, for example, the government imposed a moratorium on aerial culling after photographs of a cull sparked public outcry. In European countries, free-roaming horses are valued for their potential roles in wolf conservation (Galicia, Spain: Lagos 2013; López-Bao *et al.* 2013), fire risk reduction (Spain: Celaya *et al.* 2012; Rigueiro-Rodríguez *et al.* 2012), and the preservation of steppe ecosystems (Vodnyi Island, Russia: Spasskaya 2008, 2009). In addition, horses can have cultural significance (Portugal and Wales: Santos and Ferreira 2012; Welsh Government 2012) and touristic value (United Kingdom: Baker 1993; Verderers of the New Forest 2013). In some areas, free-roaming horses are an important component of local tradition; the horse captures (or roundups) conducted in Galicia and in Doñana National Park, Spain, for example, are honored local events that can attract visitors from around the world (Iglesia 1973; Murphy and González Faraco 2002). Regardless of where they occur, it is likely that some will view feral horses as an important part of an area's history and ecology while others will view them as destructive pests. Often the challenge of any management scheme is to strike a balance between these two opposing viewpoints.

Horse Management

Free-roaming horse management can take many forms depending upon its ultimate objective. In many areas, the goal of management is to control populations; in others, it is to reduce conflict with human activity; and still in others, it is to maintain or preserve populations. In some situations, the distinctions between these aims can be subtle and may not be mutually exclusive; conversely, these aims can directly conflict with one another. Such differences in the primary goals of managers can result in different management practices.

Management for Population Control

The control of feral horse populations can be achieved in a variety of ways, including lethal culling, contraception, and the live capture (or collection) and nonlethal removal of individuals via their sale, adoption, or maintenance in holding facilities. The feasibility and efficacy of different methods vary with the areas in which feral horses occur, both within and between continents.

Lethal Culling. Lethal culling involves the killing of individual animals; individuals can be removed and subsequently killed off site or killed on site from the ground or air. Depending upon the objective, animals may be culled indiscriminately, or they may be taken systematically according to age, sex, or reproductive state. In Argentina, the presence of feral horses in natural protected areas is of particular concern. The management of feral horses in these areas has been sporadic and occasional, however, consisting primarily of culling populations occurring in national parks (C. Zoratti, personal communication). An official governmental position toward invasive species management was taken in 1995: feral horses needed to be managed in a systematic manner, but political will and financial support were lacking. It was not until 2006 that the National Parks Administration conducted a diagnosis of the "feral horse problem" in the Natural Monument Petrified Woods in Santa Cruz province. To date, no further systemic management actions have been taken. In practice, the organized management of feral horses has been conducted at only one site, Tornquist Park, an area under Buenos Aires provincial jurisdiction, and therefore not related to the National Parks Administration. In 2006 and 2007, 220 feral horses (50% of the population) were live-trapped with mobile corrals (A. Scorolli, unpublished data) (fig. 10.3). The horses were transported to the Equine Breeding Division of the Argentine Army facilities for quarantine and were examined for parasites and contagious diseases. More than 80 horses were sent to slaughter by army equine veterinary decision. A Buenos Aires Province governor's decree declared an eradication objective, but no written strategy document or official scientific advice was provided. No efforts were made to monitor the efficacy of the removal in terms of long-term reduction in horse numbers or impacts to the park. In 2008, governmental authorities changed, and no further management was authorized.

In the Danube Delta, Romania, and Vodnyi Island, Russia, horse populations grew largely unchecked for years, but they were controlled to some extent by local

Figure 10.3 Feral horses being collected in a mobile corral in Tornquist Park, Argentina. Photo by María Fernanda Menvielle

villagers with occasional culls (Minoransky and Uzdenov 2011; O. Rosu, unpublished data). Until recently (2011), increases in horse numbers in the Danube prompted authorities to conduct extensive roundups to capture and slaughter horses (O. Rosu, unpublished data). On Vodnyi Island, culls have been used to ensure that horse populations are beneficial to the local ecology. Geobotanical studies showed that populations of up to 120 individuals can help maintain steppe ecosystems (Prishutova 2010). To maintain populations at this level, the government instituted an adaptive management strategy: each year a variable number of horses are to be removed, with the precise number determined on the basis of regular monitoring of both the population and ecosystem. In a removal conducted in 2012, for example, an equal number of young males and females were removed and used as bush meat (N. Spasskaya, unpublished data).

In Australia's Northern Territory, aerial culling from helicopter has proven to be an effective management strategy. In the early 1980s, aerial surveys indicated that the feral horse population had grown rapidly. Feral horses were damaging the environment and reducing the productivity of the cattle industry (Dobbie *et al.* 1993), necessitating a reduction in feral horse numbers. But government officers were cognizant of the potential public opposition if a major culling program were commenced prematurely. The concern was that Australia would follow the example set by the United States, where the Wild Free-Roaming Horses and Burros Act of 1971 placed constraints upon the available control options (US Congress 1971). In response, the Northern Territory government commenced the first major feral horse research program in Australia to help seek a solution to the problem (Bowman 1985; Berman 1991). The program set out to determine the Northern Territory feral horse distribution and abundance, economic and environmental impact, and to improve the understanding of feral horse ecology and methods for management (Bowman 1985; Berman 1991; Dobbie *et al.* 1993; Bryan 2001).

A fixed-wing aerial survey provided an estimate of 82,000 feral horses in the Alice Springs pastoral district (388,000 km^2) (Graham *et al.* 1986). A large proportion of the population (52,000) lived in the mountain ranges, where the rugged terrain and the abundance of natural watering points make mustering or trapping difficult (Berman 1991). A concerted, four-year control effort, combined with drought-induced mortality, resulted in a 77% reduction in the feral horse population living in the mountain ranges of Alice Springs (Low and Hewett 1990). Control activities continued, and in other areas, feral horses were completely removed. At Finke Gorge National Park, for example, a combination of live capture and aerial culling removed all feral horses. Indigenous and nonindigenous landholders, animal welfare groups, conservation groups, research scientists, and government officers were all actively involved in the program (Bryan 2001). The majority of the horses were removed within seven years. The park has now been virtually horse-free for over 20 years, and native wildlife have responded positively. Rock wallabies, for example, have expanded their distribution within the park (Bryan 2001). Feral horses were also almost completely removed from West MacDonnell Ranges National Park and from the majority of pastoral properties held by nonindigenous landholders. Benefits to native fauna and flora and reduced soil erosion were reported, as was a reduction in the number of visitor complaints about environmental damage in the National Parks (Bryan 2001).

Though lethal culling can be effective in reducing horse numbers, its use remains distasteful to much of the public. Successful management of feral horses, particularly when lethal methods are required, may be better achieved by following the steps listed below, which are based on those from Dobbie *et al.* (1993).

1. Determine distribution, abundance, rate of increase, and movement patterns.
2. Define the problem. What damage are the horses causing? What risks do they pose?
3. Identify suitable management techniques.
4. Engage interest groups to decide on management actions.
5. Implement agreed actions.
6. Assess the value of management by determining change in distribution and abundance and reduction in damage or risk.
7. Repeat steps 3-6 until an acceptable level of damage or risk is achieved.
8. Maintain this acceptable level of damage or risk indefinitely with minimal effort or cost.

The first two steps are crucial, though often bypassed. These steps require an adequate amount of scientific investigation. A good understanding of local ecology is essential for quantifying feral horse impact and improving methods for management. These steps must be followed if aerial culling is to be used, and engaging interest groups is vitally important. Aerial culling should only be conducted upon approval from these groups.

Where horses cannot be caught, or where it is impossible to treat a sufficient number with fertility control agents, the only current alternative for reducing their numbers in Australia is shooting from a helicopter. While shooting from helicopter may be relatively humane (Hampton 2013) and can rapidly reduce feral horse density, it is expensive, wasteful, distasteful to much of the public, and by itself may not provide a long-term solution. Managers tend to use shooting from helicopter as a "quick-fix" option, and because it alleviates a short-term problem, there appears to be no need to plan for long-term management to avoid the need for future aerial culling. By applying the steps listed above and using a combination of the most appropriate control options, management results could be longer lasting.

Capture and Use as Food. Traditional populations of Atlantic ponies in the Iberian Peninsula and the British Islands are considered free-roaming native horses (or *garranos*) (Bárcena 2012) that are privately owned and traditionally harvested by their owners. These harvests serve to control horse numbers, though population control is not the primary objective. Horses are collected annually to remove and sell most of the foals in Galicia (Iglesia 1973), Portugal (Portas *et al.* 1998), the New Forest (Tyler 1972; Ivey 2009; Verderers of the New Forest 2013), Dartmoor (Dartmoor Commoners' Council 2013), Exmoor (Baker 1993), and Wales (Stanley and Shultz 2012; Winton *et al.* 2013).

In Galicia, these horse harvests, called *curro* or *rapa das bestas* (fig 10.4) are an important local event in which traditions like mane and tail shearing are continued. Historically, the animals' manes and tails were used for making rope (Iglesia 1973); today, the shearing is custom. In Galicia and Portugal, the foals collected during these harvests are typically sold for meat and thus provide an incentive for the owners to maintain these populations (J. Calderón, personal communication). Individuals removed from the population were once used for transportation, haulage, or work in the farms and mines (Iglesia 1973; Baker 1993; Bárcena and Hermida 2003; Dartmoor National Park Authority 2006; Ivey 2009; Peckover and the Carneddau Pony Society 2009). These uses have all but disappeared, leading to the decline of these populations over the last 50 years. Any maintenance of these animals

Fig. 10.4 Inside the *curro*. Garrano foals are separated before adults are captured one at a time to be sheared. The method of capture varies from using a stick with a rope tied in a special knot to the manual restraint of horses on the ground. Photo by Laura Lagos

now serves only to conserve the breed and maintain tradition.

Capture and Nonlethal Removal. Capture and nonlethal removal has been used around the world to control feral horse populations with varying degrees of success. In the United States, the mandate by Congress to protect and also manage feral horse populations such that "a thriving natural ecological balance and multiple-use relationship" is maintained into perpetuity (US Congress 1971) likely limits the ability of the BLM and Forest Service (FS) to most effectively control western feral horse populations. The BLM and FS have achieved population limits primarily through the collection and removal of feral horses from the wild. Several thousand individuals (usually ranging from ~7,000 to 11,000) are gathered each year and are either adopted by private citizens, held in adoption centers, or placed into long-term holding pastures where they may be kept for the remainder of their natural lives. From 1996 to 2006, an average of ~8,500 horses per year were removed from the wild (BLM 2013*a*); the annual percentage of these animals that were adopted decreased dramatically (from 92% to 50%) during that time period. Given the high costs of owning a horse and changing economics, many more animals continue to be gathered than adopted, and the number of animals in holding has become unsustainable (National Research Council 2013). In 2013, the BLM spent $46.2 million—64% of the total budget for its Wild Horse and Burro Program—solely to maintain the animals currently in holding (BLM 2013*b*). These costs included not only the feeding and watering of these animals, but also the maintenance of their health (box 10.1). Moreover, according to BLM, horse populations continue to increase by ~20% each year (BLM 2013*b*). Clearly the capture, removal, and subsequent maintenance of feral horses in the western United States does not constitute a sufficient strategy on its own.

Capture and nonlethal removal has also been used on Vodnyi Island to maintain population levels that benefit the local ecology (see "Lethal Culling" above). In 2013, several foals and yearlings were captured and transferred to either stud farms or farmers to control population growth (N. Spasskaya, unpublished data). The efficacy of this removal regarding population control has not been established.

Contraception. Lethal culling and capture/removal methods are not accepted by much of the public, particularly in the United States, where it is illegal to hunt feral horses or to slaughter them. Such opposition has led to the development of contraceptive agents with an emphasis on developing vaccines that are effective, safe, reversible, and that do not result in significant behavioral effects (Kirkpatrick and Turner 1991). Research has largely focused on two methods of action when attempting to control fertility in females: suppression of gonadotropin secretion (gonadotropin-releasing hormone, or GnRH, vaccination) (Dalin *et al.* 2002) and stimulation of the immune system to block egg fertilization (porcine zona pellucida, or PZP, vaccination)

BOX 10.1 MANAGING THE HEALTH OF FERAL HORSES AND BURROS

Health and welfare are closely intertwined in feral horses and burros. By their nature, feral horses and burros are robust and hardy animals, but they are still susceptible to many of the same diseases and conditions as domestic horses. Feral horses and burros living on the range are not typically affected by the ailments most commonly associated with intensively managed and housed domestic horses (e.g., obesity, early-onset osteoarthritis, hoof abscesses, laminitis, colic), but injuries and disease do occur. Apart from the effects of starvation when forage or water resources are inadequate, poorly healed old injuries, and manifestations of developmental orthopedic disease seem to be the most common serious conditions seen on the range.

One of the myths about feral horses and burros is that they are somehow inherently more resistant to disease than their domestic counterparts. While it is true that the effective contact rate for disease spread between individuals may be low when they live in small, widely dispersed bands, infectious diseases do occur in free-roaming animals as well as those removed into captivity. Internal parasites (ascarids, *Oxyuris equi*, and large strongyles) are common in horses fresh off the range. Parasite burdens can be quite heavy in adolescents and adults, such as heavy ascarid burdens seen in recently removed feral horses of all ages, including adults that one might expect to have developed immunity.

Without question, the most common and challenging problem facing feral horses once they are removed from the range and held in corrals is infectious upper respiratory disease. Viral agents (equine herpes and influenza viruses) are problems, but upper respiratory infections caused by *Streptococcus equi* and *S. zooepidemicus* are most common. Exposure levels and disease spread seem low in the wild, but the real problems occur when animals are removed from the wild and congregated in captivity. Most of the animals are immunologically naïve to these infectious agents when they are captured. Combined with the stress of adapting to captivity, they become particularly susceptible to infection. Complete isolation is difficult in large dry-lot facilities, and exposure to the agents that typically cause infectious upper respiratory disease in the domestic world seems inevitable. While vaccines for these agents are commonly used, they generally are not highly efficacious in this situation. The best approach to controlling disease outbreaks seems to be to isolate new arrivals as much as possible when first removed from the wild, to use a broad vaccination protocol as soon as possible following capture, and to allow many of the respiratory conditions to "run their course" once they start to occur. Unfortunately, to some extent, such outbreaks seem inevitable given the ecology of the disease agents involved and the susceptibility of the large numbers of animals that must be handled during the transition from free-roaming to domestic life. Fortunately, while morbidity may be high, these conditions are usually self-limiting without treatment, and serious complications and mortality due to pneumonia or internal abscesses are rare. Treatment of individual animals housed in a group setting is not without added risk and stress associated with handling, so this type of treatment is usually reserved for those individuals who appear to be suffering from complications of the upper respiratory infections.

Following the transition to domestic life through adoption or sale to private ownership, the health problems faced by feral horses and burros are largely the same as those of domestic horses and burros. Some of the desirable hoof shape characteristics of the feral horse change when removed from the range into a captive environment, but mustangs used as saddle horses seem to be as athletic as any similarly built domestic horse. They also tend to be long lived and remain sound for many years when they make the transition to domestic life.

Photo by V. V. Vita

(Sacco 1977). GnRH vaccines are generally effective at achieving infertility in domestic mares; however, several applications can be necessary to induce ovarian inactivity (Dalin *et al.* 2002; Stout and Colenbrander 2004), and revaccination is necessary to maintain infertility (Killian *et al.* 2008). In the year following treatment, GnRH contraception only minimally affected time budget behaviors and did not significantly affect the reproductive behavior or physical condition of treated females in a population of feral horses living in Theodore Roosevelt National Park in North Dakota, USA (Ransom *et al.* 2014*a*). Similarly, PZP has been shown to be safe, effective, and theoretically has minimal side effects (Turner *et al.* 2001, 2002). PZP use is not practical in all populations (Killian *et al.* 2008), however; annual revaccination is necessary to ensure infertility (Turner *et al.* 2001), the time required to reach desired population levels can be substantial (Conservation Breeding Specialist Group 2006), it can indirectly affect animal behavior and physiology (Nuñez *et al.* 2009, 2010; Ransom *et al.* 2010), and its population-level efficacy may be reduced owing to local ecological feedback mechanisms (Ransom *et al.* 2014*b*).

The newly developed GonaCon and SpayVac vaccines prevent pregnancy by targeting GnRH and the zona pellucida, respectively, but single applications are thought to be effective over multiple years. Killian *et al.* (2008) showed that rates of conception for control mares were significantly higher than for SpayVac- and GonaCon-treated mares. In addition, SpayVac was more effective at achieving infertility than GonaCon in all years subsequent to treatment. Average rates of contraception over four years of study were 87.25% (range = 83–100%) and 63.50% (range = 40–93.75%) for SpayVac and GonaCon, respectively. Gray *et al.* (2011) did not find such differences between SpayVac and GonaCon-B in a population of feral horses occurring on the Virginia Range, southwest of Reno, Nevada, USA. In addition, though treated mares were less likely to conceive than control mares, contraception rates for animals treated with SpayVac were lower than in Killian *et al.*'s (2008) study, averaging 59.66% (range = 50–63%) over three years; contraception rates were similar with GonaCon-B, averaging 62.66% (range = 58–69%). Contraceptive treatment did not significantly affect foal sex ratio, foal survival, or birthing season (Gray *et al.* 2011).

The BLM has administered both liquid and pelleted PZP to various horse herds in Montana, Wyoming, and Colorado. The ability to sufficiently control population growth with PZP has not been studied systematically in open populations (Ransom *et al.* 2014*b*). In three closed populations where it was possible to scientifically observe population-level effects, however, apparent annual population growth rate was only 4–9% lower in posttreatment years when compared to pretreatment years. Moreover, population growth was positive and still increased steadily every year that a management removal was not conducted (Ransom 2012). Nevertheless, managers potentially saved the US government ~\$1.4 million during the study by combining horse removals with PZP contraception (Ransom 2012).

In the eastern United States, the National Park Service (NPS), the Foundation for Shackleford Horses, and the Corolla Wild Horse Fund currently use PZP agents to manage four feral horse herds occurring on barrier islands. These populations include the horses living on Assateague Island, off the coast of Maryland, and those occurring on Shackleford Banks, Carrot Island, and the Currituck Outer Banks, off the coast of North Carolina. Population numbers have largely been controlled to desired levels at these sites (Kirkpatrick 1995; S. Stuska, unpublished data). Despite PZP's potential drawbacks (Nuñez *et al.* 2009, 2010; Madosky *et al.* 2010; Ransom *et al.* 2010, 2013) and the uncertainty regarding its efficacy in large, open populations (Ransom *et al.* 2014*b*), successes in eastern populations, the potential fiscal implications, and the promising results with SpayVac (Killian *et al.* 2008) have made zona pellucida–directed contraception of feral mares an attractive option for the management of feral horses in the western United States.

Management for the Reduction of Human–Horse Conflict. In several European countries, the high occurrence of horse–vehicle collisions has influenced the management of free-roaming horse populations. For example, Galicia registered an average of 75 collisions per year between 2006 and 2010 (Valero *et al.* 2013). Similarly, in the New Forest, 69 and 64 accidents were registered in 2011 and 2012, respectively (Verderers of the New Forest 2013). In Galicia, increased fencing and regulations ensuring compensation to drivers (by horse "owners") are the only measures in place to mitigate this issue. These policies fail to acknowledge the importance of combined approaches, including increased public awareness and drivers' education, to the reduction of wildlife-related traffic accidents (Langbein *et al.* 2011). Conversely, managers in New Forest are taking several steps to reduce horse-related accidents (New Forest National Park Authority 2013; Verderers of the New Forest 2013). They encourage the police to enforce a speed limit of 40 miles per hour for all unfenced roads and collaborate with the administration of the county highway to develop more effective signage. In addition, there is an effort to increase pony visibility; roadside vegetation is trimmed, and high-risk horses that commonly graze on the road side are fitted with reflective collars.

Management to Maintain and Preserve Populations. Despite the need to control feral horse numbers in some areas, those same animals are sometimes fed to supplement their diet in poor environmental conditions. Feral horse populations on Vodnyi Island have been fed during particularly severe winters (N. Spasskaya, unpublished data), against scientific recommendations (Minoransky and Uzdenov 2011). Similarly, Four Paws, an international animal welfare organization, has been providing winter supplements to the Danube Delta population since 2010 (O. Rosu, personal communication). Supplementing animals in this way can lead to population increases, further necessitating population control measures; however, the public's notion that horses are a domestic species and therefore need to be cared for (Slaukstins 2002; Vera 2009) can result in this type of conflicting management practice.

Horses as Conservation Management Tools

Herbivore grazing can be used as a conservation tool to decrease habitat management intensity; it is inexpensive, environmentally friendly, and produces mosaic-like patches, creating more natural landscapes (Kampf 2002). Throughout Europe, several programs have introduced horses to increase biodiversity (Murray 2008; Rewilding Europe 2012) and to maintain (Vulink *et al.* 2010), improve (Forder 2006), or preserve various habitats (Cosyns *et al.* 2001), thereby making them more suitable for other species. Perhaps the most well-known introduction is at the Oostvaardersplassen, a 56-km^2 fenced area in the Netherlands. Koniks, a hearty breed of horses originating in Poland, were introduced to the area in 1984 (Heck cattle had been introduced in 1983) to help maintain greylag goose (*Anser anser*) and barnacle goose (*Branta leucopsis*) habitat. Grazing by horses and cattle shortens the grasslands and slows vegetation succession, making the area suitable for the geese and a number of other bird species. In addition, the horses may attract tourism, resulting in social and economic benefits for local communities, landowners, and stakeholders (Rewilding Europe 2012). A similar program has been instituted in Latvia at Lake Pape (Slaukstins 2002; Schwartz 2005), as have several in England by organizations such as the Royal Society and Protection of Birds, the National Trust, the Forestry Commission, and the Wildlife Trusts (Murray 2008).

Their history of domestication makes it difficult for some members of the public to truly consider these horses "wild" animals. Several groups believe that if horses are to be introduced to an area, they ought to be provided with food, water, and shelter (Slaukstins 2002; Vera 2009). Such public concern has largely driven the management policies regarding these populations. For example, initially, a minimal intervention policy, including the culling of dying animals to simulate predation (van Dierendonck 2012), was adopted in the Oostvaardersplassen. Later, the International Committee on Management of the Oostvaardersplassen (ICMO) adopted new management policies to satisfy the public's animal welfare concerns. Research was to be conducted on the population, the population was to be more closely monitored, animals were provided shelter access, and those unlikely to survive the winter were culled to prevent suffering (ICMO 2006).

In 2010, the ICMO reevaluated the culling policy to more accurately predict the likelihood of animal survival, creating the Early Reactive Culling Protocol (ICMO 2010). This protocol considers individual behavior, individual condition score, population density, food availability, time of year, and environmental conditions to make more informed culling decisions; animals unlikely to survive are thus culled in an earlier stage of decline to minimize suffering (ICMO 2010; van Dierendonck 2012).

More recent horse introduction projects, such as those instituted in the United Kingdom and Spain, have yet to encounter such issues. In other areas where endangered breeds have been introduced, increases in population numbers are integral to the reestablishment of the breed (Rewilding Europe 2012). In areas instituting rewilding projects, more natural, density-dependent processes, including predation and intraspecific competition, are expected to control populations. Regardless, when semidomestic animals are released into the wild, population control may become necessary (Foundation for Restoring European Ecosystems 2013), especially in areas that do not contain predators.

Current Developments

Today, the management of free-roaming horses continues to be a "work in progress." In 2011, for example, the BLM approached the National Academy of Sciences to request an independent technical evaluation of the science, methodology, and technical decision making of their Wild Horse and Burro Program. The panel recommended the reconciliation of management practices with feral equid population ecology;

the use of more comprehensive population models to forecast program costs and evaluate management alternatives for animals on the range and in holding facilities; and suggested that the most promising fertility control methods for feral horses and burros are chemical vasectomy for males and the GonaCon and PZP vaccines for females (National Research Council 2013). But ~95% of horse herds managed by the BLM consist of open populations, where immigration and emigration cannot be controlled. In such populations, the identification of individual animals for monitoring or treatment is not possible with the technologies currently available to the BLM (National Research Council 2013). These limitations make the panel's recommendations exceedingly difficult to carry out. In addition, contraceptive agents are unlikely to effectively control such large, open populations (Ransom *et al.* 2014*b*). Yet the fact remains that current management practices are unsustainable and may actually promote higher population growth rates in some circumstances. The BLM removals typically keep feral horse numbers below levels that would be affected by food limitation and density dependence. Adverse effects to pregnancy rates, female fecundity, and survival rates associated with food competition and density dependence are therefore prevented, potentially resulting in increased population growth (National Research Council 2013). Legislation that would put the task of managing feral horse populations in the hands of states rather than the federal government has been considered by the US Congress (US Congress 2014). It is difficult to predict how such a policy shift might affect feral horse management in the United States; however, it will likely make an already piecemeal policy even less cohesive, and as with most policy regarding feral horses, it is likely to result in considerable controversy.

In Argentina, collaborations between governmental agencies and university researchers are developing more viable solutions for their feral horse populations (A. Scorolli, unpublished data). A long-term research project (1995 to present) has been investigating the demography and social organization of the feral horse population in Tornquist Park. A model has been developed in which the proportion of adult females with different body condition scores (collected at winter's end, when animals are typically in the poorest condition) is used to predict the population's likelihood of reaching carrying capacity, the potential population growth, and foaling rates (Scorolli 2012). Outcomes of the model may help inform management strategies, as the aforementioned factors can be used to make assumptions about the horses' environmental impact. Moreover, both the time and funding necessary to collect the relevant data are considerably less than that typically needed to collect data for traditional demographic models. This new model has been partially validated with data from 2011 (Scorolli 2012) and is currently being refined. In addition to these developments, the problem of invasive alien species (IAS), including feral domestics, was considered at the Argentine National Expert Workshop on Alien Fauna in 2011 and 2012. A national strategy on the IAS project is currently under way, promising further progress toward achieving viable solutions for the control of feral horse populations in Argentina.

In different areas of Spain, regional governments seek to mitigate human–horse conflicts with traditional populations through the tracking of horse "owners." This tracking ensures that those affected by the horses can seek damage compensation. Spanish regional governments are applying existing European domestic horse regulations, which stipulate that all equids have an identification document (or passport) and that they be implanted with a microchip linked to the document and to a database containing details about the animals' owners (European Commission 2008). This measure is incompatible with the existence of feral horse populations and puts traditional populations and the customary methods used to manage them at risk of extinction (Lagos 2013; López-Bao *et al.* 2013). Article 7 of the regulation, an exemption to microchip measure, takes free-roaming equids into account, stating that "the competent authority of the EU countries may decide that equidae constituting defined populations living under wild or semi-wild conditions in certain areas, including nature reserves, to be defined by that authority, shall be identified only when they are removed from such areas or brought into domestic use." This exemption has been applied in Wales for their free-roaming horse populations in Carneddau (Welsh Government 2012) and in England, in accordance with the Horse Passports Regulation of 2009 (UK Secretary of State 2009), for the Exmoor, Dartmoor, and New Forest populations, but it has not been applied in Spain or Portugal. Fertility control methods are currently being considered in Europe as well. Four Paws has developed an action plan to manage the horses in the Danube Delta, which includes the contraception of females to help control population numbers (Rosu 2013).

In Australia, aerial culling was considered by many to be the most humane method for reducing horse numbers by the end of the 1980s. The animals were often found in remote and rugged areas, and provided that highly trained, experienced officers conducted the culling, adhering to strict operational guidelines (Bryan 2001), this method was (and continues to be) highly supported. Aerial culling has continued in central Australia, with 16,254 feral horses being shot from helicopter between 2002 and 2013 (G. Edwards, personal communication). Little management of feral horses has occurred on land held by indigenous owners in central Australia, and large populations of feral horses remain in these areas.

Despite the general acceptance of culling, conflict can and does arise in response to lethal management methods. Community conflict most often arises when management actions occur without supporting local research and public consultation. In October 2000, for example, a culling operation was conducted in Guy Fawkes River National Park after a bushfire destroyed the primary food source for a large population of feral horses. Six hundred feral horses were shot from helicopter, resulting in significant community opposition, and as a result the New South Wales minister for the environment commissioned an independent review to examine the Guy Fawkes operation. While the operation was found to be well planned, "it would have been prudent for the Service to have sought the involvement and cooperation of the RSPCA in planning and carrying out the operation, with emphasis on both the welfare of the horses and the significant ill-effects of the large horse herd on native flora and fauna if nothing was done, and that local land owners should also have been involved in some way from the outset, to ensure that they knew what was happening and why" (English 2000).

Studies required to improve feral horse management at Guy Fawkes River National Park have been conducted since the culling operation (Schott 2005; Dawson *et al.* 2006; Vernes *et al.* 2009; Lenehan 2011), and a successful capture and rehoming program has been established. The Guy Fawkes cull in 2000 provided the impetus for more thorough planning of feral horse population management in many parts of Australia. Preparation of formal management plans involving community interest groups may help overcome the problems experienced at Guy Fawkes and help ensure that all appropriate preparations are conducted before action is taken. While these developments are encouraging, long-term solutions have yet to be achieved.

Conclusion

There are ~1.5–2 million free-roaming horses around the world. The extinction or extirpation of their natural predators in nearly all of the regions where feral horses occur necessitate their management. From the data presented here, it is clear that there is no single best management strategy. In the western United States, taxpayers spend tens of millions of dollars each year to maintain the nearly 50,000 feral horses in holding; in the eastern United States, horses are controlled primarily with contraception, potentially altering their behavior and reproductive physiology. In South American countries, there is little to no organized management of feral horses, save for those occurring in Argentina, where new collaborations seem promising. In European countries, the situation is different, as free-roaming horses are sometimes considered a native species; nevertheless, people have different attitudes regarding traditional, feral, and introduced horse populations, ranging from cultural icon and essential keystone species to problem animal; these diverse attitudes have resulted in inconsistent management regimens across nations. Finally, in Australia, feral horses adversely affect the native ecosystems for which they are not adapted. Although scientists have developed seemingly humane ways to reduce populations, it is clear from the information presented here that lethal culls are not a panacea and need to be conducted in conjunction with other measures to ensure consistent population-level declines. Overall, managers have limited tools with which to control feral horse populations; politics and public opinion limit these options even further. The history of horse domestication complicates the issue, blurring the line between wild and domestic animal.

The management of free-roaming horses is and will continue to be a difficult task. With every decision, managers and policymakers need to consider the efficacy, feasibility, costs, and benefits of different strategies; the potential effects to horse health and welfare; and the impacts of public opinion on management implementation. Without natural predators to control feral horse populations, almost any method we choose will likely necessitate ongoing human intervention. A balance between the costs and benefits, to the surrounding ecology and both feral horse and human populations, is difficult to strike, but with sufficient study of horse behavior and demographics, and with clear, precise definitions of the problems that horses pose to a region, we can increase the likelihood of effective and ethical management in diverse geographic regions.

ACKNOWLEDGMENTS

The authors sincerely thank all those who generously contributed both their time and knowledge to this effort—there are too many to name here. Their help undoubtedly made this chapter a more complete and accurate review of feral horse management around the world, both past and present. The authors also want to thank Jason Ransom and Petra Kaczensky, whose initiative, hard work, and tenacity made the meeting in Vienna a success and this book possible. Thanks are also owed to Jason and Petra, in addition to two anonymous reviewers, for their thoughtful consideration of an earlier version of this chapter.

REFERENCES

Achilli, A., A. Olivieri, P. Soares, H. Lancioni, B.H. Kashani, U.A. Perego, S.G. Nergadze, V. Carossa, M. Santagostino, and S. Capomaccio. 2012. Mitochondrial genomes from modern horses reveal the major haplogroups that underwent domestication. Proceedings of the National Academy of Sciences of the USA 109:2449–2454.

Administraţi Rezervaţiei Biosferei Delta Dunări. 2012. The situation of abandoned horses in Letea Forest Area of the Danube Delta biosphere reserve. Available at www.self-willed-land.org.uk/articles/Horses_from_Letea.pdf.

Baker, S. 1993. Survival of the fittest: A natural history of the Exmoor pony. Exmoor Books, Exeter, UK.

Bárcena, F. 2012. Garranos: Os póneis selvagens (*Equus ferus sp.*) do norte da Península Ibérica. Pages 75–96 *in* Livro de Atas: I Congresso Internacional do Garrano. Arcos de Valdevez, Portugal.

Bárcena, F., and R. Hermida. 2003. Mamíferos de Galicia. Pages 432–579 *in* R. Otero Pedrayo (Ed.), Enciclopedia Galega. Editorial Novos Vieiros, A Coruña, Spain.

Bento Gonçalves, A., A. Vieira, F. Ferreira Leite, L. da Vinha, and P. Alexandra Malta. 2011. Evaluation and implications of free-ranging garrano horses in the risk of forest fires: The study case of Vieira do Minho Municipality (Portugal). Pages 53–77 *in* D.A. Boehm (Ed.), Forestry: Research, ecology and policies. Nova Science, New York, USA.

Berman, D. 1991. The ecology of feral horses in central Australia. Thesis, University of New England, Armidale, NSW, Australia.

BLM. Bureau of Land Management. 2013*a*. Wild horse and burro statistics. Washington, DC, USA. Available at www.wildhorseandburro.blm.gov/statistics/index.htm.

BLM. Bureau of Land Management. 2013*b*. Wild horse and burro quick facts. Washington, DC, USA. Available at www.blm.gov/wo/st/en/prog/whbprogram/history_and_facts/quick_facts.html.

Bowman, A. 1985. Analysis of aerial survey data from Alice Springs district and implication for feral horse management. B Natural Resources Project, University of New England, Armidale, NSW, Australia.

Bryan, R. 2001. Humane control of feral horses in central Australia. Paper presented at the 12th Australasian Vertebrate Pest Conference, Melbourne, VIC, Australia.

Celaya, R., L.M.M. Ferreira, U. García, R. Rosa García, and K. Osoro. 2012. Heavy grazing by horses on heathlands of different botanical composition. Pages 219–226 *in* M. Saastamoinen, M.J. Fradinho, A.S. Santos, and N. Miraglia (Eds.), Forages and grazing in horse nutrition. EAAP Publication No. 132. Wageningen Academic, Wageningen, Netherlands.

Conservation Breeding Specialist Group. 2006. Horses of Assateague Island population and habitat viability assessment workshop, Berlin, MD. National Park Service, Washington, DC, USA.

Cosyns, E., T. Degezelle, E. Demeulenaere, and M. Hoffmann. 2001. Feeding ecology of Konik horses and donkeys in Belgian coastal dunes and its implications for nature management. Belgian Journal of Zoology 131:111–118.

Dalin, A.M., Ø. Andresen, and L. Malmgren. 2002. Immunization against GnRH in mature mares: Antibody titres, ovarian function, hormonal levels and oestrous behaviour. Journal of Veterinary Medicine Series A 49:125–131.

Dartmoor Commoners' Council. 2013. Pony drifts. Dartmoor Commoners' Council, Tavistock, Devon, UK. Available at www.dartmoorcommonerscouncil.org.uk/.

Dartmoor National Park Authority. 2006. The Dartmoor ponies factsheet. Dartmoor National Park Authority, Newton Abbot, Devon, UK. Available at www.dartmoor-npa.gov.uk.

Dawson, M.J., C. Lane, and G. Saunders. 2006. Proceedings of the National Feral Horse Management Workshop. Invasive Animals Cooperative Research Centre, Canberra, ACT, Australia.

de Villalobos, A.E., and S.M. Zalba. 2010. Continuous feral horse grazing and grazing exclusion in mountain pampean grasslands in Argentina. Acta Oecologica: International Journal of Ecology 36:514–519.

de Villalobos, A.E., S.M. Zalba, and D.V. Pelaez. 2011. *Pinus halepensis* invasion in mountain pampean grassland: Effects of feral horses grazing on seedling establishment. Environmental Research 111:953–959.

Dobbie, W.R., D. Berman, and M. Braysher. 1993. Managing vertebrate pests: Feral horses. Australian Government Publishing Service, Canberra, ACT, Australia, Available at www.feral.org.au/managing-vertebrate-pests-feral-horses-2.

English, A. 2000. Report on the cull of feral horses in Guy Fawkes River National Park in October 2000. University of Sydney, Sydney, NSW, Australia.

European Commission. 2008. Commission regulation no. 504/2008 of June 6, 2008, implementing Council Directives 90/426/EEC and 90/427/EEC as regards methods for the identification of equidae. Brussels, Belgium. Available at http://ec.europa.eu/environment/life/index.htm.

Forder, V. 2006. Conservation grazing: Konik horse, European beaver and wild boar. Wildwood Trust, Herne Bay, Kent, UK. Available at www.wildwoodtrust.org/files/conservation-grazing.pdf.

Foundation for Restoring European Ecosystems. 2013. Natural grazing: Wild and semi-wild animals as landscape architects. Foundation for Restoring European Ecosystems, Beuningen, Netherlands. Available at www.freenature.eu/free/download/documenten/natuurlijke-begrazing_uk.pdf

Furbish, C.E., and M. Albano. 1994. Selective herbivory and plant community structure in a mid-Atlantic salt marsh. Ecology 75:1015–1022.

Graham, A., K. Johnson, and P. Graham. 1986. An aerial survey of horses and other large animals in the Alice Springs and

Gulf regions. Conservation Commission of the Northern Territory, Alice Springs, NT, Australia.

Gray, M.E., D.S. Thain, E.Z. Cameron, and L.A. Miller. 2011. Multi-year fertility reduction in free-roaming feral horses with single-injection immunocontraceptive formulations. Wildlife Research 38:475–481.

Hampton, J.O. 2013. Assessment of the humaneness of feral horse helicopter shooting operations in the Northern Territory: Tempe Downs. Ecotone Wildlife Veterinary Services, Canberra, ACT, Australia.

Hunt, K. 1992. Horse evolution. Page 205 *in* B.J. MacFadden (Ed.), Fossil horses: Systematics, paleobiology, and evolution of the family equidae. Cambridge University Press, New York, USA.

ICMO. International Committee on the Management of Large Herbivores in the Oostvaardersplassen. 2006. Reconciling nature and human interests. The Hauge, Netherlands.

ICMO. International Committee on the Management of Large Herbivores in the Oostvaardersplassen. 2010. Natural processes, animal welfare, moral aspects and management of the Oostvaardersplassen. The Hague, Netherlands.

Iglesia, P. 1973. Los caballos gallegos explotados en régimen de libertad o caballos salvajes de Galicia. Thesis, Facultad de Veterinaria, Universidad Complutense de Madrid, Madrid, Spain.

Ivey, J. 2009. Report on New Forest traditions: Our New Forest. A living register of language and traditions. New Forest Centre, Lyndhurst, Hampshire, UK. Available at www.newforestcentre.org.uk/uploads/publications/65.pdf

Kampf, H. 2002. Nature conservation in pastoral landscapes: Challenges, chances and constraints. Pages 15–38 *in* B. Redecker, P. Finck, W. Hardtle, U. Riecken, and E. Schroder (Eds.), Pasture landscapes and nature conservation. Springer-Verlag, Berlin, Germany.

Killian, G., D. Thain, N.K. Diehl, J. Rhyan, and L. Miller. 2008. Four-year contraception rates of mares treated with single-injection porcine zona pellucida and GnRH vaccines and intrauterine devices. Wildlife Research 35:531–539.

Kirkpatrick, J.F. 1995. Management of wild horses by fertility control: The Assateague experience. Monograph NRSM-95/26. US National Park Service, Washington, DC, USA.

Kirkpatrick, J.F., and J.W. Turner.1991. Reversible contraception in nondomestic animals. Journal of Zoo and Wildlife Medicine 22:392–408.

Lagos, L. 2013. Ecología del lobo (*Canis lupus*), del poni salvaje (*Equus ferus atlanticus*) y del ganado vacuno semiextensivo (*Bos taurus*) en Galicia: Interacciones depredador-presa. Thesis, Universidad de Santiago de Compostela, Santiago de Compostela, Spain.

Langbein, J., R. Putman, and B. Pokorny. 2011. Traffic collisions involving deer and other ungulates in Europe and available measures for mitigation. Pages 215–259 *in* R. Putman, M. Apollonio, and R. Andersen (Eds.), Ungulate management in Europe: Problems and practices. Cambridge University Press, Cambridge, UK.

Leblond, A., A. Sandoz, G. Lefebvre, H. Zeller, and D.J. Bicout. 2007. Remote sensing based identification of environmental risk factors associated with West Nile disease in horses in Camargue, France. Preventive Veterinary Medicine 79:20–31.

Lenehan, J. 2011. Ecological impacts of feral horses in grassy woodland and open-forest gorge country in a temperate-subtropical wilderness. Thesis, University of New England, Armidale, NSW, Australia.

López-Bao, J.V., V. Sazatornil, L. Llaneza, and A. Rodríguez. 2013. Indirect effects on heathland conservation and wolf persistence of contraditory policies that threaten traditional free-ranging horse husbandry. Conservation Letters 6:448–455.

Low, W.A., and M.R. Hewett. 1990. Aerial survey of horses and other large animals in the central ranges of the Alice Springs district. Unpublished report to the Conservation Commission of the Northern Territory, Alice Springs, NT, Australia.

Loydi, A., and R.A. Distel. 2010. Floristic diversity under different intensities of large herbivore grazing in mountain grasslands of the Ventania System, Buenos Aires. Ecologia Austral 20:281–291.

Loydi, A., and S.M. Zalba. 2009. Feral horses dung piles as potential invasion windows for alien plant species in natural grasslands. Plant Ecology 201:471–480.

Loydi, A., R.A. Distel, and S.M. Zalba. 2010. Large herbivore grazing and non-native plant invasions in montane grasslands of central Argentina. Natural Areas Journal 30:148–155.

Madosky, J.M., D.I. Rubenstein, J.J. Howard, and S. Stuska. 2010. The effects of immunocontraception on harem fidelity in a feral horse (*Equus caballus*) population. Applied Animal Behaviour Science 128:50–56.

Magdalena, R., and T. Vidal. 1984. Los garranos galaicos. Quercus 15:35–38.

Matos Vieira Leite, J. 2012. A raça Equina Garrana. Pages 30–49 *in* N. Vieira de Brito and G. Candeiras (Eds.). Proceedings of the Livro de Atas, I Congresso Internacional do Garrano [in Portuguese]. Arcos de Valdevez, Portugal.

Minoransky, V.A., and A.M. Uzdenov. 2011. Feral horses of Vodnyi Island (Manich-Gudilo Lake, Rostov Region, Russia). News Biosphere Reserve. [In Russian.] Askania Nova 13:135–145.

Murphy, M.D., and J.C. González Faraco. 2002. Las yeguas marismeñas de Doñana: Naturaleza, tradición e identidades sociales en un espacio protegido. Revista de Dialectología y Tradiciones Populares 57:5–40.

Murray, D.A. 2008. Conserving Britain's last wild ponies. Geographical Magazine 80:48–53.

National Park Service. 2013. Assateague Island National Seashore. Washington, DC, USA. Available at www.assateagueisland.com/island_info/assateague_info.htm.

National Research Council. 2013. Using science to improve the BLM Wild Horse and Burro Program: A way forward. National Academies Press, Washington, DC, USA.

New Forest National Park Authority. 2013. Animal accidents. Lymington, Hampshire, UK. Available at www.newforestnpa.gov.uk/info/20094/commoning/40/animal_accidents/3.

Nimmo, D.G., and K.K. Miller. 2007. Ecological and human dimensions of management of feral horses in Australia: A review. Wildlife Research 34:408–417.

North Carolina Coastal Reserve and National Estuarine Research Reserve. 2013. Rachel Carson. Available at www.nccoastalreserve.net/web/crp/rachel-carson.

Nuñez, C.M.V., J.S. Adelman, C. Mason, and D.I. Rubenstein. 2009. Immunocontraception decreases group fidelity in a feral horse population during the non-breeding season. Applied Animal Behaviour Science 117:74–83.

Nuñez, C.M.V., J.S. Adelman, and D.I. Rubenstein. 2010. Immunocontraception in wild horses (*Equus caballus*) extends reproductive cycling beyond the normal breeding season. PLoS ONE 5:e13635.

Outram, A.K., N.A. Stear, R. Bendrey, S. Olsen, A. Kasparov, V. Zaibert, N. Thorpe, and R.P. Evershed. 2009. The earliest horse harnessing and milking. Science 323:1332–1335.

Pacheco, M.A., and E.A. Herrera. 1997. Social structure of feral horses in the llanos of Venezuela. Journal of Mammalogy 78:15.

Peckover and the Carneddau Pony Society. 2009. First annual sale by public auction of approx 100 Carneddau Welsh Mountain pony foals, youngstock and draft mares. Grazing Animals Project, Stoneleigh Park, Warwickshire, UK. Available at www.grazinganimalsproject.org.uk/news/first_public_auction_of_carneddau_ponies_tyn_llwyfan_5th_december_2009.html.

Portas, M.C.P., J.M. Vieira Leite, and J.J.D. Oliveira e Sousa. 1998. A raça Garrana: Um contributo para o seu estudo. Veterinaria Técnica Dezembro 98:18–26.

Prieditis, A. 2002. Impact of wild horses herd on vegetation at Lake Pope, Latvia. Acta Zoologica Lituanica 12 (4):392–396.

Prishutova, Z.G. 2010. Feral horses (*Equus caballus*) as a component of protected grassland ecosystems in the Rostovsky Reserve. Russian Journal of Ecology 41:55–59.

Ransom, J.I. 2012. Population ecology of feral horses in an era of fertility control management. Thesis, Colorado State University, Fort Collins, USA.

Ransom, J.I., B.S. Cade, and N.T. Hobbs. 2010. Influences of immunocontraception on time budgets, social behavior, and body condition in feral horses. Applied Animal Behaviour Science 124:51–60.

Ransom, J.I., N.T. Hobbs, and J. Bruemmer. 2013. Contraception can lead to trophic asynchrony between birth pulse and resources. PLoS ONE 8:e54972.

Ransom, J.I., J.G. Powers, H.M. Garbe, M.O. Oehler Sr., T.M. Nett, and D.L. Baker. 2014*a*. Behavior of feral horses in response to culling and GnRH immunocontraception. Applied Animal Behaviour Science 157:81–92.

Ransom, J.I., J.G. Powers, N.T. Hobbs, and D.L. Baker. 2014*b*. Ecological feedbacks can reduce population-level efficacy of wildlife fertility control. Journal of Applied Ecology 51:259–269.

Rewilding Europe. 2012. Wild horses released in Western Iberia rewilding area. Nijmegen, Netherlands. Available at http://rewildingeurope.com/news/articles/wild-horses-released-in-western-iberia-rewilding-area.

Rigueiro-Rodríguez, A., R. Mouhbi, J.J. Santiago-Freijanes, M.D.P. González-Hernández, and M.R. Mosquera-Losada. 2012. Horse grazing systems: Understory biomass and plant biodiversity of a Pinus. Scientia Agricola 69:38–46.

Rodgers, R.B. 1985. Assateague Island National Seashore feral pony management plan. Assateague Island National Seashore, Berlin, MD, USA.

Rosu, O. 2013. Research plan: Evaluation of immunocontraception on the Letea feral horses following the active immunization with porcine zona pellucida. Page 16 *in* Vier Pfoten—Four Paws. Universitatea de Ştiinţe Agronomice Şi Medicină Veterinară Bucureşti, Vier Pfoten, Romania.

Roztocze National Park. 2013. Roztocze National Park, Poland. Available at www.roztoczanskipn.pl.

Rubenstein, D.I., and C.M.V. Nuñez. 2009. Sociality and reproductive skew in horses and zebras. Pages 196–226 *in* R. Hager and C.B. Jones (Eds.), Reproductive skew in vertebrates: Proximate and ultimate causes. Cambridge University Press, Cambridge, UK.

Russell, V. 1976. New Forest ponies. David & Charles, Newton Abbot, Devon, UK.

Sacco, A.G. 1977. Antigenic cross-reactivity between human and pig zona pellucida. Biology of Reproduction 16:164–173.

Santos, A.S., and L.M.M. Ferreira. 2012. The Portuguese Garrano breed: An efficient and sustainable production system. Forages and Grazing in Horse Nutrition 132:481–484.

Schott, C. 2005. Ecology of free-ranging horses in Northern Guy Fawkes River National Park NSW, Australia. Thesis, University of New England, Armidale, NSW, Australia.

Schwartz, K.Z.S. 2005. Wild horses in a "European wilderness": Imagining sustainable development in the post-Communist countryside. Cultural Geographies 12:292–320.

Scorolli, A.L., and C. Zoratti. 2007. Feral horse management in Argentina: Two case studies of joint work. Second Latin-american National Parks and Protected Areas Congress. Bariloche, Argentina.

Scorolli, A.L., and A.C. Lopez Cazorla. 2010. Feral horse social stability in Tornquist Park, Argentina. Mastozoologia Neotropical 17:391–396.

Scorolli, A.L. 2012. Feral horse body condition: A useful tool for population management? Presented at the International Wild Equid Conference, 18–22 September 2012, Vienna, Austria.

Slaukstins, V. 2002. The Lake Pape: Grazing of coastal grasslands. WWF Latvia project. Pages 197–207 *in* B. Redecker, P. Finck, W. Härdtle, U. Riecken, and E. Schröeder (Eds), Pasture landscapes and nature conservation. Springer-Verlag, Berlin, Germany.

Southern States. 2013. Running wild—The feral ponies of Mount Rogers, Virginia. Available at www.southernstates.com/articles/the-wild-ponies-of-mount-rogers-virginia.aspx

Spasskaya, N.N. 2008. Feral horses are not strangers in the steppe. [In Russian.] Steppe Bulletin 25:52–56.

Spasskaya, N.N. 2009. Steppe and horses: Conflict or cooperation? (Rostovsky Reserve). Reserve management: Problems in protection and ecological restoration of grassland ecosystems. Pages 130–134 *in* Proceedings of the international scientific-practical conference dedicated to the 20th anniversary of the organization of GPP "Orenburg." [In Russian.] Orenburg, Russia.

Stanley, C.R., and S. Shultz. 2012. Mummy's boys: Sex differential maternal–offspring bonds in semi-feral horses. Behaviour 149:251–274.

Stout, T.A.E., and B. Colenbrander. 2004. Suppressing reproductive activity in horses using GnRH vaccines, antagonists or agonists. Animal Reproduction Science 82–83:633–643.

Tupac-Yupanqui, I., S. Dunner, B. Sañudo, A. González, S. Argüello, *et al.* 2011. Genetic characterization of the caballo Monchino breed and its relationships with other Spanish local equine breeds [in Spanish]. Archivos de Zootecnia 60:425–428.

Turner, J.W., I.K.M. Liu, D.R. Flanagan, A.T. Rutberg, and J.F. Kirkpatrick. 2001. Immunocontraception in feral horses: One inoculation provides one year of infertility. Journal of Wildlife Management 65:235–241.

Turner, J.W., I.K.M. Liu, D.R. Flanagan, K.S. Bynum, and A.T. Rutberg. 2002. Porcine zona pellucida (PZP) immunocontraception of wild horses (*Equus caballus*) in Nevada: A 10 year study. Reproduction Supplement 60:177–186.

Tyler, S.J. 1972. The behaviour and social organization of the New Forest ponies. Animal Behaviour Monographs 5:85–196.

UK Secretary of State. 2009. The horse passports regulations 2009. No. 1611. London, UK. Available at www.legislation.gov.uk/uksi/2009/1611/contents/made.

US Congress. 1971. The Wild Free-Roaming Horses and Burros Act, Public Law 92-195, 92nd Congress.

US Congress. 2014. Wild Horse Oversight Act. 113th Congress.

Valero, E., J. Picos, F. Lago, and L. Lagos. 2013. Los accidentes de tráfico relacionados con fauna salvaje en Galicia: Incidencia, patrón y soluciones. Presented at the 6th Congreso Forestal Español, Vitoria, Spain.

van Dierendonck, M. 2012. Early reactive culling protocol in the Oostvaardersplassen Nature Reserve, the Netherlands. Presented at the International Wild Equid Conference, 18–22 September 2012, Vienna, Austria.

Vera, F. 2009. Large-scale nature development—The Oostvaardersplassen. British Wildlife 20:28–36.

Verderers of the New Forest. 2013. Available at www.verderers.org.uk.

Vernes, K., M. Freeman, and B. Nesbitt. 2009. Estimating the density of free-ranging wild horses in rugged gorges using a photographic mark-recapture technique. Wildlife Research 36:361–367.

Vulink, J.T., M.R. van Eerden, and R.H. Drent. 2010. Abundance of migratory and wintering geese in relation to vegetation succession in man-made wetlands regimes: The effects of grazing. Ardea 98:319–327.

Walton, I. 2008. Mamíferos de Chile. Lynx Edicions, Barcelona, Spain.

Warmuth, V., A. Eriksson, M.A. Bower, J. Cañon, G. Cothran, O. Distl, M.-L. Glowatzki-Mullis, H. Hunt, C. Luís, and M. do Mar Oom. 2011. European domestic horses originated in two Holocene refugia. PLoS ONE 6:e18194.

Warmuth, V., A. Eriksson, M.A. Bower, G. Barker, E. Barrett, *et al.* 2012. Reconstructing the origin and spread of horse domestication in the Eurasian steppe. Proceedings of the National Academy of Sciences of the USA 109:8202–8206.

Welsh Government. 2012. Equine identification and the semi-feral ponies of Wales. Cardiff, Wales. Available at http://wales.gov.uk/topics/environmentcountryside/ahw/horses/equineidsemiferalponies/?lang=en.

Winton, C.L., M.J. Hegarty, R. McMahon, G.T. Slavov, N.R. McEwan, M.C.G. Davies-Morel, C.M. Morgan, W. Powell, and D.M. Nash. 2013. Genetic diversity and phylogenetic analysis of native mountain ponies of Britain and Ireland reveals a novel rare population. Ecology and Evolution 3:934–947.

Zalba, S.M., and N.C. Cozzani. 2004. The impact of feral horses on grassland bird communities in Argentina. Animal Conservation 7:35–44.

11

Wild Equid Captive Breeding and Management

MANDI WILDER SCHOOK,
DAVID M. POWELL, AND
WALTRAUT ZIMMERMANN

Wild equids are iconic symbols of Africa and Asia, making them widely exhibited in zoological collections worldwide. But beyond their crowd appeal, wild equids housed in zoological institutions, reserves, and related conservation organizations serve as a resource for conservation, education, research, and, perhaps most significantly, a source for reintroduction (Saltz and Rubenstein 1995). As populations of wild equids continue to dwindle, we find it is imperative to meet the challenge of maintaining healthy insurance populations. The major goal of captive breeding programs is to create sustainable populations. This means breeding animals at a rate to either maintain or increase current animal numbers and in a manner that maintains genetic diversity. Inbreeding (and associated loss in genetic diversity) is associated with increased risk of disease and infertility that can ultimately compromise an entire population (Gomendio *et al.* 2000; Ruiz-Lopez *et al.* 2010). The success of efforts to achieve sustainable breeding goals relies on expertise and coordination in such diverse disciplines as population genetics, animal husbandry, veterinary medicine, nutrition, behavior, ecology, reproductive biology, and physiology. While some challenges exist, there are significant opportunities emerging to increase the contribution of captive breeding to the conservation of wild equids.

The Role of Studbooks and Management Plans in Captive Breeding

The most basic tool for population management of captive wild equids is a detailed pedigree, or studbook. The studbook captures the population's genetic and demographic history and can be analyzed to determine age- and sex-specific fecundity, mortality, and genetic status. Studbooks may be regional (e.g., North America) or international in scope. Currently there are international studbooks for Somali wild ass (*Equus africanus somalicus*), Przewalski's horse (*E. ferus przewalskii*), Grevy's zebra (*E. grevyi*), Hartmann's mountain zebra (*E. zebra hartmannae*), Asiatic wild ass (*E. hemionus*) separately for the subspecies Persian onager (*E. h. onager*) and Turkmen kulan (*E. h. kulan*), and kiang (*E. kiang*). Every animal in the managed population receives a unique studbook number and is tracked throughout its lifetime by recording all transfers between institutions. Parentage and sex of all animals are also recorded, and additional data such as identifying marks, genetic origin, contraception status, medical notes, and necropsy results can be recorded as well.

Currently, captive wild equids are primarily managed at a regional scale through zoological associations (e.g., the Association of Zoos and Aquariums, or AZA, in North America; the European Association of Zoos and Aquariums, or EAZA; and the Zoo and Aquarium Association in Australasia, or ZAA), though over time it will be necessary to move toward global management to maintain genetic diversity (Lees and Wilcken 2011; Lacy 2013). Within regions, populations are managed through coordinated programs called Species Survival Plans (SSPs) and European endangered species programs (EEPs) in the United States and Europe, respectively. At regular intervals (one to three years), the species coordinator and studbook keeper produce a breeding and transfer plan based on demographic and genetic analyses of the population. Each animal receives a recommendation of whether to breed, or whether to be moved between institutions for breeding or exhibition. The animal's mean kinship (MK), which is a measure of each individual's relatedness to the population as a whole (Ballou and Lacy 1995) and therefore a measure of its genetic importance, is an important factor in determining which animals are recommended for breeding. Additionally, the demographic and genetic status of the population, space available for the species in the region, needs of holding institutions, and programmatic population goals all help to shape the plan. Periodic analyses are conducted to determine whether a regional population could benefit genetically or demographically from an exchange of animals between regions. Appropriate animals for import or export are identified, as well as their intended mates in the receiving region, and the transfers are executed. Maintaining excellent records on population status and pedigree relationships is the foundation for genetic management, but day-to-day management of wild equids in captivity requires science and husbandry-based solutions to a number of challenges.

Challenges

For all captive equids, maintenance of healthy and viable breeding populations is predicated on knowledge of basic husbandry, including social, environmental, and nutritional requirements. Husbandry guidelines (Rademacher and Winkler 2000; Zimmermann and Kolter 2016) are established to minimize hazards and health risks for captive animals and to promote animal welfare, but husbandry depends on adequate knowledge of species biology, or science that evaluates how particular husbandry protocols affect animals, and the experience of animal care personnel. In some cases it may take years of experience and research to understand the needs of a species in captivity. Proper identification and parentage verification are critical in order to manage on the population level and make recommendations for individuals, but they can also be challenging in large herds of animals that look more or less similar to one another. Many polygynous species, as all wild equids are, have an adult sex ratio that is approximately equivalent (see chap. 6). Captive facilities must therefore develop protocols for managing same-sex groups of animals, which at times can be difficult for groups of male equids. Related to this is the challenge of breeding animals regularly enough to maintain reproductive viability while managing limited spatial and financial resources in captive facilities.

Husbandry and Individual Management

Husbandry requirements are more or less similar across species of wild equids. What does differ between some species is their social system, and thus holding conditions must comply with species needs. All Asiatic equids are cold adapted, and the wild horse and wild ass have partly overlapping distribution areas (Heptner and Naumov 1966; Nobis 1971; Bannikov 1981; Groves 1986; Eisenmann 1992). They can tolerate extreme temperatures, at least for a short while, in heat up to 50°C/122°F and cold down to –40°C/–40°F, so in zoos they can be maintained with minimal needs for shelter. In contrast, African equids need temperature-controlled housing during winter months in higher latitudes.

Routine preventive medical practices for nondomestic equids usually include vaccinations and fecal parasite surveillance as well as enteric parasite treatments, hoof trims, and dental procedures when needed. Practices are dependent upon geographic location, individual animal and collection, medical histories, enclosure designs, and management practices. Vaccinations commonly administered include tetanus toxoid, Eastern equine encephalitis, rabies, Western equine encephalitis, and West Nile virus. Equine rhinopneumonitis vaccination is sometimes also administered, especially in breeding situations. Fecal parasite screening is generally performed twice annually, but it can be conducted more or less often. Anthelmintic treatments can be administered to a particular individual when excessive parasites are identified. Hoof trims may need to be performed to prevent hoof overgrowth depending on the exhibit design, diet, and geographic location, and dental procedures (filing) are sometimes required in aged individuals.

With similar nutritional needs as domesticated equids, wild equids can be fed a variety of diets formulated for domestic horses. These diets are formulated to meet the various needs individual animals might have, including weight gain or loss, fiber and energy requirements, and so on. Most captive wild equids are fed a diet composed of a pelleted ration and grass or legume hay. These diets are largely supplemented by providing additional opportunities for natural grazing or browsing in exhibit enclosures. Horses and plains zebra (*E. quagga* and all recognized subspecies) may require a richer diet than the desert-adapted wild asses and the Grevy's zebra. While the Asiatic equids, African wild ass, and Grevy's zebra may enrich their diet with browse, plains zebra seem to be strict grazers (see chap. 4). Wild equids may also require supplementation with additional vitamin E (tocopherol, 1,000 mg per os per day), especially during winter, when hay instead of green fodder is fed, a diet that can be lower in vitamins. A deficiency in vitamin E may result in ataxia or white muscle disease and myopathy (Dierenfeld *et al.* 1997), especially in situations that can be considered stressors, such as transport.

All equids are fast runners; enclosures—whether dry lots or pastures—should be large enough to allow running, water permeable, and structured such that the animals have to move between food, water, and resting places to abrase their hooves in a natural way. Sandy ground for rolling and big tree trunks for scratching are important considerations for comfort and natural behaviors. Only a few wild animal parks worldwide can keep wild equids on pastures year-round. The grass is quickly destroyed by grazing and hoof beat, especially during wet seasons. Although feeding naturally on grass all day would be ideal for all equid species, this usually cannot be achieved. Similar to domestic equids, however, a good alternative to pasture is a dry lot with several feeding stations placed as far apart as possible to encourage movement. Designing exhibits for exercise is important for maintaining a healthy body condition. As a rule of thumb, only young equids move enough of their own accord. Older animals do not retain this motivation and would not move enough if food, water, and comfort places were all located in close vicinity. Offering a percentage of the exhibit with a hard surface such as crushed limestone will aid in hoof abrasion and minimize the need for hoof trims.

Because equid behavior has been studied intensively in situ and ex situ (see chap. 3), holding conditions can be adapted to meet different needs according to the social system of each species. Usually, equid species living in harem groups (horses, plains zebra, and mountain zebra) can and should be kept as a harem, although with some harem species—namely, the mountain zebra—a strong hierarchy and mate choice can make the exchange of herd members for breeding a challenge. Although wild asses are permanently (Africa) or temporarily (Asia) territorial in the wild, in zoos the stallion can be kept together with females. It is easier with wild asses to exchange group members, as they do not have a strict hierarchy like the harem-living equids. Should a situation arise that requires separation of an individual for a short timespan, extra space must be available out of sight of the other group members. Equid stallions can become aggressive when prevented from approaching a mate. Individual temperament also periodically requires adjustments to husbandry in wildlife (Powell and Gartner 2010). Exceptions to stallion-harem housing should be considered for Grevy's zebra. In the wild, adult stallions are permanently territorial and mate with females passing through. In zoos, Grevy's stallions may not tolerate being in close vicinity to females year-round, which either requires a large enclosure or a separate enclosure for the male (Rademacher 1997).

Identification and Parentage Testing

Identification

The International Zoo Yearbook (Mohr 1968) states that unambiguous individual identification of zoo animals is required for management. Otherwise, managing health, recording exact life data, and monitoring breeding and reproduction are impossible. Research on reproduction and behavior also depends on clear recognition of individuals (Zimmermann and Kolter 2015). Furthermore, unequivocal identification becomes an essential need beyond the purposes of an individual zoo in order to maintain studbooks and develop species survival plans. All efforts to manage species according to sophisticated programs based on genetic and demographic analyses are in vain when basic individual data are erroneous owing to the inability to distinguish between individuals. A look into the Przewalski's horse studbook shows hundreds of parentage entries with individual misidentification (Bowling *et al.* 2003). Legal authorities that regulate the trade of endangered species also insist on a permanent marker. In Europe, the mandatory marking is a microchip or transponder. In North America, markings can include a variety of permanent identifiers.

All African equids are marked individually and conspicuously enough by their coat color pattern, so that recognition using photographs is quite easy. In contrast,

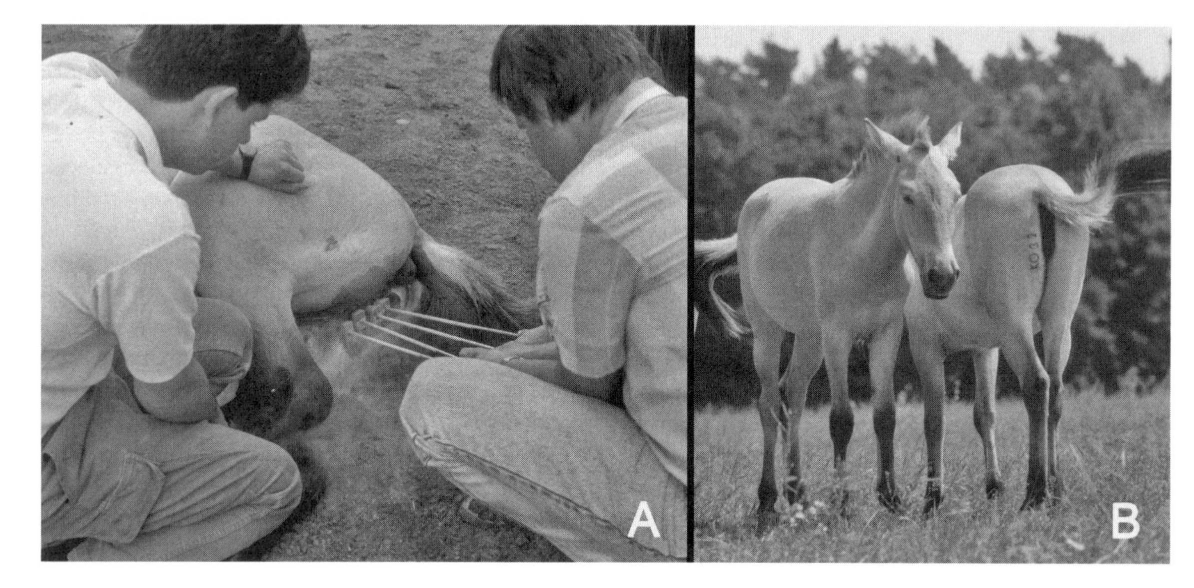

Fig. 11.1 Freeze branding with liquid nitrogen works well to provide a permanent means of identification in Asian equids that are not readily identifiable by natural markings alone. The brand is easily spotted from long distances in large pastures, such as in semireserves. Photos by Waltraut Zimmermann

Przewalski's horses and hemiones have a relatively uniform color. Variations may occur in striped or nonstriped legs, white or dark nose or belly, and shoulder stripe or marking. Photo documentation for Asiatic equids thus must also focus on small details, such as number and location of crowns (cowlicks) between the eyes, shape and location of permanent scars, location of one or more warts, and location of black and white markings. Yet Asiatic equids are also sent to semireserves and reintroduction projects, where these characteristics might not be sufficient for proper identification.

In other mammals (e.g., antelopes, caprines), both ear notches and tags are used for identification, but these are not recommended for equids because of ear biting during fighting and social encounters, which may remove the tags or render the notching code unreadable. In Europe, the freeze brand was recommended and turned out to be useful. If immobilization is required for transport, freeze branding can be done at that time. The procedure uses liquid nitrogen and aluminum types (fig. 11.1A). Any animal can then be correctly identified any place in the world, with permanent identification that is unmistakable (e.g., studbook number) and visible with binoculars in semireserves or in the wild (fig. 11.1B).

Parentage Testing

The rapid development of deoxyribonucleic acid (DNA) methods in the field of criminal forensics has provided opportunities to apply these techniques to the identification of animals when parentage or other identity questions arise. Tests for domestic horses are also widely available to prevent fraud in racing sports. In the 1980s, the genetics laboratory at the University of California, Davis, USA, became one of the first institutions worldwide to focus on equine genetics. Today, such institutions are abundant around the globe. While in the beginning only blood or tissue samples delivered enough DNA to analyze, today labs are able to get results from minimal traces (e.g., hair roots) that are comparatively easy to obtain and can be stored for years.

Przewalski's horses became the first wild equid subjects for such DNA tests, as it had been assumed that the identities of several individuals had been switched accidently. Dr. H.C. Erna Mohr (1894–1968) started the Przewalski's horse studbook, and it is to her credit that the early history of Przewalski's horses in captivity was documented (Mohr 1959). After a stagnation in population growth as a result of two world wars, numbers of Przewalski's horses increased steadily, with almost 2,150 living animals to date. Bigger groups, staff replacement at zoos, and more frequent exchanges around the world caused increased misidentifications owing to the similar phenotypes of the horses. Thus questionable individuals have found entry into the studbook. The only possibility to solve this problem was to DNA-type the animals in question for parentage verification and studbook corrections (see also chap. 7). Today, genetic testing is used to supplement and verify data for many captive equid studbooks.

Housing Excess Males and Maintaining Bachelor Groups in Limited Spaces

Almost every mammal species where several females form permanent or temporary groups led by one male proves to be a challenge for zoos: both sexes are often produced in equal numbers, but only one adult male is usually be tolerated per breeding group, and holding space is usually limited. In Europe, holding space for males did not exist before 1985, the founding year of the EEP. Even so, regular exchange of harem stallions was necessary to reduce inbreeding in the population. Yearling males that were too young for breeding also needed to be placed, as eventually they would be exposed to aggression from their sires. It was therefore imperative that places be found to keep excess males, and in Europe today, stallion groups exist for all equids where young males can grow up and develop social skills (Hoffmann 1985).

Over the years, experience has shown that it is not the size of an enclosure that influences aggression between the stallions, but rather the number of stallions per enclosure. Contrary to logical thought, including more stallions does not lead to more aggression. When there are many stallions together (>10), they split into groups that result in less agonistic behavior overall (Zharkikh 2009). Bachelor groups of mountain and plains zebra are quite stable, and severe fighting in captivity has not occurred. Grevy's zebra and Asiatic wild ass bachelor groups also do well because their social system does not include a strong hierarchy (Ginsberg 1988). Instead, pairs often form close bonds; therefore it seems recommendable to keep groups with even numbers only (Heuschkel *et al.* 1999). A mixture of equid species or subspecies is also possible. In Europe, stallions of Przewalski's horse and onager, onager and kulan, and even Grevy's zebra and Somali wild ass have been housed together with varying degrees of success.

Some pharmaceutical tools have been used to manage aggression in male equids. Zehnder *et al.* (2006) documented use of a synthetic progestogen, altrenogest, for controlling aggression in a male Grevy's zebra. A dose of 40 mg per day proved sufficient to control aggressive behaviors when initially introduced to female conspecifics. The altrenogest was later discontinued until the next introduction, and no further aggression issues were documented. Long-acting neuroleptics such as perphenazine decanoate (Flugger and Jurczynski 2005) have been used to reduce aggression in Persian onagers and plains zebra, while fluphenazine decanoate (B. Wolfe, unpublished data) or perphenazine enanthate in combination with haloperidol has been used when introducing bachelor groups of Przewalski's horse stallions and Persian onagers (Atkinson and Blumer 1997). The agents work for two to three weeks with a single injection and allow herd hierarchies to be established without severe aggression. These pharmaceutical agents should be used with caution and under the supervision of a veterinarian, as serious side effects can occur from overdose.

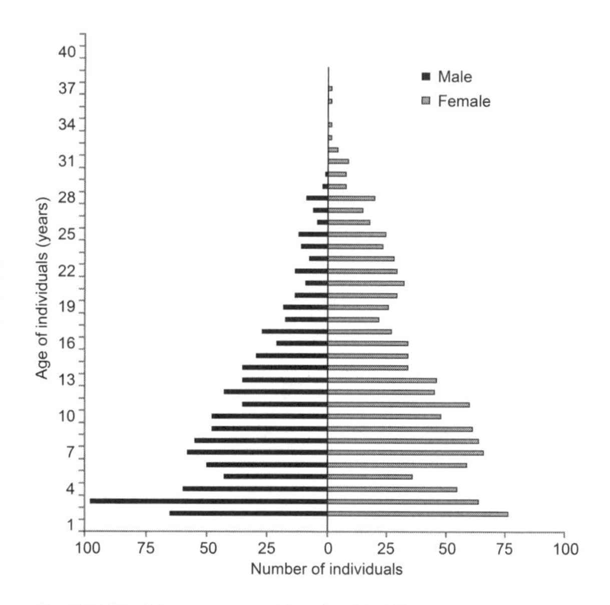

Fig. 11.2 Healthy age pyramid of the 2013 European Endangered Species Program Przewalski's horse population.

Population-Level Considerations

Population Sustainability: Challenges in Maintaining Genetic Diversity and Associated Risks

Demographic and genetic management of populations is necessary for long-term sustainability, but often the immediate challenge in keeping populations robust is a demographic rather than genetic one. Healthy populations are characterized by an abundance of young individuals that represent the future reproductive capacity of the population; this is often visualized in the form of an age pyramid that should generally be triangular in shape, with the narrow point at the top representing older animals (fig. 11.2). A population age structure of this sort is often realized through allowing most of the sexually mature population to breed regularly. This presents obvious challenges in captivity, where the availability of space is limited and populations are relatively long-lived without natural threats to survival.

Captive animals often experience better health, nutrition, and body condition and may therefore start reproducing earlier, have higher reproductive success, and

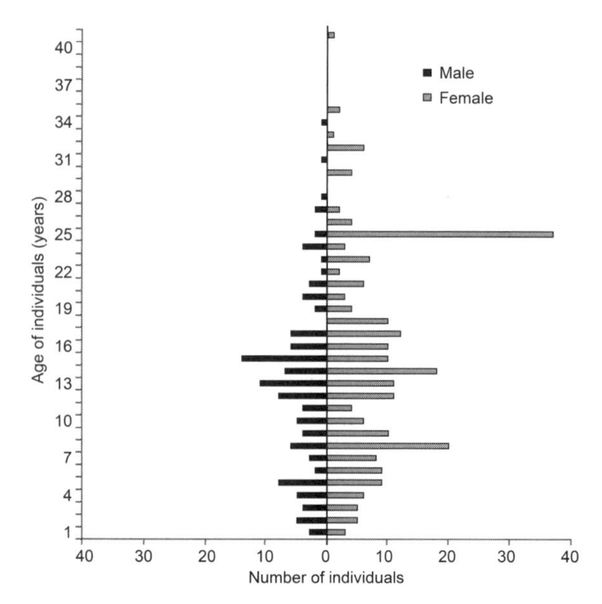

Fig. 11.3 The population structure of plains zebra held in member zoos of the Association of Zoos and Aquariums (AZA) is roughly columnar in shape. *Source*: Association of Zoos and Aquariums 2013

live longer than free-ranging counterparts. Limitations on space and the need to maintain genetic diversity often mean that only selected pairings or groups of individuals are allowed to breed at a given time. In many cases, this leads to a population structure wherein all age classes are roughly equal in number, and the resulting age pyramid is columnar in shape (fig. 11.3). The challenge at this stage in population management is to balance genetic goals and spatial constraints while having animals breed at least semiregularly to maintain reproductive health and possibly social skills.

In a worst-case scenario, the population becomes aged, with more individuals in the older age classes than in the younger age classes, and the resulting age structure resembles an inverted triangle. If this occurs, reproductive capacity and genetic diversity in the population are compromised, as few individuals are of prime reproductive age, with even fewer immature animals to be the next generation of breeders. This age structure also represents a possible genetic bottleneck, depending on the genetic constitution of the reproductive and prereproductive age classes. If genetic diversity is already low and inbreeding already high, then the population is also susceptible to inbreeding depression, which can take a variety of forms, including compromised survival and reduced reproductive success (see National Resource Council 2013 for a review). In Przewalski's horses, a loss in gene diversity has been correlated with abnormal reproductive function (Collins *et al.* 2012). A combination of poor demographic and genetic health can easily put a population on the trajectory toward extinction (box 11.1).

How can poor demographic and genetic health best be improved in captivity, where there are true limitations on space? Obtaining new bloodlines from the wild and exchanging animals between zoos and breeding centers in different regions of the world is often challenging because of significant costs, political difficulties, and in some cases disease considerations. As is the case when struggling to conserve populations in situ, the answer lies in having as diverse a toolbox as possible to address challenges. A closer look at captive space and how it is used is a starting point.

According to the International Species Information System, in 2013 there were ~3,000 nondomestic equids in ISIS member facilities around the world. An overwhelming 47.5% of the captive space currently allocated to equids is occupied by the plains zebra, which is the least endangered and most abundant wild equid on the planet (fig. 11.4). Plains zebra occupy 55% of the total space devoted to African wild equids. Zebra in general are highly recognizable by zoo visitors and considered an attractive icon of Africa. Plains zebra are also straightforward to manage in captivity and mix well with other species, making them the African equid of choice to exhibit in zoos (Powell 2001). In North America, Grevy's zebra can be considered more difficult to manage, and some have reported that they do not mix well with other species; however, there have been exceptions in North America, and this experience has not been reported in European holding institutions. As populations of mountain zebra and Somali wild ass increase in zoos, there is increasing evidence to indicate that these species are also suitable for mixed-species exhibits. Thus there is justification for considering an intentional reduction of the plains zebra population in captivity to make significant space available for other equids. It is not always easy to substitute an Asian equid for an African one in zoos because exhibits are sometimes designed to approximate geographic regions of the world (e.g., the African plains). A case could be made for zoos to consider mixed-species Asian semiarid grassland exhibits in addition to or instead of the ubiquitous African savannah. One possible advantage the Asian equids have over their African counterparts is that they are winter hardy, a particularly desirable characteristic for temperate-zone zoos.

Contrary to Europe, North American zoos are seeing a trend toward decreasing amounts of space allocated to ungulates. It is estimated that nearly 1,000

BOX 11.1 PRZEWALSKI'S HORSE A- AND B-LINE MANAGEMENT

The Przewalski's horse population in Europe is a good example of how the preservation of an endangered species can fail if basic principles of demographic and genetic management are ignored over decades. Twelve wild-caught Przewalski's horses and a domestic horse constitute the founders of the today's world population. As mentioned by eyewitnesses, not all foals captured by Hagenbeck seemed to be pure Przewalski's horses (Mohr 1959). The colt studbook #17 Bijsk 7, "Sultan," and the filly studbook #18 Bijsk 8, bought by the Bronx Zoo, were such individuals.

After the individuals reached adulthood, the Bronx Zoo Director William T. Hornady complained about this pair in a letter addressed to Hagenbeck: "one of your horses turned out to be a Mongolian pony, with the other animal too small to properly represent the species." The pair was finally passed over to the Cincinnati Zoo. Both horses showed an odd phenotype: the stallion lacked black pigment in its coat and had a white star on its forehead, while the mare had a long, hanging mane and atypical tail.

Because of suspicion of domestic horse genes in the founder population, Przewalski's horses were split into two lines for breeding management: the so-called A- and B-lines. Breeders believed the A-line (also called the Munich line) was the "purebred" line, despite evidence that the great-grandchild of studbook #17 and #18 (studbook #121, "Roma") had phenotypic evidence of domestic genes, and it was shipped to Munich in 1937 and became one of the breeding mares (Zimmermann *et al.* 2009). The B-line (also called the Prague line) was known to include six founders plus one domestic founder. The A-line, with a maximum of nine founders, became highly inbred over time. At least 15 stallions turned out to be infertile (no sperm production or bad sperm quality), and 8 mares had no offspring despite being in a breeding situation. In some females, atrophic ovaries were found (Hegel *et al.* 1990; Zimmermann 1997). In 2012, it was no longer possible to improve population demographics or genetics, and the A-line was finally given up. The A-line was nearly managed to extinction because breeders did not want to mix the so-called purebred animals with the B-line. In Europe, only 46 A-line horses had no blood of the dubious pair (#17 and #18), and 10 (22%) of them were infertile, making the potential breeding pool even smaller.

In the B-line (Prague line), inbreeding was kept at an acceptable limit by an early addition of A-line stallions, which increased the founder group to 13 horses. The resulting line has been referred to as the M-line (mixture of all) since the 1970s. A genetic breeding strategy combined with some phenotypic selection prevented domestic traits from manifesting in the population. No infertility has been reported, and the population has increased worldwide up to 2,114 horses in 2014. Today, the breeding pool of the Przewalski's horses is represented by the M-line (all founders) only.

Bijsk 7 and Bijsk 8 at the Cincinnati Zoological Garden, ca. 1913–1914.

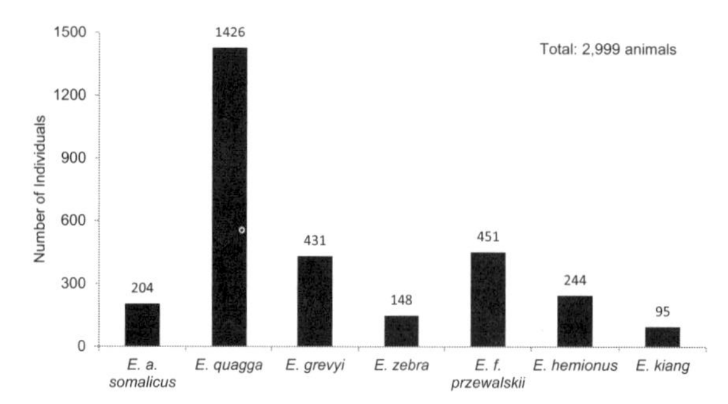

Fig. 11.4 Total number of individuals of African and Asian wild equid species held globally in member zoological institutions of the International Species Information System.

spaces for ungulates have been lost in the last 10 years. The reasons range from ungulates being perceived as difficult to manage, being less charismatic than other megafauna, requiring too much space to manage social behavior and grazing pressure, experiencing loss of interest in certain species (e.g., Przewalski's horses and Arabian oryx), and undergoing a push to create larger mixed-species exhibits (e.g., the African savannah) or to use the space for other revenue-generating activities. The AZA, which oversees coordinated breeding programs in North American zoos, considers a sustainable population to be one that retains 90% or more of its original genetic diversity for a minimum of 100 years and/or 10 generations (Boyle *et al.* 2011). As of September 2013, the Grevy's zebra was the only wild equid population considered sustainable in the AZA, while populations of plains zebra, Przewalski's horse, Somali wild ass, Hartmann's zebra, and Persian onager did not meet the criteria for being considered sustainable. The EAZA assessed studbook populations in its member institutions using a sustainability score card (see Leus *et al.* 2011 for details) and reported that only populations of Asiatic wild ass and Grevy's zebra met all five criteria for sustainability, while Przewalski's horse, Somali wild ass, and Hartmann's zebra populations met only three or four. The challenge is finding ways to maximize use of existing space through mixed-species exhibits, alternative management strategies, breeding centers and private sector partnerships, aggression management and control, time sharing of exhibit space, and management of single-sex groups to allow production of offspring without creating additional social groups (Houston *et al.* 2011). It is thus imperative that zoos and other captive breeding centers consider the options presented by Houston *et al.* (2011) to help maximize the use of captive space for building sustainable wild equid populations.

Population Management of Surplus Individuals

Contraception and euthanasia are tools that can contribute to more efficient use of space and populations that are more genetically and demographically healthy. Separating males from females to control breeding produces a greater demand on space and may lead to some individuals being housed alone; permanent sterilization (e.g., castration) of individuals can have unintended genetic and demographic consequences and behavioral effects, and the sterilized individuals continue to take up space that could be used for more offspring. In contrast, targeted contraception and culling may better achieve the simultaneous goals of genetic and demographic population health, efficient use of space, and animal welfare.

Contraception of feral horses and asses for population control has been intensively studied over the past two decades, and the details and mechanisms of contraception in equids are discussed in chapter 10. Contraception to control captive breeding is somewhat less documented, however (Kirkpatrick *et al.* 1995; Frank *et al.* 2005). Contracepting individuals, rather than separating to prevent breeding, allows equids to be maintained in social groups, which allows for maximum available space usage and likely improves welfare. Despite only limited published data, the PZP contraceptive vaccine has been used for management purposes in at least 135 wild equid females in European and US zoological populations, including zebras, Asian wild asses, and Przewalski's horses (K. Frank, unpublished data). A booster dose of the contraceptive vaccine, delivered by intramuscular injection, is recommended about 6 weeks after the primer and then every 12 months (every 8 months for zebra species). But contraception for longer than three years is not recommended if reversibility and future conception are desired (Ransom *et al.* 2013). Therefore this approach to contraception presents a reliable but short-term tool for targeted population control. Administration of progestins, such as altrenogest, can be used for short-term suppression of estrus and ovulation in female equids. The oral form must be administered daily and long-acting intramuscular injections must be given every 10–12 days, however, both methods can be difficult to achieve in intractable animals.

It may seem counterintuitive to consider culling (euthanasia) as a tool for aiding in the preservation of animal populations, especially in zoos and breeding centers, which many members of the public feel are "sanctuaries" for animals. The mission of these institutions, however, is to maintain populations of species for the long term so that animals are available for public education and enjoyment, research, and conservation (World Association of Zoos Aquariums 2005). This means that managers must take a long-term view of their populations and make decisions that will support the species in perpetuity while ensuring the highest standards of welfare for animals in the population.

Most animals have evolved to remain reproductive throughout their lifetimes to maximize production of offspring. Thus mammalian females regularly go through cycles of sexual receptivity, pregnancy, and parturition. These cycles involve profound changes in uterine structure and function, and it is becoming increasingly evident in wildlife that preventing regular reproduction can compromise subsequent fertility, particularly when females experience long interbirth intervals (Penfold *et al.* 2014). The link between reproduction and fertility has been known in the livestock industry for some time; however, it is not often documented or studied because subfertile animals are usually slaughtered. In some taxa, animal health may also be affected; for example, in some canids, consecutive nonconceptive ovulatory cycles are associated with higher incidence of pyometra (uterine infection; Asa *et al.* 2014). Contraceptives that prevent repeated nonconceptive cycles (e.g., many of the hormone-based contraceptives) should support higher posttreatment fertility and uterine health, but they do carry a nonzero risk to subsequent fertility and in some cases can result in dramatic changes in behaviors supported by circulating hormone levels (Adams *et al.* 1978; Henderson and Shively 2004; Asa and Porton 2010). It may be the case that females should be breeding more regularly than is needed to maintain or grow the captive population. In these cases, euthanasia of surplus offspring is an alternative that allows for the maintenance of fertility and reproductive capacity in a way that is unaffected by constraints on space. Euthanasia can be used on postreproductive individuals to maximize space for breeding animals and also allows managers to control sex ratio of the population and maximize genetic diversity through selective removal of overrepresented individuals. Finally, animals that are regularly breeding engage in a wider range of behaviors (e.g., courtship, mating, parental care) than nonbreeding animals, which may be important for maintaining critical behavioral skills in animals destined to be future breeders or prepared for reintroduction.

A third option that would allow for captive facilities to work with larger, more sustainable populations is to consider new partnerships with individuals and organizations outside of the traditional zoo industry. Accrediting bodies exist in a number of regions (e.g., AZA, EAZA, Australasian Regional Association of Zoological Parks and Aquaria), but there are additional nonaccredited partners in the private sector that could provide valuable space and perhaps expertise in managing larger herds of wild equids for conservation purposes. Some examples are discussed below (see Semireserves for Asiatic Equids). Philosophy and approach toward working with the private sector will vary regionally and will also likely be guided by institutional and regulatory considerations, but there are models that have been successful.

As zoos and other accredited captive breeding centers move forward with efforts to achieve sustainable populations of animals in the face of rapid loss of biodiversity and global climate change, it seems prudent to consider as many approaches as possible in the tool kit for sustainable population management, even if they appear difficult at the outset. A better understanding of philosophies and approaches across regions and guidance on implementation should lay the foundation for successful, sustainable population management.

Regional Isolation and the Importance of Working Together

Zoos have evolved from primarily entertainment institutions to ones that focus on education and preservation of endangered species with a scientific foundation. The SSPs in the United States and EEPs in Europe and others (e.g., Australasian Species Management Program in Australia) came into being between 1980 and 1985. The idea of species population management was first limited to the particular continent, but global, scientifically based cooperation quickly became essential.

Differences in culture, regional public acceptance, technical progress, and scientific methods were and still are discussed during annual workshops. Today, most equid programs cooperate on an intercontinental basis in order to manage populations globally and preserve genetic diversity.

Opportunities

In striving to meet the sometimes daunting challenges of maintaining sustainable captive populations of wild equids, population managers and researchers have created unique solutions that offer significant opportunities. The creation of semireserves and breeding centers, for example, provides significant land holdings to manage greater numbers of wild equids in naturalistic and semifree ranging habitats. Other opportunities have arisen from significant advances in scientific disciplines of genetics and reproductive biology, creating new tools to effectively monitor populations and augment captive breeding efforts. Finally, zoos have evolved into conservation organizations and now make a significant contribution to conservation and reintroduction efforts of wild equids. These are discussed in detail below.

Semireserves for Asiatic Equids

The first semireserve on the European continent for Asiatic equids was Askania Nova, Ukraine (formerly Russia). The German founder, Friedrich von Falz-Fein (1863–1920), set aside about 2,000 hectares of private land, which was primarily steppe habitat, for wild equids. In 1921, his property was nationalized. Today deer, bison, antelope, and equid species live together in this reserve. About 120 Przewalski's horses and 90 Asiatic wild asses graze there intermittently. Askania Nova is now a world heritage site, as is the Hortobágy National Park in Hungary, where the largest captive population of Przewalski's horses (283 in 2014) roam in the Pentezug Puszta (2,400 hectares; Zimmermann *et al.* 2009).

Smaller semireserves are also of high importance: they are often parts of former military areas, which became valuable flora-fauna-habitats (FFH) in the European community. Together with ruminants (e.g., cattle, deer, sheep, goats), they keep the areas open and thus preserve rare plant and animal species typical for open grasslands. There are currently 17 semireserves in Europe with almost 500 Przewalski's horses and 2 reserves with about 95 Asiatic wild asses. Today, the EEP cannot imagine the Przewalski's horse population without them. With the exception of the Askania Nova and Pentezug reserves, they serve as holding places for single-sex reserve groups—the most important horses from the genetic point of view (low mean kinship values) are maintained until they are needed to be brought back to zoos for breeding. Some places also take less genetically valuable young females to be reared and then shipped to other breeding programs in either the United States or Australia, or particularly to Mongolia and China for reintroduction if they fulfill the conditions for being released (healthy, nonpregnant, socialized, acclimated to grazing year-round, genetically suitable).

The semireserve concept is not well developed in the United States, possibly because the wild equid species present today were never found in North America, and introducing "exotic" or "nonnative" species to public land is not recommended. Even so, North American zoos are developing the concept of breeding consortia with abundant land holdings for large numbers of animals. The Conservation Centers for Species Survival (www.conservationcenters.org) is an example of a collaborative effort to use institutions with large acreage and state-of-the-art facilities for animal handling to intensively study and propagate Persian onagers and Przewalski's horses.

Captive Research

In addition to captive management, zoological institutions take an active role in both research and in situ conservation of wild equids. The better a wild animal species is understood, the easier it is to preserve it successfully in situ and ex situ. Today, all equid species can be kept successfully in zoological gardens. Much of recent research related to captive management is thus concentrated on genetics and artificial insemination.

When SSPs (1980) and EEPs (1985) came into existence, not much was known about the validity of subspecies designations in zebras and asses, and the Przewalski's horse was still a constant subject of debate, as hybridization with domestic horses had occurred in the past. Studies revealed that the domestic horse had only 64 chromosomes while the Przewalski's horse had 66 (Benirschke *et al.* 1965), but only the F1 generation (first generation bred in captivity) of Przewalski's horses could clearly be identified as having 65 chromosomes (indicating some evidence of hybridization with domestic equids). Yet offspring in the F2 generation onward again had 66 chromosomes, so evidence of hybridization was no longer present. A long-lasting and unique cooperation between the SSP, EEP, and the involved institutions (Center for Reproduction of Endangered Species or CRES, San Diego, California, USA; the Veterinary Genetics Laboratory, Davis, California, USA; Cologne Zoo, Cologne, Germany) helped answer many questions and solved numerous problems over the years (Ryder 1986; Ryder *et al.* 1988). Still, today, DNA material is shipped from Europe to the United States and vice versa to continue to unlock the relationship between domestic and Przewalski's horses (Ryder 1994; Goto *et al.* 2011).

Two subspecies of Asiatic wild ass, *E. h. onager* and *E. h. kulan*, were regarded as a single species but have now been recognized and bred as separate subspecies on the basis of morphological characteristics (Groves and Mazak 1967). Recent studies have demonstrated that the chromosomal polymorphism was detected in both subspecies, however, and findings on mitochondrial DNA molecules speak more for a common ancestor (see chap. 7). More research is needed to decide their taxonomic position (Nielsen *et al.* 2007). Genetic research was also conducted to correctly classify the plains zebra (Groves 1986; Groves and Bell 2004). The origins of founder animals in the captive population have often been unknown; stripes and other phenotypic characteristics were extremely variable and thus

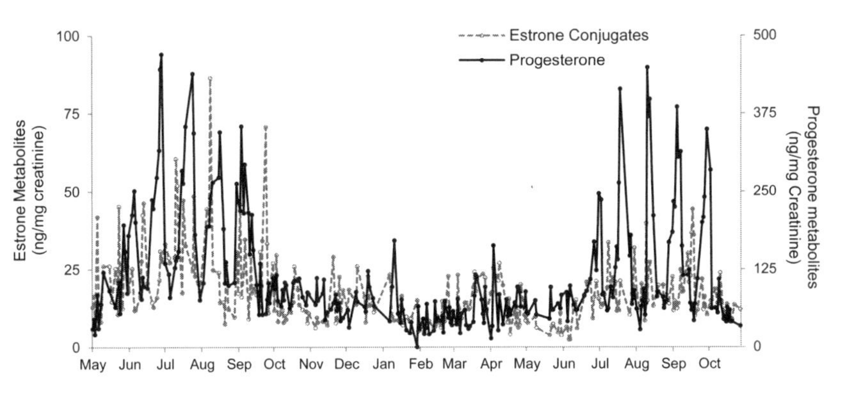

Fig. 11.5 Urinary hormone monitoring of estrogen and progesterone metabolites reveals the seasonal nature of reproductive activity in female Persian onagers. During the breeding season from June to November, females have an estrus cycle every ~25 days, characterized by an estrogen peak and then a progesterone peak.

not reliable in identifying correct subspecies. As genome sequencing has become an established technique in genetic laboratories globally, relationships between species and the evolution of species remain the focus of many research projects.

Despite cooperation between regions and sharing of information, the exchange of genetics between continental regions can be slower than ideal because of the cost of transporting an animal overseas and quarantine requirements. One possible solution is transporting frozen semen between regions and using artificial insemination to augment breeding programs and increase genetic exchange. Artificial insemination (AI) presents possible benefits to global genetic management by permitting collection and banking of genetic material (sperm) not only from captive but also from free-ranging individuals. There is reduced cost and risk to the animals when transporting frozen semen compared to live animals, and semen cryopreservation allows population managers to bank material to preserve the genetic diversity in the captive population. Sperm (and theoretically eggs) can be frozen indefinitely, and then used to reintroduce genes from founder animals decades later, as they are periodically lost from the population (Comizzoli *et al.* 2009). Banking genetic material, in effect, acts as an insurance policy against loss of genetic diversity.

Establishing methods for cryopreserving genetic material and assisted reproductive techniques requires an in-depth understanding of the unique reproductive physiology of each species. Scientists and veterinarians have studied AI and semen cryopreservation for decades in domestic equids, yet the success rates for both techniques are still variable. For example, the success rate of AI with frozen semen can vary between 20% and 60%, depending on the breed and individual stallion. While domestic horses and asses can serve as a model for developing these tools for wild equids, it is important to ascertain the aspects of reproductive biology that differ among species.

What we know of domestic equid reproductive biology stems largely from measuring hormones in blood samples, semen collection using dummy mounts, and serial transrectal ultrasonography of the ovaries (Ginther 1974; Ginther *et al.* 1987; Blanchard *et al.* 2003). Such hands-on techniques are rarely possible or practical when it comes to studying wild equids, even in a captive environment. For wild equids, noninvasive techniques, such as monitoring hormone metabolites in urine or feces, have proven invaluable and provided the bulk of current knowledge about the estrous cycle. Two years of monitoring urinary hormone metabolites in female onagers, for example, revealed details about their seasonality and the length of each estrous cycle (fig. 11.5). Knowledge of ovarian follicle development and ovulation in wild equids has come from facilities with specialized handling devices that allow for brief restraint (~3 minutes) and transrectal ultrasonic examination of the ovaries (Schook *et al.* 2013) as well as necropsy findings (Westlin-van Aarde *et al.* 1988).

All equids, domestic and wild, are monovulatory, resulting in the birth of one offspring, though both do exhibit varying degrees of double ovulations and rare twinning events, where the offspring are not likely to survive. Domestic and Przewalski's horses as well as Asian wild asses are typically seasonal breeders that exhibit estrous cycles and ovulate approximately every 21–25 days during the 6 months surrounding summer in temperate climates, and slightly longer in tropical climates (Ginther 1974; Collins *et al.* 2012; Schook *et al.* 2013) (see chap. 6). Domestic and African wild asses are less seasonal (Ginther *et al.* 1987; Pagan *et al.* 2009), showing cycle lengths of 25–28 days for most or all months of the year, and zebras cycle year-round (Joubert 1974; Asa *et al.* 2001; Nuñez *et al.* 2011). For African species, knowledge of seasonality in each species is important to animal managers in colder climates so that unexpected births do not occur during weather conditions that a newborn foal would be ill equipped

Fig. 11.6 This onager foal born in 2010 was the first wild equid to be produced by artificial insemination. Photo by Mandi Wilder Schook

to handle. In such situations, managers may choose to pull the male during the coldest winter months so that breeding does not occur. Domestic horses and asses as well as Przewalski's horses and zebras ovulate when follicles reach 30–45 mm in diameter, while Asian wild asses ovulate when follicles reach 20–30 mm in diameter (Ginther *et al.* 1972; Westlin-van Aarde *et al.* 1988; Tibary *et al.* 2006; Taberner *et al.* 2008; Collins *et al.* 2012; Schook *et al.* 2013). Knowledge of these differences in seasonality, estrous cycle length, and follicle size is vital when working out the specific timing necessary for artificial insemination to be successful.

Using knowledge gained from basic reproductive biology and noninvasive hormone monitoring techniques, the first wild equids were produced from artificial insemination using fresh ($n = 1$ foal) or frozen ($n = 1$ foal; fig. 11.6) semen in 2010 (Schook *et al.* 2013). These onager births represented a 66% success rate with AI (two of three females inseminated became pregnant and carried to term). The timing of the insemination can be gauged using noninvasive hormone monitoring techniques, thus requiring minimal handling of a wild species. The success of this procedure demonstrates that AI can be a practical tool for the occasional exchange of genetics between regions for global population of wild equid species.

The Role of Zoos in Equid Conservation and Reintroduction

Zoos have millions of visitors annually and can raise both awareness and conservation funds to support in situ programs. In addition, zoos often have staff with considerable expertise and passion for a particular species. Staff can be given both time and resources to participate in or coordinate particular conservation efforts. European and North American zoos coordinate Przewalski's horse reintroduction projects, for example, and they support conservation programs for Grevy's and mountain zebra as well as for Asian and African wild ass. Many zoos also offer competitive grants that specifically fund in situ research and conservation efforts. In a myriad of ways, zoos support in situ programs that monitor and protect current populations, promote habitat preservation, engage local communities, provide sustainable employment, and seek to understand sources of human–wildlife conflict.

Some specific examples of conservation support for African equids include Marwell Wildlife (UK) and Saint Louis Zoo (USA), each with a long-standing involvement in the conservation of Grevy's zebra, including research focused on conservation and anthropological impacts in Kenya (Langenhorst and Davidson 2012). In 1968, when the Yotvata Hai-Bar Nature Reserve (12 km^2) was founded, several kulan and onager specimens from zoos were introduced to form protected populations in the north of Eilat, Israel. Since 1992, zoos have also contributed by sending Przewalski's horses back to the countries of origin for reintroduction. Such projects are expensive and need a continued support over decades. Without technical and personal assistance and a long-lasting cooperation with the organizations and staff on site, such projects are not successful (see chap. 14).

Along with zoos' support of conservation efforts, a sound program of worldwide captive population management of endangered African and Asian wild equids is of vital importance until each species has a stable population in their countries of origin. Zoos attract the interest of the public by reporting about conservation and reintroduction projects, collecting money, and convincing more organizations to sponsor these projects in order to make the long-lasting captive breeding of an endangered species a success story.

Conclusion

Maintaining captive populations of wild equids has proven to be a valuable tool for conservation of these species. Whether through public education, fundraising, research, technology transfer, or the provision of animals for reintroduction, zoos have demonstrated that the goals of sound husbandry, excellent animal welfare, and in situ conservation are compatible and achievable. Challenges remain, as the long-term sustainability of these populations requires further innovation and scientific research. New approaches and philosophies related to captive propagation must be explored and tested while simultaneously maintaining rigorous standards of care and management that are time-tested. Zoos must also continually find ways to generate funds to support in situ programs. The groundwork has begun as zoos have explored management options like semireserves, breeding consortia, and assisted reproduction. Similarly, zoos are engaging their audiences in new ways with technology and social media to extend their reach and inspire conservation support. The future of wild equids in situ is uncertain but will most certainly require intensive management in the wild, possibly resembling the kinds of management used in captivity. Zoos and other captive breeding centers have much to offer in this regard in developing integrated, sustainable conservation programs for wild equids.

ACKNOWLEDGMENTS

The authors thank Paul Calle for providing information on routine veterinary care for wild equids, Dan Beetem for sharing his expertise in equid management, and Nilda Ferrer for assisting with gathering ISIS data on wild equids in captivity.

REFERENCES

Adams, D.B., A.R. Gold, and A.D. Burt. 1978. Rise in female-initiated sexual activity at ovulation and its suppression by oral contraceptives. New England Journal of Medicine 299:1145–1150.

Asa, C.S., and I.J. Porton. 2010. Contraception as a management tool for controlling surplus animals. Pages 469–482 *in* K.V.T. Devra, G. Kleiman, and C.K. Baer (Eds.), Wild mammals in captivity: Principles and techniques for zoo management. University of Chicago Press, Chicago, USA.

Asa, C.S., J.E. Bauman, E.W. Houston, M.T. Fischer, B. Read, C.M. Brownfield, and J.F. Roser. 2001. Patterns of excretion of fecal estradiol and progesterone and urinary chorionic gonadotropin in Grevy's zebras (*Equus grevyi*): Ovulatory cycles and pregnancy. Zoo Biology 20:185–195.

Asa, C.S., K.L. Bauman, S. Devery, M. Zordan, G.R. Camilo, S. Boutelle, and A. Moresco. 2014. Factors Associated with uterine endometrial hyperplasia and pyometra in wild canids: Implications for fertility. Zoo Biology 33:8–19.

Association of Zoos and Aquariums. 2013. Plains zebra studbook. Equid Taxon Advisory Group, USA.

Atkinson, M.W., and E.S. Blumer. 1997. The use of a long-acting neuroleptic in the Mongolian wild horse (*Equus przewalskii przewalskii*) to facilitate the establishment of a bachelor herd. Proceedings of the Annual Meeting of the American Association of Zoo Veterinarians:199–200.

Ballou, J.D., and R.C. Lacy. 1995. Identifying genetically important individuals for management of genetic variation in pedigreed populations. Pages 76–111 *in* J.D. Ballou, M. Gilpin, and T.J. Foose (Eds.), Population management for survival and recovery: Analytical methods and strategies in small population conservation. Methods and cases in conservation science. Columbia University Press, New York, USA.

Bannikov, A.G. 1981. The Asiatic wild ass. Institute of Ecological and Evolutionary Problems, Moscow, Russia.

Benirschke, K., N. Malouf, R.J. Low, and H. Heck. 1965. Chromosome complement: Differences between *Equus caballus* and *Equus przewalskii*, Poliakoff. Science 148:382–383.

Blanchard, T.L., D.D. Varner, J. Schumacher, C.C. Love, S.P. Brinsko, and S.L. Rigby. 2003. Semen collection and artificial insemination. Pages 131–142 *in* Manual of equine reproduction, 2nd ed. Mosby, Saint Louis, MO, USA.

Bowling, A.T., W. Zimmermann, O. Ryder, C. Penado, S. Peto, L. Chemnick, N. Yasinetskaya, and T. Zharkikh. 2003. Genetic variation in Przewalski's horses, with special focus on the last wild caught mare, 231 Orlitza III. Cytogenetic and Genome Research 102:226–234.

Boyle, P., B. Andrews, C. Dorsey, M. Fouraker, D. Pate, M. Reed, and B. Wiese. 2011. Building sustainable zoo populations. Connect, Association of Zoos and Aquariums, Silver Spring, MD, USA.

Collins, C.W., N.S. Songsasen, M.M. Vick, B.A.Wolfe, R.B. Weiss, C.L. Keefer, and S.L. Monfort. 2012. Abnormal reproductive patterns in Przewalski's mares are associated with a loss in gene diversity. Biology of Reproduction 86:28.

Comizzoli, P., A.E. Crosier, N. Songsasen, M.S. Gunther, J.G. Howard, and D.E. Wildt. 2009. Advances in reproductive science for wild carnivore conservation. Reproduction in Domestic Animals 44 Supplement 2:47–52.

Dierenfeld, E.S., P.P. Hoppe, M.H. Woodford, N.P. Krilov, V.V. Klimov, and N.I. Yasinetskaya. 1997. Plasma alpha-tocopherol, beta-carotene, and lipid levels in semi-free-ranging Przewalski horses (*Equus przewalskii*). Journal of Zoo and Wildlife Medicine 28:144–147.

Eisenmann, V. 1992. Origins, dispersals, and migrations of *Equus* (*Mammalia, Perissodactyla*). Courier Forschungsinstitut Senckenberg 153:161–170.

Flugger, M., and K. Jurczynski. 2005. Experiences with the use of the long-acting-tranquilizer perphenazine in onagers (*Equus hemionus onager*) and Chapman's zebras (*Equus burchelli antiquorum*) in Hagenbecks Tierpark. Erkrankungen der Zootiere 42:30–35.

Frank, K.M., R.O. Lyda, and J.F. Kirkpatrick. 2005. Immunocontraception of captive exotic species. Zoo Biology 24:349–358.

Ginsberg, J.R. 1988. Social organization and mating strategies of an arid adapted equid: The Grevy's zebra. Thesis, Princeton University, Princeton, NJ, USA.

Ginther, O.J. 1974. Occurrence of anestrus, estrus, diestrus, and ovulation over a 12-month period in mares. American Journal of Veterinary Research 35:1173–1179.

Ginther, O.J., H.L. Whitmore, and E.L. Squires. 1972. Characteristics of estrus, diestrus, and ovulation in mares and effects of season and nursing. American Journal of Veterinary Research. 33:1935–1939.

Ginther, O.J., S.T. Scraba, and D.R. Bergfelt. 1987. Reproductive seasonality of the jenney. Theriogenology 27:587–592.

Gomendio, M., J. Cassinello, and E.R.S. Roldan. 2000. A comparative study of ejaculate traits in three endangered ungulates with different levels of inbreeding: Fluctuating asymmetry as an indicator of reproductive and genetic stress. Proceedings of the Royal Society of London B: Biological Sciences 267:875–882.

Goto, H., O.A. Ryder, A.R. Fisher, B. Schultz, S.L. Kosakovsky Pond, A. Nekrutenko, and K.D. Makova. 2011. A massively parallel sequencing approach uncovers ancient origins and high genetic variability of endangered Przewalski's horses. Genome Biology and Evolution 3:1096–1106.

Groves, C.P. 1986. The taxonomy, distribution, and adaptation of recent equids. Pages 11–51 *in* M. Meadow and R.H. Uerpmann (Eds.), Equids in the ancient world. Reichert-Verlag, Wiesbaden, Germany.

Groves, C.P., and C.H. Bell. 2004. New investigations on the taxonomy of the zebra's genus *Equus*, subgenus *Hippotigris*. Mammalian Biology: Zeitschrift für Säugetierkunde 69:182–196.

Groves, C.P., and V. Mazak. 1967. On some taxonomic problems of Asiatic wild asses with the description of a new subspecies (Perissodactyla; Equidae). Mammalian Biology: Zeitschrift für Säugetierkunde 32:321–355.

Hegel, G., H. Wiesner, and T. Hänichen. 1990. Fertilitätsstörungen bei Przewalski-Urwildpferden. Presented at Internationales Symposion zur Erhaltung des Przewalskipferdes, Zoologischer Garten Leipzig, Leipzig, Germany.

Henderson, J.A., and C.A. Shively. 2004. Triphasic oral contraceptive treatment alters the behavior and neurobiology of female cynomolgus monkeys. Psychoneuroendocrinology 29:21–34.

Heptner, V.G., and N.P. Naumov. 1966. Die Säugetiere der Sowjetunion. Jena, Fischer, Germany.

Heuschkel, B., A. Kroehne, and W. Zimmermann. 1999. Die Haltung von Grevy-zebras im Kölner Zoo. Zeitschrift des Kölner Zoo 42:103–120.

Hoffmann, R. 1985. On the development of the social behavior in immature males of a feral horse population (*Equus Przewalskii F caballus*). Zeitschrift Fur Saugetierkunde: International Journal of Mammalian Biology 50:302–314.

Houston, B., D. Powell, B. Huffman, and M. Fischer. 2011. Striving for sustainability: The ungulate manager's toolbox. Animal Keeper's Forum 38:362–367.

Joubert, E. 1974. Notes on the reproduction in Hartmann zebra *Equus zebra hartmannae* in south west Africa. Madoqua 1:31–35.

Kirkpatrick, J.F., W. Zimmermann, L. Kolter, I.K.M. Liu, and J.W. Turner. 1995. Immunocontraception of captive exotic species. 1. Przewalski's horses (*Equus Przewalksii*) and banteng (*Bos javanicus*). Zoo Biology 14:403–416.

Lacy, R.C. 2013. Achieving true sustainability of zoo populations. Zoo Biology 32:19–26.

Langenhorst, T., and Z. Davidson. 2012. Grevy's zebra conservation: 2013 report and funding proposal prepared for supporters and members of the Grevy's zebra EEP conservation projects. Marwell Wildlife, Marwell, UK.

Lees, C.M., and J. Wilcken. 2011. Global programmes for sustainability. WAZA Magazine 12:2–10.

Leus, K., L.B. Lackey, W. van Lint, D. de Man, S. Riewald, A. Veldkam, and J. Wijmans. 2011. Sustainability of European Association of Zoos and Aquaria bird and mammal populations. WAZA Magazine 12:11–14.

Mohr, E. 1959. The primitive wild horse, *Equus przewalskii Poljakoff* 1881. Die neue Brehm-Bucherei, Wittenberg Lutherstodt, A. Ziemsen Verlog, Germany.

Mohr, E. 1968. Studbooks for wild animals in captivity. International Zoo Yearbook 8:159–166.

National Research Council. 2013. Using science to improve the BLM Wild Horse and Burro Program: A way forward. National Academies Press, Washington, DC, USA.

Nielsen, R.K., C. Pertoldi, and V. Loeschcke. 2007. Genetic evaluation of the captive breeding program of the Persian wild ass. Journal of Zoology 272:349–357.

Nobis, G. 1971. Vom Wildpferd zum Hauspferd: Zur Phylogonie pleistozäner Equiden Eurasiens und das Domestikationsproblem unserer Hauspferde. Böhlau-Verlag, Cologne, Germany.

Nuñez, C.M.V., C.S. Asa, and D.I. Rubenstein. 2011. Zebra reproduction. Pages 2851–2865 *in* A.O. McKinnon, E.L. Squires, W.E. Vaala, and D.D. Varner (Eds.), Equine reproduction. Wiley-Blackwell, West Sussex, UK.

Pagan, O., F. Von Houwald, C. Wenker, and B.L. Steck. 2009. Husbandry and breeding of Somali wild ass *Equus africanus somalicus* at Basel Zoo, Switzerland. International Zoo Yearbook 43:198–211.

Penfold, L.M., D. Powell, K. Traylor-Holzer, and C. Asa. 2014. "Use it or lose it": Characterization, implications, and mitigation of female infertility in captive wildlife. Zoo Biology 33:20–28.

Powell, D. 2001. Zebra. Pages 1371–1375 *in* C.E. Bell and L.E. Fisher (Eds.), Encyclopedia of the world's zoos. Fitzroy Dearborn, London, UK.

Powell, D.M., and M.C. Gartner. 2010. Applications of personality to the management and conservation of nonhuman animals. Pages 185–199 *in* M. Inoue-Murayama, M. Kawamura, and A. Weiss (Eds.), From genes to behavior: Social structure, personalities, communication by color. Springer, Tokyo, Japan.

Rademacher, U. 1997. Sozialverhalten von Grevy zebras in Zoologischen Gärten. University of Bielefeld, Bielefeld, Germany.

Rademacher, U., and A. Winkler. 2000. EEP husbandry guidelines for African equids. Der Zoologisch-Botanische Garten Stuttgart, Wilhelma, Germany.

Ransom, J.I., N.T. Hobbs, and J. Bruemmer. 2013. Contraception can lead to trophic asynchrony between birth pulse and resources. PLoS ONE 8:e54972.

Ruiz-Lopez, M.J., D.P. Evenson, G. Espeso, M. Gomendio, and E.R.S. Roldan. 2010. High levels of DNA fragmentation in spermatozoa are associated with inbreeding and poor sperm quality in endangered ungulates. Biology of Reproduction 83:332–338.

Ryder, O.A. 1986. Genetic investigations: Tools for supporting breeding programme goals. International Zoo Yearbook 24/25:157–162.

Ryder, O.A. 1994. Genetic studies of Przewalski's horses and their impact on conservation. Pages 75–92 *in* L. Boyd and K.A. Houpt (Eds.), Przewalski's horse: The history and biology of an endangered species. State University of New York Press, Albany.

Ryder, O.A., J.H. Shaw, and C.M. Wemmer. 1988. Species, subspecies and ex situ conservation. International Zoo Yearbook 27:134–140.

Saltz, D., and D.I. Rubenstein. 1995. Population dynamics of a reintroduced Asiatic wild ass (*Equus hemionus*) herd. Ecological Applications 5:327–335.

Schook, M.W., D.E. Wildt, R.B. Weiss, B.A. Wolfe, K.E. Archibald, and B.S. Pukazhenthi. 2013. Fundamental studies of the reproductive biology of the endangered persian onager (*Equus hemionus onager*) result in first wild equid offspring from artificial insemination. Biology of Reproduction 89:41.

Taberner, E., A. Medrano, A. Peña, T. Rigau, and J. Miró. 2008. Oestrus cycle characteristics and prediction of ovulation in Catalonian jennies. Theriogenology 70:1489–1497.

Tibary, A., A. Sghiri, and M. Bakkoury. 2006. Reproductive patterns in donkeys. Proceedings of the 9th Congress of the World Equine Veterinary Association, Marrakech, Morocco:311–319.

Westlin-van Aarde, L.M., R.J. van Aarde, and J.D. Skinner. 1988. Reproduction in female Hartmann's zebra, *Equus zebra hartmannae*. Journal of Reproduction and Fertility 84:505–511.

World Association of Zoos Aquariums. 2005. Building a future for wildlife—The World Zoo and Aquarium conservation strategy. Bern, Switzerland.

Zehnder, A.M., J.C. Ramer, and J.S. Proudfoot. 2006. The use of altrenogest to control aggression in a male Grant's zebra (*Equus burchelli boehmi*). Journal of Zoo and Wildlife Medicine 37:61–63.

Zharkikh, T.L. 2009. Behaviour of bachelor males of the Przewalski horse (*Equus ferus przewalskii*) at the reserve Askania Nova. Der Zoologische Garten 78:282–299.

Zimmermann, W. 1997. Das Erhaltungszuchtprogramm Przewalskipferd, eine zehnjährige Zusammenarbeit in Europa. Pages 189–199 *in* A.U. Schreiber and J. Lehmann (Eds.), Populationsgenetik im Artenschutz. Landesanstalt für Ökologie, Bodenordnung und Forsten, Nordrhein-Westfahlen, Germany.

Zimmermann, W., and L. Kolter. 2016. EEP Asiatic equids husbandry guidelines. Zoologischer Garten Köln, Cologne, Germany.

Zimmermann, W., K. Brabender, and L.A. Kolter. 2009. A Przewalski's horse population in a unique European Steppe reserve—The Hortobágy National Park in Hungary. Pages 257–284 *in* Equus. Zoo Praha, Prague, Czech Republic.

PART III CONSERVATION

12

Status and Conservation of Threatened Equids

PATRICIA D. MOEHLMAN,
SARAH R.B. KING, AND
FANUEL KEBEDE

During the Pleistocene, equids were one of the most abundant and ecologically important grazing animals on the grasslands of Africa, Asia, and the Americas. Today there remain only seven species of wild equid—three species of ass, three of zebra, and one wild horse—and most of these are threatened with extinction. African wild ass (*Equus africanus*), Grevy's zebra (*E. grevyi*), mountain zebra (*E. zebra*), Asiatic wild ass (*E. hemionus*), and Przewalski's horse (*E. ferus przewalskii*) have all experienced significant range reductions and population declines, both historically and over the last 100 years. Only the plains zebra (*E. quagga*) and kiang (*E. kiang*) persist in large numbers and are categorized as Least Concern by the International Union for Conservation of Nature (IUCN) Red List. Zebras, asses, and horses can serve as "flagship" species for the conservation of biodiversity in their ecosystems (e.g., Liu *et al.* 2014). Many threatened wild equids live in arid habitats where access to water and forage is critical. Arid habitats are also home to human populations that are at risk from the same climatic extremes that threaten wild equids. Conservation of wildlife is therefore closely linked to local people actively participating in and benefiting from the conservation management of their areas (Low *et al.* 2009; Kebede *et al.* 2012).

In most populations of large, long-lived herbivores, drought years and low offspring survival have the potential to be compensated by recruitment over the long term, but when individuals are also killed for meat, medicine, and money, it is difficult for a population to recover (Moehlman 2002, 2005). Moreover, competition with herders for food and water may reduce available range and resources, and have a negative effect on juvenile survival and recruitment (Williams 2002). During drought conditions, pastoralists may limit wildlife access to water, and forage may be excessively grazed at greater distances from permanent water sources (Williams 2002; Kebede *et al.* 2012). Additionally, wildlife, including equids, may become an alternative source of protein. The combination of human-induced adult mortality and a low reproductive rate in equids may increase their vulnerability to extinction compared to other ungulates.

In the conservation of species, it is useful to differentiate between "declining populations" and "small populations" (Caughley 1994). Most declining populations of wild equids face external threats to their survival, including illegal and nonsustainable hunting, loss of habitat due to increasing herds of livestock, the complete loss of pastures to agriculture, and reduced access to forage and water. But some species and populations face internal threats from small population size. Specifically, these factors include slow population growth owing to density-dependent social interactions, inbreeding, hybridization, and vulnerability to stochastic factors such as disease, droughts, and extreme winters (see chap. 6). Small populations are also more vulnerable to natural predation by carnivores. Additionally, climate change has the potential of increasing the variability of rainfall and temperature as well as frequency of negative stochastic events.

Why are so many wild equids threatened with extinction? Does basic equid biology—that is, a slow reproductive rate, dependence on water, and hindgut digestion—limit their options and flexibility? Or does potential competition with humans and their livestock lead to declines? Does a combination of restricted access to water and forage limit foal survival and recruitment? Or can overhunting alone send equid populations into a downward spiral? In this chapter, we provide data and insight as to how these factors singly or in concert can threaten wild equid populations. Understanding how environmental and anthropogenic factors affect wild equid populations is critical for determining why most equid species are threatened. Many believe the demise of equids results from a combination of habitat loss and overhunting by humans. Additionally, equid biology may make them more vulnerable and less likely to cope with long-term habitat and climate change (Coughenour 2002; Moehlman 2002).

It is important to evaluate the role of equids in their ecosystems. For the past two decades, data sets from Africa indicate that in protected areas where equids occur, their population sizes have tended to remain stable when most other medium- to large-sized herbivores have fluctuated significantly (Moehlman *et al.* 1997; Ogutu and Owen-Smith 2003; Grange *et al.* 2004). Equids can potentially outcompete ruminants when resources are limiting, which means that they will be less affected by the effect of climate on primary productivity (Ogutu and Owen-Smith 2003), but they can potentially be more affected by predation (Grange *et al.* 2004, 2015). The role of equid social organization may also mediate the impact of resource availability and predation as well as their fundamental demographic traits. It is important to determine whether equid life history strategies are significantly different from ruminant herbivores of a similar body size.

The main factor affecting wild equid survival and reproduction is access to water and adequate forage. During severe droughts, lack of water and forage can lead to high mortality in all age and sex classes. Research on Grevy's zebra has shown that even when rainfall is normal, if females and their foals have to forage far from water, foals do not survive and recruitment is low (Williams 1998). When populations are forced to the edge of their normal range, water and forage resources may not be adequate for the sustenance of a stable population; this may have been a factor in the decline of the Przewalski's horse. In large ecosystems like the Serengeti, plains zebra can move seasonally for water and forage, and thus have maintained a stable population for decades (Ogutu *et al.* 2011). Similarly, Asiatic wild asses in the Gobi Desert have been able to move seasonally over large distances (Kaczensky *et al.* 2011*b*) and appear to have a large stable population. But roads, railroads, and fences have the potential to fragment large ecosystems and adversely affect equid populations and other large ungulates (Ito *et al.* 2013).

In addition to direct impacts, these human impacts can indirectly affect equids by eliminating access to water and forage via fencing, agriculture, and the presence of large herds of domestic livestock. These have affected Grevy's zebra and Asiatic wild ass in the western portions of their range. Domestic livestock can also affect wild equids by the transmission of disease, either directly or through vectors (Robert *et al.* 2005), which may have a disproportionate effect on populations that are already stressed as a result of poor habitat or reintroduction efforts. Human development can also influence species' distribution such that populations become more isolated and suffer low genetic diversity (Hill 2009). Species that were formerly allopatric may develop overlapping ranges, and the probability of hybridization increases (Cordingley *et al.* 2009). Both legal and illegal hunting for meat, medicine, and commercial gain can adversely affect wild equid populations. When such increased—and often unsustainable—mortality is added to poor recruitment, the result can be devastating and drive small populations to extinction.

The Przewalski's horse provides a dramatic story of how a successful equid can be compromised by competition for water and forage and how, in combination with hunting and severe weather, can be driven to extinction in the wild. It is worrying that a similar fate may lie ahead for the Asiatic wild ass in most of its range; it also had a Eurasian distribution but has been reduced to mainly small and isolated populations.

Status of Wild Equids in the IUCN Red List of Threatened Species

The current state of knowledge of the seven wild equids and of the environmental and anthropogenic threats to their long-term conservation provides insight as to what research and conservation actions are needed to secure their future. An important factor in successfully conserving wild equids is obtaining accurate information on their population status, distribution, and threats to survival. The IUCN Red List of Threatened Species was developed to provide science-based information on the current threatened status of global biodiversity (IUCN 2015). To be assessed under one of the threatened categories (Critically Endangered, Endangered, or Vulnerable), the best available evidence must indicate that the species meets the criteria for this listing. The IUCN Equid Specialist Group strives to keep species assessments up to date, with the whole Red List being published approximately every four years. Current status and full details from the most recent assessments of equid species are available at www.iucnredlist.org.

Assessments are made at the species level for the global range, although for species such as the mountain zebra, where there is taxonomic dispute about whether the two subspecies should be species, assessments are also made at a subspecific level. Domestic species are not considered to be wild equid species, so feral horses and feral donkeys are not assessed. Furthermore, for the Red List criteria, the term "population" refers to the total number of mature individuals in the species rather than the normal biological usage, with subpopulations being geographically distinct groups that have little genetic exchange with the rest of the population (IUCN 2013). Mature individuals of wild species are those that are capable of reproduction (IUCN 2013), so for most species of equid, that means individuals that are >2 years old, and sometimes 5 years old. For reintroduced species such as Przewalski's horse, however, the population consists of both wild- and captive-born individuals. Wild-born individuals are considered mature as for any other species, but captive-born individuals can only be included as mature individuals in the population when they have reached the age of maturity and produced offspring that have proven viable (i.e., the reintroduced individual is a grandparent).

Twenty-five percent of all mammals that had adequate data for assessment ($n = 1{,}139$) are threatened with extinction (Schipper *et al.* 2008). As a family, equids are highly endangered, with >70% of the species assessed as threatened (five of the seven equid species). To put this in context, nearly one-quarter of all mammal species are threatened, with Perissodactyls being the second most threatened order (>80% of Perissodactyl species are threatened or recently extinct). Quantifying this level of threat under the objective IUCN Red List of Threatened Species allows conservationists and policymakers to develop an appropriate response to prevent further decline. The Red List can be used to develop strategies for prioritizing species and areas for conservation action. An analysis of the impact of conservation actions on the status of the world's vertebrates yielded evidence that targeted conservation action can reverse declines in biodiversity and species threatened status (Hoffman *et al.* 2010). For example, Przewalski's horse has been down-listed from Critically Endangered to Endangered (King *et al.* 2015), and the mountain zebra was down-listed from Endangered to Vulnerable in 2008 (Moehlman *et al.* 2008*a*).

All equid species are threatened in varying degrees by (1) range reduction, (2) reduction in water and forage accessibility, (3) predation and illegal hunting, (4) fragmented and small population size, and (5) reduced gene flow. Climatic extremes and stochastic events like drought and severe winters can exacerbate these factors.

Least Concern

Plains zebra and kiang are currently classified as Least Concern by the IUCN, meaning that the species are considered widespread and abundant. These two species are the only wild equids that do not qualify for an IUCN threatened category. Both species have large populations and have not experienced severe reduction in their historic range (figs. 12.1–12.4). Their status indicates that equids can flourish, and they provide insight as to the conditions needed for sustaining viable populations. These two species represent the two different social systems seen in equids: plains zebra have a harem/family mating system and occur in relatively mesic habitats; kiang social organization is a "territorial defense" system similar to that of other wild equids living in arid conditions (Shah *et al.* 2008; St-Louis and Côté 2008). They also demonstrate the extremes at which equids can live: from high-elevation grasslands to semideserts. Both species have ranges spanning large parts of a continent with contiguous grassland and areas with little human population pressure. Both species also have a large abundance within protected areas that are supporting the population, although outside these areas populations are often declining. The fact that these species continue to be widespread and abundant provides hope for the conservation of other equids in an increasingly anthropocentric environment.

Plains Zebra

Plains zebra are the most abundant wild equid. Their current range is from South Sudan and Ethiopia to South Africa. In some locales, however, overhunting

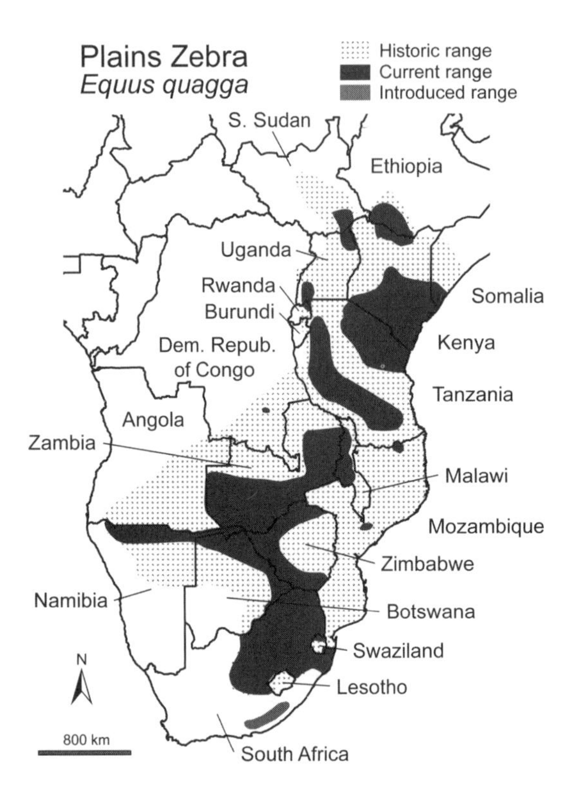

Fig. 12.1 Plains zebra (*E. quagga*) historic and current range. Based on Hack *et al.* 2002

and habitat loss have resulted in declining populations. The quagga (*E. quagga quagga*), a subspecies/subpopulation of the plains zebra, was driven to extinction by overhunting and loss of resources owing to competition with livestock (Higuchi *et al.* 1984; Klein and Cruz-Uribe 1999; Leonard *et al.* 2005), and plains zebra are now extinct in two countries in which they formerly occurred: Burundi and Lesotho. In Angola, only one or two small populations persist (P. vaz Pinto, personal communication), with one of the populations being supported by transboundary dispersal from Namibia.

In South Africa, plains zebra are at risk resulting from loss of habitat from agricultural and livestock ranching and decrease in genetic diversity in small populations (<100 zebra) on private property, as well as spread of pathogens and parasites through translocations into new areas. Furthermore, competition with livestock for food and water has resulted in hunting and fencing off zebra from rangelands (Stears *et al.* 2015). Information from South Africa provides insight as to the future threats to plains zebra in other parts of its range. As human populations increase and agriculture and ranching become more developed, there will be increased competition for land and resources, and fencing of ranches will isolate populations and reduce or eliminate gene flow.

The existence of substantial protected areas—for example, the Serengeti-Mara ecosystem—helps to ensure the continued existence of this species, but populations outside parks and on private land are more at risk.

Fig. 12.2 Plains zebra.
Photo © Patricia D. Moehlman

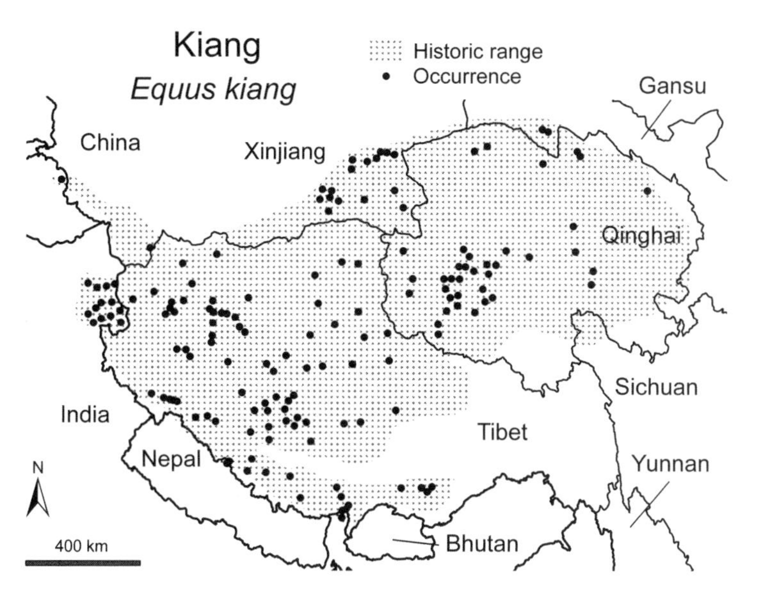

Fig. 12.3 Kiang (*E. kiang*) historic and current range. Black dots are central locations of current populations. Based on Shah 2002

Because of the extensive range of plains zebra and their local occurrence in both protected and nonprotected areas, they present a challenge in terms of frequency and accuracy of monitoring. Hack *et al.* (2002) recommended that, in addition to more frequent monitoring, conservation action was needed for assessing risk, improving knowledge of basic biology, and investigating economic alternatives of various utilization strategies.

Fig. 12.4 Kiang. Photo © Patricia D. Moehlman

Kiang

Most kiang are found on the Tibetan Plateau between 2,700 and 5,400 m above sea level, but they also range into India, Nepal, and Pakistan. As of 2008, the global population estimate was 60,000 to 70,000 animals (Shah *et al.* 2008). This estimate is based on multiple surveys with different methodology over multiple years, however. The areas to be surveyed are vast, and kiang populations are fragmented and move seasonally. In some areas, populations appear to be increasing, such as in the Chang Tang Nature Reserve (Schaller *et al.* 2005), but they may concentrate in areas where the human density is low (Fox and Bårdsen 2005). In Ladakh, kiang density appears to be stable (Bhatnagar *et al.* 2006).

Access to water during dry periods is critical. Kiang are potentially threatened by overhunting and conflicts with humans and livestock. In the past, local populations have been decimated and are no longer present or are scattered in what were former ranges. Better protection and enforcement of hunting regulations have resulted in population increases in some areas (Schaller *et al.* 2005). In the future, increasing populations of people and livestock have the potential to negatively affect kiang both in terms of loss of habitat and forage and the introduction of disease. Management on a local scale is needed to temper and resolve issues concerning habitat, forage, and water use (e.g., Tsering *et al.* 2006). The increasing use of fencing on the Tibetan Plateau endangers both the kiang and pastoral livelihoods. Mining for gold, both by commercial companies and opportunistic individuals, is also a threat to the viability of kiangs and their habitat.

Similar to the plains zebra, the numbers and range of kiang make their numbers difficult to assess. They often occur in remote and isolated mountainous regions, where surveys are difficult. Like other animals that are perceived as common, it is important that regular

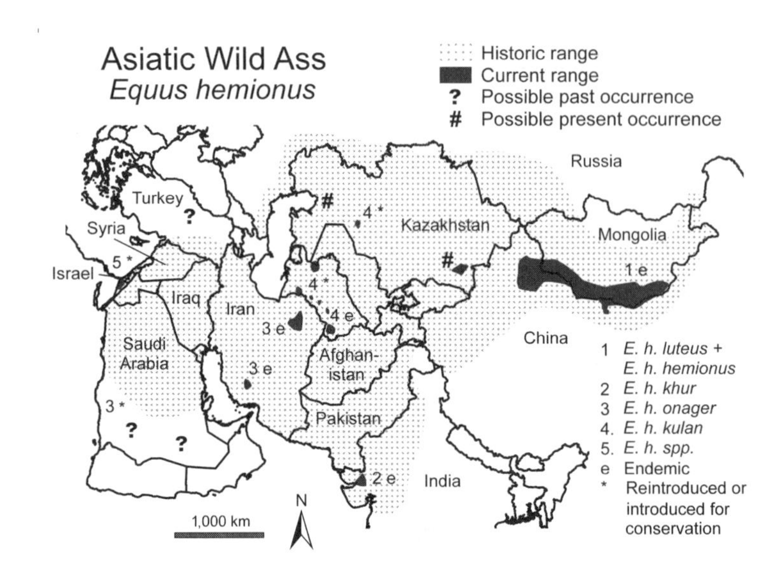

Fig. 12.5 Asiatic wild ass (*E. hemionus*) historic and current range. Based on Feh et al. 2002

population assessments are made in both local and range wide areas so that any decline is noticed before it becomes irreversible.

Vulnerable

Asiatic wild ass and mountain zebra are classified as Vulnerable by the IUCN Red List, as they face a high risk of extinction in the wild. Like the plains zebra and kiang, the Asiatic wild ass and mountain zebra represent the two different social systems in equids, illustrating that social system is not directly associated with endangered status. Both Asiatic wild ass and mountain zebra have experienced major reductions in range over the last 100 years, and both have fragmented populations within their historic range. They each have one more or less major contiguous population of a subspecies that is currently stable or increasing—khulan (*E. h. hemionus*) in Mongolia and China, and Hartmann's mountain zebra in Namibia (*E. z. hartmannae*)—but the other subspecies are found in small and isolated populations. The large populations in Mongolia / China and Namibia are robust and appear to be stable. Protected areas and community-based conservation development have played a significant role in sustaining these populations, but some types of economic development (mining, transportation infrastructure) pose a threat by partitioning their range and reducing seasonal nomadic movements. In the past, illegal hunting, especially for commercial purposes, has decimated formerly robust populations within short periods of time. Focused and appropriate management actions (better protection and reintroductions) have allowed some of the populations to recover.

Asiatic Wild Ass

Historically, Asiatic wild ass had one of the largest known ungulate ranges, found from Mongolia, China, and Russia west to Turkey, Syria, and the Arabian Peninsula (figs. 12.5 and 12.6). By the nineteenth century, the range of the Asiatic wild ass had declined significantly. Today the range is discontinuous, and most populations are small and isolated. The most abundant subpopulation of the species occurs in the southern part of Mongolia and adjacent northern China. Indigenous isolated populations survive in the Rann of Kutch (India), the Badkhyz Strictly Protected Area (Turkmenistan), and Touran National Park and Bahram-e-goor Reserve (Iran). Populations have been reintroduced in Turkmenistan, Kazakhstan, Uzbekistan, and Israel. The reestablished populations in Ukraine and Israel are not of the subspecies that originally occurred there, and the reintroduced population in Israel is of hybrid origin (*E. h. onager* and *E. h. kulan*) (Feh *et al.* 2002).

The Asiatic wild ass was listed by the IUCN as Endangered in 2008 because it was estimated to have declined by >50% over the previous 16 years, and Wingard and Zahler (2006) reported that the illegal trade in Mongolian khulan was removing ~3,000 individuals per year from the population (Lkhagvasuren 2007; Moehlman *et al.* 2008c). But recent surveys in Mongolia indicate that the population has remained relatively stable (Ransom *et al.* 2012; Norton-Griffiths *et al.* 2013), and illegal poaching is now minimal (Kaczensky *et al.* 2015).

There are five recognized subspecies of Asiatic wild ass (Oakenfull *et al.* 2000; Grubb 2005): the khulan in Mongolia and China; the khur in India (*E. h. khur*);

Fig. 12.6. Khulan. Photo © Anne-Camille Souris 2008

the Turkmen kulan, native to Turkmenistan and reintroduced in other areas of Turkmenistan, Kazakhstan, Uzbekistan, and Ukraine, and extirpated in Afghanistan (*E. h. kulan*); the onager in Iran (*E. h. onager*); and the Syrian wild ass, which is extinct but formerly from Syria south into the Arabian Peninsula (*E. h. hemippus*). As a species, the Asiatic wild ass is legally protected in Mongolia, Turkmenistan, Iran, and India. The khulan and khur are included in CITES Appendix I, with the other subspecies being in CITES Appendix II. The species is included in Appendix II of the Convention on the Conservation of Migratory Species of Wild Animals.

The largest population of Asiatic wild ass is the khulan in Mongolia and China, with around 75% of the global population of the species found in Mongolia. Recent surveys indicated that the population is ~42,000 individuals in Mongolia (Reading *et al.* 2001; Ransom *et al.* 2012; Norton-Griffiths *et al.* 2013), with ~5,000 in China (Q. Cao, personal communication, based on Bi 2007 and Chu *et al.* 2008). Unfortunately, there is little chance of migration between Mongolia and China because of a border fence (Kaczensky *et al.* 2011*a*; P. Kaczensky, personal communication). The population in Mongolia is threatened by competition for water and pasture with livestock. The improved economy in Mongolia as a result of mineral resource extraction has reduced the level of offtake of khulan by illegal hunting, but now represents a new threat as more linear infrastructure such as roads, railways, and fences are built, with the potential to prevent migrations and dispersal of the population (Batsaikhan *et al.* 2014; Kaczensky *et al.* 2015).

The Indian khur population in the Little Rann of Kutch went from 3,000 to 5,000 in 1946 to a low of 360 in 1967 (Clark and Duncan 1992). Since then, with increased protection, the population has recovered to ~4,000 individuals (in 2009; N. Shah, personal communication). This area is threatened by hydrological and agricultural development, and the current population has limited area for expansion (Shah and Qureshi 2007), resulting in the subspecies also being persecuted by local people as they raid crops.

The Turkmen kulan had sharply declined until the establishment of Badkhyz Strictly Protected Area, bordering Iran and Afghanistan, in 1941. With this protection, the population increased to 6,000 animals in 1993. There was a precipitous decline to 646 by 2000, which was attributed to illegal commercial hunting (Feh *et al.* 2002; V. Lukarevskyi, personal communication). The estimate in 2013 was 420 individuals (N. Khudaykuliyev, personal communication). In the 1980s, when the population was robust, it provided animals for reintroduction efforts elsewhere in Turkmenistan and to Kazakhstan (see chap. 14). The current population in Turkmenistan, including kulan that have spread across the border to Uzbekistan (Kuznetsov 2014; N. Marmazinskaja, unpublished data 2012/2013), is around 920 animals in five locations (Kaczensky *et al.* 2015). By comparison, the reintroduced kulan (32 in 1982–1984) in Kazakhstan are increasing, and current estimates are ~3,000 in three populations (Meldebekov *et al.* 2010; Plakhov *et al.* 2012; Kaczensky *et al.* 2015; R. Habibrakhmanov, personal communication). The government is planning new reintroductions to recently established protected areas (O. Pereladova, personal communication). With the exception of Altyn Emel National Park in Kazakhstan, populations of the kulan subspecies are small and isolated. The populations in Turkmenistan have suffered severe declines due to illegal hunting and limited seasonal nomadic movements arising from conflicts with agricultural development.

The onager, or Persian wild ass, suffered a drastic decline due to habitat loss and direct and indirect conflicts with people and livestock. Before 1950, this subspecies was widely distributed in the central and eastern arid plains of Iran. Recently, populations have been restricted to two protected areas (Bahram-e-Goor Protected Area and Touran Biosphere Reserve). In 1996, there were an estimated 96 onagers in Bahram-e-Goor and 471 animals in Touran Biosphere Reserve (Feh *et al.* 2002), but while the Bahram-e-Goor population has increased to around 600 onagers, the Touran population has declined to 145 individuals (in 2014; M. Hemani, personal communication). The core area of Bahram-e-Goor Protected Area has become Qatruiyeh National Park, and the increased management support appears to have benefitted the population. The two populations in Iran are small, isolated, and vulnerable to local extinction, and they remain threatened by competition with livestock and poaching. Should the populations increase and expand their ranges, there is the potential that this threat will be exacerbated.

Hartmann's Mountain Zebra
Equus zebra hartmannae
A
Historic range
Current range
Angola
Namibia
Botswana
South Africa
N
250 km

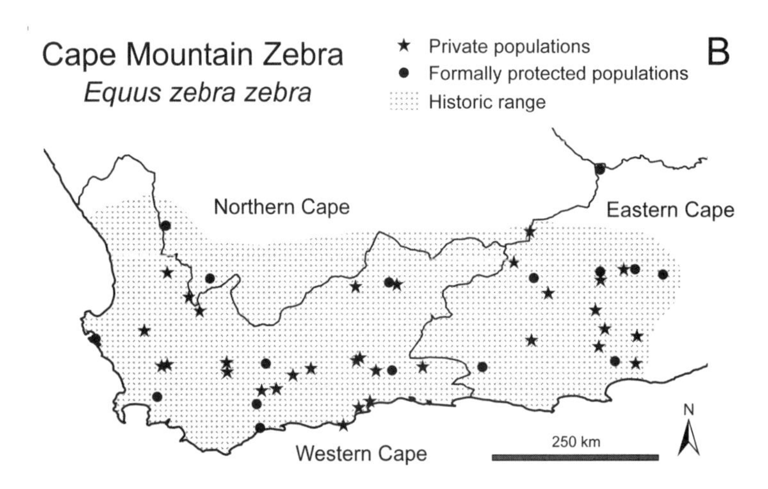

Fig. 12.7 Mountain zebra (*E. zebra*) historic and current range of Hartmann's mountain zebra (A) and Cape mountain zebra (B). *Sources*: A based on Novellie *et al.* 2002; B based on Hrabar *et al.* 2015

Fig. 12.8 Cape mountain zebra. Photo © H. Hrabar

With the exception of the Mongolian khulan population, Asiatic wild ass populations are small and isolated, and are therefore demographically and genetically vulnerable. Disease and drought are "stress events" that are a constant threat to small, isolated wild ass populations. A disease outbreak of African horse sickness in the 1960s, for example, resulted in a major decline and the extinction of small khur populations (Gee 1963). Continued fragmentation and marginalization of the smaller populations could result in similar extinctions.

Mountain Zebra

Historically, mountain zebras occurred from southern Angola through Namibia to southern South Africa (Novellie *et al.* 2002; Novellie 2008; Penzhorn 2013). Two subspecies are recognized: Hartmann's mountain zebra, which occurred in the mountains of Namibia between the Namib Desert and the central plateau and extended into southern Angola (fig. 12.7A), and Cape mountain zebra (*E. z. zebra*), which occurred in the mountains in the Western and Eastern Cape Provinces (figs. 12.7B, 12.8). The Cape mountain zebra is listed in CITES Appendix I. Hartmann's mountain zebra is listed in CITES Appendix II, and there is a commercial trade in sport hunting and skins (2,000–3,000 per year). This trade can be sustainable but needs to be carefully monitored and managed on a per population basis.

Similar to the Asiatic wild ass, the mountain zebra has one subspecies that is flourishing (Hartmann's mountain zebra in Namibia): the population trend is positive and current estimates as of 2000 are over 25,000 (J. Muntifering, personal communication). The Hartmann's mountain zebra had an estimated population of 50,000 in 1973 (Joubert 1973), and there was a decline to about 7,500 in 1989 (Novellie *et al.* 2002). The decline was attributed to droughts and fencing that prevented migration and access to water and forage, plus the impact of legal and illegal hunting (Novellie *et al.* 1992). Since the 1980s, Hartmann's mountain zebra has recovered significantly owing to better protection and community-based and private enterprise development of conservation for this subspecies. Hartmann's mountain zebra occur in four key protected areas in Namibia (15% of the population), with around 25% of the national population on conservancies in communal lands and the remaining 60% on commercial livestock and game farms (Novellie *et al.* 2002). In Namibia, the establishments of artificial water points have allowed Hartmann's mountain zebra to occupy previously unsuitable habitat, such that their present range differs from that in historical times. Genetic analyses indicate that current heterogeneity is reasonably robust (Moodley and Harley 2005).

By contrast, the Cape mountain zebra was almost extirpated in the 1950s, and the population bottlenecked at 30 individuals (Hrabar *et al.* 2015). The surviving natural populations of Cape mountain zebra occur only in Mountain Zebra National Park, Gamka Mountain Reserve, and the Kammanassie Mountains of South Africa (Novellie *et al.* 2002). An extensive reintroduction program established populations in parts of the historic range, but removal of too many animals may have led to the demise of the Onteniqua and Biviaanskloof populations (Lloyd 2002; Novellie *et al.* 2002; Penzhorn 2013). Cape mountain zebra are now found throughout South Africa in private game ranches and reserves. As a result of these conservation actions, the population increased to about 1,200 individuals in 1998, and since then there has been a steady increase in the population. As of 2013, the population was estimated to be 3,100 on protected and private lands (Hrabar *et al.* 2015). This increase is attributed to an increased size in the key national parks and a trend toward wildlife rather than domestic livestock on private lands (Hrabar and Kerley 2013). The 52 known subpopulations are isolated because of fencing; however, metapopulation management is critical for the future of this subspecies. Genetic analyses of the mitochondrial deoxyribonucleic acid (DNA) control region of Cape mountain zebra provide further evidence that this subspecies experienced a severe bottleneck and exhibits a very low nucleotide (Craddock $h = 0.0$, Kammanassie $h = 0.545$) and haplotype diversity (Craddock $\pi = 0.0$, Kammanassie $\pi = 0.006$; Moodley and Hartley 2005). While the population now appears to be increasing, threats to this subspecies remain: loss of genetic diversity, availability of suitable habitat, hybridization, lack of management capacity, disease, and hunting (Hrabar *et al.* 2015).

Endangered

Grevy's zebra and Przewalski's horses are classified as Endangered under the IUCN Categories and Criteria, as they are facing a high risk of extinction in the wild. Again, the two species represent the two different types of social system represented by equids, but their population trajectories appear to be headed in different directions. Przewalski's horses have been brought back from the brink of extinction through reintroduction efforts to its former range, while Grevy's zebra have suffered a range and population decline, only ameliorated by extensive conservation efforts and introduction to an area south of its historic range. Threats to these species are largely competition for forage and water with livestock, and the dangers of stochastic events affecting small and isolated populations. Conservation efforts for both species hinge on efforts with the local community to support their presence and mitigate competition with livestock.

Grevy's Zebra

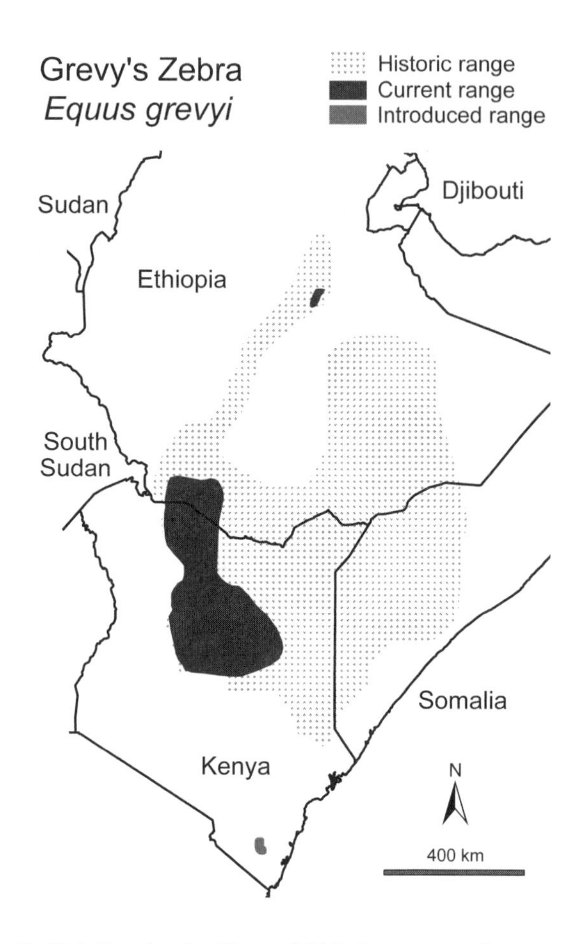

Fig. 12.9 Grevy's zebra (*E. grevyi*) historic, current, and introduced range. *Source*: Kebede *et al.* 2014

Grevy's zebra have undergone a significant range reduction (figs. 12.9 and 12.10) and population decline, and they are listed in CITES Appendix I. Grevy's zebra are legally protected in Ethiopia, and although official protection has been limited, community-based conservation has been more effective. In Kenya, they have been protected by a hunting ban since 1977. While under the Wildlife Conservation and Management Act No. 376 of 1976 (Part II of the First Schedule), the Grevy's zebra was listed as a "Game Animal" (Williams 2002); they have been up-listed to a legally "Protected Animal" in Kenya.

Grevy's zebra are confined to the Horn of Africa, specifically Ethiopia and Kenya, and may persist in South Sudan. Historically, they ranged from the Alledeghi Plain in Ethiopia through the Ogaden, to southern Somalia and into northern Kenya (Bauer *et al.*1994). Currently, Grevy's zebra have a discontinuous range. The northernmost population is small and isolated in the Alledeghi Wildlife Reserve, Ethiopia (Kebede *et al.* 2014). The next population to the south is at Lake Ch'ew Bahir in southern Ethiopia, and this population extends just north of Mount Kenya, although a few animals are found farther southeast along the Tana River. A small, introduced, population survives in and around Tsavo East National Park in Kenya. Grevy's zebra have been extirpated from southern Somalia, where the last confirmed sightings date to 1973 (Rowen and Ginsberg 1992). There are no confirmed records that the species ever occurred in Eritrea or Djibouti (Yalden *et al.* 1986; Bauer *et al.* 1994). Sightings from South Sudan require verification (Moehlman *et al.* 2008*b*).

In the 1970s, there was an estimated global population of ~15,000 Grevy's zebra (Rowen and Ginsberg 1992) that declined to about 4,600 in the late 1980s, then to ~2,200 in 2008 (B. Low, personal communication). The global population in 2011 was estimated as 2,837 (Kenya, 2,546; Ethiopia, 281) (Kenya Wildlife Service 2012), yielding a total population estimate of 1,100 mature individuals, with the largest subpopulation estimated at 432 mature individuals (Kenya Wildlife Service 2012). The decline from the 1970s to the present is roughly 80%. Since the early 2000s, there has been a small but positive increase in both the Kenyan and Ethiopian populations. Kenya has 90% of the global population.

Fig. 12.10 Grevy's zebra.
Photo © Patricia D. Moehlman

Grevy's zebra are dependent on access to water; conservation actions that allow them to share water and forage with pastoralists and their livestock are thus critical. Community-based conservation in northern Kenya has led to the establishment of conservation conservancies and improved survivorship (Low *et al.* 2009; Kenya Wildlife Service 2012). On the Laikipia Plateau, protection and reduced competition with domestic livestock have seen Grevy's zebra numbers increasing since they first expanded into this area in the early 1970s (Williams 2002, 2013). Similar to other territorial equid species, the solitary behavior of this species may make them more vulnerable to predation by lions and spotted hyenas, especially in the Laikipia area (Mwasi and Mwangi 2007).

The density and area of occupancy of Grevy's zebra fluctuate seasonally as animals move in search of resources. During the dry season, when they are dependent on permanent water, animals tend to be more concentrated. The exception is in the Alledeghi Wildlife Reserve in Ethiopia. Contrary to expectations, Grevy's zebra are more concentrated during the wet season in order to avoid the pastoralists and livestock that move into the area in the wet season (Kebede *et al.* 2012) (see box 12.1).

The major threats to Grevy's zebra include reduction of available water sources, habitat degradation due to overgrazing, competition for resources, hunting, and disease (Rowen and Ginsberg 1992; Williams 2002; Kebede 2013). In Kenya, hunting for skins in the late 1970s may have contributed to the observed decline, although the continuing decline can be attributable to low recruitment (Williams 1998, 2002), which can result from competition for resources—both food and access to water—with pastoral people and domestic livestock (Williams 1998). A low level of hunting of Grevy's zebra for food, and in some areas medicinal use, continues (Williams 2002). Additionally, the water supply in critical perennial rivers has declined, most notably in the Ewaso Ng'iro River, where overabstraction of water for irrigation schemes has reduced dry-season river flow by 90% over the past three decades (Williams 2002). The proposed Lamy Port and Lamu Southern Sudan-Ethiopia Transport Corridor has the potential to sever migration routes and further fragment existing populations. In Ethiopia, killing of Grevy's zebra was the primary cause of the decline as of 2007 (F. Kebede, personal communication). Disease can also contribute to increased mortality. Recently, Muoria *et al.* (2007) recorded an outbreak of anthrax in the Wamba area of southern Samburu, Kenya, during which >50 animals succumbed to the disease.

Przewalski's Horse

The Przewalski's horse has recovered from being categorized as Extinct in the Wild by the IUCN in the 1960s to its current listing of Endangered. The last wild population of Przewalski's horses was found in southwestern Mongolia and adjacent Gansu, Xinjiang, and inner Mongolia (China). Wild horses were last seen in 1969 north of the Tachiin Shar Nuruu in the Dzungarian Gobi Desert of Mongolia (Paklina and Pozdnyakova 1989; Boyd *et al.* 2008) (figs. 12.11 and 12.12). When the species was assessed in 1996, it was listed as Critically Endangered because there was at least one mature Przewalski's horse in the wild, thanks to reintroduction efforts. In 2011, the reintroduced population had had at least 50 mature individuals in the wild for 5 years, qualifying Przewalski's horse for down-listing to Endangered. As of 2013, there were 427 free-ranging individuals at the three Mongolia reintroduction sites (C. Feh, personal communication; O. Ganbaatar, personal communication; D. Usukhjargal and N. Bandi, personal communication; King *et al.* 2015). Przewalski's horse is listed in CITES Appendix I as *E. przewalskii*. It is legally protected in Mongolia, and the reintroduction sites are mostly within protected areas.

Five reintroduction sites have been established. Three are in Mongolia: Hustai National Park (570 km^2) in the Mongol Daguur Steppe of central Mongolia; Takhin Tal, from which horses were released into the Great Gobi B Strictly Protected Area (9,000 km^2) in the Dzungarian Gobi Desert; and the fenced Seriin Nuruu (140 km^2) in the Khomiin Tal buffer zone of the Khar Us Nuur National Park in western Mongolia. There are two reintroduction sites in China, although both are in early stages. The largest population is at the Kalamaili Nature Reserve in Xinjiang Uighur Autonomous Region, where there were 127 Przewalski's horses in 2013 (Xia *et al.* 2014).

Historically, there were many factors that precipitated the extinction of the Przewalski's horse: cultural and political changes, hunting, military activities, climatic change, and competition with livestock and increasing land-use pressure (Sokolov *et al.* 1992; Zhao and Liang 1992; Ryder 1993; Bouman and Bouman 1994). Capture expeditions probably diminished the remaining Przewalski's horse populations by killing and dispersing the adults (van Dierendonck and Wallis de Vries 1996). The harsh winters of 1945, 1948, and 1956 probably had an additional impact on the small population (Bouman and Bouman 1994). Increased pressure on, and rarity of waterholes in, their last refuge should also be considered as a significant factor contributing

BOX 12.1 SUITABLE HABITAT ANALYSES FOR CONSERVATION MANAGEMENT

The Grevy's zebra is one of the most endangered wild equids. They have suffered significant reduction of natural range, and the loss of habitat and illegal hunting are the greatest threats to their survival in the wild. Pertinent scientific information is critical for effective conservation management that can safeguard this species. Two fundamental aspects of appropriate management are to determine suitable habitat and to identify the components of wildlife-human interactions. Wild equids are dependent on frequent access to water and normally can disperse farther from permanent water during the wet season.

Species distribution modeling can be used to determine the seasonal habitat use and geographic range of a species. The maximum entropy model (Maxent) was used to analyze the seasonal pattern of Grevy's zebra habitat use and which factors affected their distribution. Determination of suitable habitat allows managers to demarcate and prioritize protection for areas that are critical to the species survival. It also allows a prediction of where a rare species might still persist that has not been documented. It provides a tool for achieving a better understanding of where a species range might expand. Grevy's zebra have experienced a significant reduction in their historic range. The largest and most isolated population in Ethiopia occurs in the Alledeghi Wildlife Reserve. This grassland habitat is important for both wildlife and pastoralists and their livestock.

Presence of Grevy's zebra was determined on the basis of direct observation and fecal piles during the wet and dry seasons. The most important predictor variable was precipitation, followed by distance from settlement and slope. Contrary to expectations, Grevy's zebra had a larger area of utilization during the dry season (563 km^2) than during the wet season (437 km^2; see figure, opposite). Though animals are normally expected to have a wider distribution during the wet season, this behavior can be affected by human presence. During the wet season, pastoralists establish temporary settlements and bring large herds of livestock into the reserve. As a result, competition for forage and space is increased during the wet season, and Grevy's zebra behaviorally avoid people and their livestock. Primary productivity

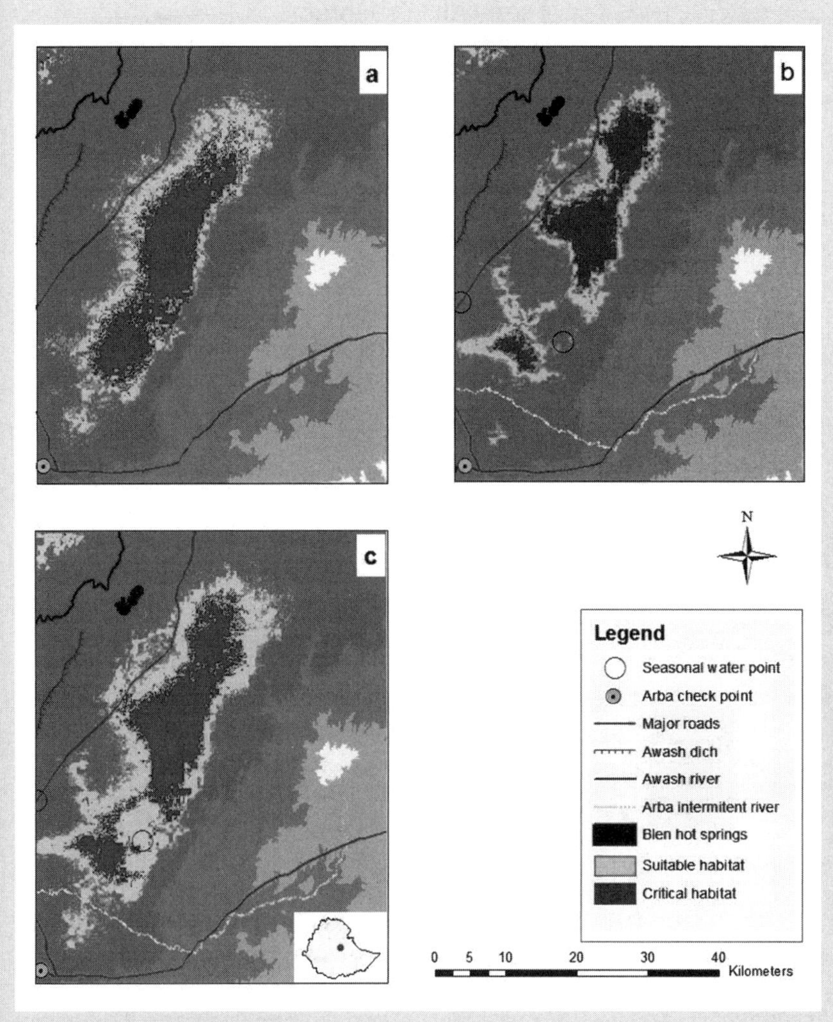

Suitable habitat for Grevy's zebra in the Alledeghi Wildlife Reserve in Ethiopia during the dry season (A), wet season (B), and throughout the year (C). Adapted from Kebede *et al.* 2012

is high, and Grevy's zebra can survive in higher densities in smaller areas. By contrast, during the dry season, people and their livestock leave the reserve, and Grevy's zebra disperse into a larger area.

Clearly, the wet-season use of the Alledeghi Wildlife Reserve by pastoral peoples has a significant impact on Grevy's zebra distribution. In addition, anthropogenic factors have greater variability than biological and topographic aspects. Anthropogenic variables can be influenced by food security, politics, and economics. In the Alledeghi Wildlife Reserve, there are conflicts between the Afar and Issa communities, and through time, their areas of settlement and grazing could change, which would have a direct effect on Grevy's zebra distribution. Hence management needs to develop adaptive strategies and to work in cooperation with local communities such that both people and wildlife can survive. The use of species distribution modeling can be an important and effective tool for assessing ecological and socioeconomic components of a species existence in the wild, and enable prioritization of appropriate conservation actions.

Fig. 12.11 Przewalski's horse (*E. f. przewalskii*) historic and reintroduced range. Based on Boyd *et al.* 2008

Fig. 12.12 Przewalski's horse. Photo © Patricia D. Moehlman

to their extinction (van Dierendonck and Wallis de Vries 1996).

Population growth analyses of the two largest reintroduced populations indicate that the Hustai Nuruu population's growth is steady and consistent (Usukhjargal and Bandi 2013; King *et al.* 2015). The release of captive-bred individuals has ceased, and established captive- and wild-born individuals are successfully reproducing. By contrast, the Gobi population is still dependent on releases. But the wild-born individuals had a higher survival than reintroduced animals during the extreme winter of 2009/2010 (Kaczensky *et al.* 2011*a*).

The reintroduced populations in Mongolia and China are susceptible to the inherent dangers faced by all small and isolated populations: loss of genetic diversity as a result of a small founder population and generations in captivity, and restricted range. Proper management of the metapopulation and the exchange of individuals between reintroduction sites can mitigate this threat, however. The multiple external threats are hybridization with domestic horses, competition for resources with domestic horses and possibly other livestock, infectious disease transmitted by domestic horses, and stochastic events such as severe winters (Kaczensky *et al.* 2011*a*). Additional threats are lack of information, knowledge, and awareness, and mining. The three Mongolian reintroduction sites currently keep detailed records on natality and survivorship, however, and are integrating local community economic development. There have been multiple workshops of stakeholders involved in the Mongolian reintroduction to advise further conservation actions that need to be taken (Boyd 2007).

Critically Endangered

Critically Endangered species are those that are determined by the IUCN Red List Categories and Criteria to be facing an extremely high risk of extinction in the wild. This is the most severely threatened category and indicates that conservation actions are essential for the species to remain extant. The only Critically Endangered equid species is the African wild ass, which has experienced a catastrophic decline in numbers and range and is the world's most endangered equid. Conservation actions are complicated by political instability and the extreme aridity of the area where it resides.

African Wild Ass

The African wild ass is listed on the IUCN Red List of Threatened Species as Critically Endangered because of its small population size. The species numbers at best ~200 mature individuals and may be undergoing a continuing decline due to climate and human/livestock impacts. No subpopulation has numbers in excess of 50 mature individuals (Moehlman *et al.* 2008*d*). The African wild ass is listed in CITES Appendix I.

African wild ass occur in Eritrea and Ethiopia, and some animals may persist in Somalia, Djibouti, Sudan, and Egypt (figs. 12.13 and 12.14). The Nubian wild ass (*E. a. africanus*) has not been observed in the wild since the 1970s. This subspecies once ranged from the Nubian Desert of northeastern Sudan, to the shores of the Red Sea, and south into northern Eritrea. The Somali wild ass (*E. a. somaliensis*) was found in the Denkelia region of Eritrea, the Danakil Desert, and the Awash River Valley in the Afar region of northeastern Ethiopia, western Djibouti, and into the Ogaden region of eastern Ethiopia. In Somalia, they historically ranged from Meit and Erigavo in the north to the Nugaal Valley, and as far south as the Shebele River (Moehlman 2002; Moehlman *et al.* 2008*d*).

In Ethiopia, there has been a severe population decline since the early 1970s. Klingel (1972) estimated a density of 18.6 wild asses per 100 km² in an area of ~10,000 km². During that survey, Yangudi-Rassa National Park had the highest density (30 wild asses per 100 km²). In 1994, no wild asses were seen during a ground survey of the Yangudi-Rassa National Park, but local Issa pastoralists reported that they were present but rare, and occurred at an approximate density well below 1 animal per 100 km² (Moehlman *et al.* 1998). In 2007, Kebede *et al.* (2007) surveyed the historic range of the African wild ass in Ethiopia and determined that they have been extirpated from Yangudi-Rassa National Park and the Somali region and that the only remaining population was in the northeastern Afar region. The total number of wild ass observed during this survey was 25 in an area of 4,000 km², yielding a rough density of 0.625 animals per 100 km². This density is higher than that of the 1994–1998 survey, which was 0.5 wild asses per 100 km² in an area of 2,000 km² (Moehlman *et al.* 2008*d*).

In the Serdo-Hillu area of Ethiopia, where there has been a research and conservation program with the Ethiopian Wildlife Conservation Authority and local Afar pastoralists, the population has remained stable. Although the number of African wild ass in this area has not declined, the population is small and under high risk of extinction. There may be <200 African wild asses left in Ethiopia.

In Eritrea, there are limited long-term data. The first successful survey was made in 1996, and since then there has been a research and conservation program with the Ministry of Agriculture and the Hamemelo Agricultural College. The main study site in the Northern Red Sea Zone has a population of roughly 47 individuals per 100 km² (Moehlman *et al.* 1998; Moehlman 2002), which is the highest population density found anywhere in the current range of this species and is similar to population densities recorded in Ethiopia in the early 1970s. The study area is limited (100 km²), but recent research indicates that African wild ass currently inhabit ~11,000 km² in the Denkeli Desert of Eritrea (Teclai 2006). Surveys are needed to determine the distribution and density of African wild ass in this larger area. A rough estimate of African wild ass in Eritrea would yield a total of possibly 400 individuals.

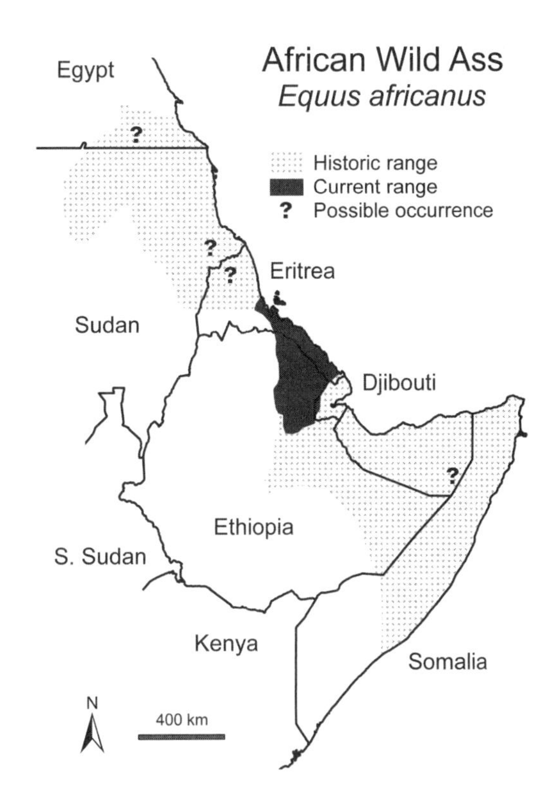

Fig. 12.13 African wild ass (*E. africanus*) historic and current range. Adapted from Moehlman 2002

Fig. 12.14 African wild ass.
Photo © Patricia D. Moehlman

In Somalia in 1997, local pastoralists said that there were <10 African wild asses left in the Nugaal Valley; an earlier ground survey in 1989 in the Nugaal Valley yielded population estimates of roughly 135–205 animals or ~2.7–4.1 per km² (Moehlman *et al.* 1998). Some animals may remain near Meit and Erigavo, but this area has not been surveyed since the 1970s (Moehlman *et al.* 2008*d*). It is not known whether the African wild ass currently persists in Somalia.

In summary, the total number of observed African wild asses in Eritrea and Ethiopia is 70 individuals; there may be as many as 600 individuals in these two countries, but this figure is a rough extrapolation from more intensely studied areas. The number of mature individuals is approximately one-third of the population (Feh *et al.* 2001); hence the minimum number of mature individuals is 23 and the maximum might be 200. In Ethiopia in the past 35 years, there has been a >95% population decline, and in the past 12 years, the African wild ass has been extirpated from a further 50% of its range (Kebede *et al.* 2007). In Eritrea, the population is stable and slowly increasing, but it is difficult to predict population trends into the future. The desert habitat of the African wild ass in both Eritrea and Ethiopia suffers from recurrent and extreme droughts (Kebede 1999; Moehlman *et al.* 2008*d*).

DNA extracted from fecal samples collected from animals in Eritrea and Ethiopia resulted in the identification of five mitochondrial DNA haplotypes: one haplotype (group of polymorphisms) specific to the Eritrean population (haplotype D), one haplotype specific to the Ethiopian population (haplotype E), and three shared haplotypes (haplotypes A, B, and C). These results suggest that there is or has been gene flow between the subpopulations (Afrera, Serdo) in Ethiopia and the population in Eritrea (Oakenfull *et al.* 2002; Rosenbom *et al.* 2015).

The major threat to the African wild ass is hunting for food and medicinal purposes (Kebede 1999; Moehlman 2002; Moehlman *et al.* 2013). Limited access to drinking water and forage (largely arising from competition with livestock) is also a major constraint, with reproductive females and foals less than three months old most at risk. Hence it will continue to be important to determine critical water supplies and basic forage requirements, allowing management authorities to determine how to conserve the African wild ass (in consultation with local pastoralists) (Moehlman 2002; Moehlman *et al.* 2008*d*, 2013; Kebede 2013; Kebede *et al.* 2014). The third potential threat to the survival of the African wild ass is possible interbreeding with the domestic donkey. To date, however, genetic analyses indicate that there are no hybrids (Moehlman *et al.* 2008*d*; A. Beja-Pereira, personal communication).

In Ethiopia, Yangudi-Rassa National Park (4,731 km²) and the Mille-Serdo Wild Ass Reserve (8,766 km²) were established in 1969. The former has never been formally gazetted, however, and both areas are utilized by large numbers of pastoralists and their livestock. These areas are remote and extremely arid, and the Ethiopian Wildlife Conservation Authority has not had sufficient funds or personnel for appropriate management (Kebede 1999). In Eritrea, the government designated the African wild ass area between the Buri Peninsula and the Dalool Depression as a high-priority area for conservation protection as a nature reserve. In both Eritrea and Ethiopia, research and conservation programs supported by the governments have been critical for sustaining African wild ass populations. The development of protected and multiple-use areas with local community involvement is needed to secure access to water and forage and to reduce the threat of illegal hunting.

Recommended research and conservation actions include:

1. Ecosystem/landscape-based research and management on the African wild ass in Eritrea and Ethiopia, including interactions among pastoralists, livestock, wildlife, and the environment.
2. Active involvement of local pastoralists in the preparation and management of long-term action plans.
3. Postgraduate training of personnel in Eritrea and Ethiopia.
4. Surveys in northern Eritrea, Djibouti, Sudan, and Egypt to determine whether African wild ass populations survive in these areas.
5. Genetic research on the African wild ass and local domestic donkey populations to clarify whether hybridization has occurred.

Conclusion

Threatened wild equid species have both harem-forming and territorial types of mating systems and occur in both arid and mesic habitats, indicating that limited access to water and forage in combination with unsustainable hunting and potentially disease are the major causes for wild equid declines, rather than any cause relating to social system, digestive mechanism, or environment.

Human cultural beliefs can often help protect equid species. Religious beliefs historically limited hunting of kiang on the Tibetan Plateau; in the Danakil of Ethiopia, the Koran and Muslim religious leaders prohibit the eating of African wild ass (St-Louis and Côté 2008; Kebede 2013). Thus, in some areas, mortality due to human hunting and consumption can be reduced. Wild equids have experienced amazing recoveries as a result of human intervention. Committed management programs have helped save both Przewalski's horses and Cape mountain zebra from extinction. Improved awareness and the support of local communities have aided the conservation of Grevy's zebra, African wild ass, Asiatic wild ass, and Hartmann's mountain zebra. The continued commitment of wildlife conservation authorities, research personnel, and local communities will be critical for sustaining wild equid populations into the future.

ACKNOWLEDGMENTS

The authors would like to express their gratitude to all those who have contributed to equid conservation. Without their efforts, the grasslands of the world would be less diverse. In particular, they thank all of the assessors, contributors, and reviewers of the IUCN Red List equid assessments, both present and past. Without their knowledge and expertise, the assessments would be much less accurate. The text cites personal communication with many individuals; both those mentioned and many other biologists and conservationists who are equid specialists or are working in a relevant region have been incredibly generous in sharing data, comments, and information. The authors appreciate their essential contribution. P.D.M. is particularly grateful to Olga Pereladova, Mahmoud Hemami, and Lhagvasuren Badamjav for their help, and cannot thank Randy Boone enough for all of his expertise and diligence in preparing the historic and current range maps. P.D.M. is also grateful to the donors who have supported so much training, research, and conservation for wild equids: EcoHealth Alliance, Thye Foundation, Model Foundation, Basel Zoo, Liberec Zoo, Saint Louis Zoo Conservation Program, the Whitley Awards, Plock Zoo, SeaWorld Busch Gardens Conservation Fund, and Gilman International Conservation Fund.

REFERENCES

Batsaikhan, N., B. Buuveibaatar, B. Chimed, O. Enkhtuya, D. Galbrakh, *et al.* 2014. Conserving the world's finest grassland amidst ambitious national development. Conservation Biology 28:1736–1739.

Bauer, I.E., J. McMorrow, and D.W. Yalden. 1994. The historic ranges of three equid species in north-east Africa: A quantitative comparison of environmental tolerances. Journal of Biogeography 21:169–182.

Bhatnagar, Y.V., R. Wangchuk, H.H.T. Prins, S.E. Van Wieren, and C. Mishra. 2006. Perceived conflicts between pastoralism and conservation of the kiang *Equus kiang* in the Ladakh Trans-Himalaya, India. Environmental Management 38:934–941.

Bi, J.H. 2007. A study on the status of Asiatic wild ass (*Equus hemionus hemionus*) and its ecological problems. [In Chinese.] Thesis, Beijing Forestry University, Beijing, China.

Bouman, I., and J. Bouman. 1994. The history of Przewalski's horse. Pages 5–38 *in* L. Boyd and K.A. Houpt (Eds.), Przewalski's horse: The history and biology of an endangered species. State University of New York Press, Albany, NY, USA.

Boyd, L. 2007. Workshop to discuss the future of the takhi (*Equus ferus przewalskii*) in Mongolia: A report. Mongolian Journal of Biological Sciences 5:53–54.

Boyd, L., W. Zimmermann, and S.R.B. King. 2008. *Equus ferus przewalskii*. IUCN red list of threatened species. Version 2009.2. International Union for Conservation of Nature, Gland, Switzerland. Available at www.iucnredlist.org.

Caughley, G. 1994. Directions in conservation biology. Journal of Animal Ecology 63:215–244.

Chu, H.-J., Z.-G. Jiang, W.-X. Lan, C. Wang, Y.-S. Tao, and F. Jiang. 2008. Dietary overlap among kulan *Equus hemionus*, goitered gazelle *Gazella subguttorsa* and livestock. [In Chinese.] Current Zoology 54:941–954.

Clark, B., and P. Duncan. 1992. Asian wild asses. Pages 17–21 *in* P. Duncan (Ed.), Zebras, asses and horses: An action plan for the conservation of wild equids. International Union for Conservation of Nature, Gland, Switzerland.

Cordingley, J.E., S.R. Sundaresan, I.R. Fischhoff, B. Shapiro, J. Ruskey, and D.I. Rubenstein. 2009. Is the endangered Grevy's zebra threatened by hybridization? Animal Conservation 12:505–513.

Coughenour, M. 2002. Ecosystem modelling in support of conservation of wild equids—The example of the Pryor Mountain Wild Horse Range. Pages 154–162 *in* P.D. Moehlman (Ed.), Equids: Zebras, asses and horses. Status survey and conservation action plan. International Union for Conservation of Nature, Gland, Switzerland.

Feh, C., B. Munkhtuya, S. Enkhbold, and T. Sukhbaatar. 2001. Ecology and social structure of the Gobi khulan *Equus hemionus* subsp. in the Gobi B National Park, Mongolia. Biological Conservation 101:51–61.

Feh, C., N. Shah, M. Rowen, R. Reading, and S.P. Goyal. 2002. Status and action plan for the Asiatic wild ass (*Equus hemionus*). Pages 62–71 *in* P.D. Moehlman (Ed.), Equids: Zebras, asses and horses. Status survey and conservation action plan. International Union for Conservation of Nature, Gland, Switzerland.

Fox, J.L., and B.-J. Bårdsen. 2005. Density of Tibetan antelope, Tibetan wild ass and Tibetan gazelle in relation to human presence across the Chang Tang Nature Reserve of Tibet, China. Acta Zoologica Sinica 51:586–597.

Gee, E.P. 1963. The Indian wild ass: A survey. Journal of the Bombay Natural History Society 60:517–529.

Grange, S., P. Duncan, J.-M. Gaillard, A.R.E. Sinclair, P.J.P. Gogan, C. Packer, H. Hofer, and M.L. East. 2004. What limits the Serengeti zebra population? Oecologia 140:523–532.

Grange, S., F. Barnier, P. Duncan, J.-M. Gaillard, M. Valeix, H. Ncube, S. Périquet, and H. Fritz. 2015. Demography of plains zebras (*Equus quagga*) under heavy predation. Population Ecology 57:201–214.

Grubb, P. 2005. Order Perissodactyla. Pages 629–636 *in* D.E. Wilson and D.M. Reeder (Eds.), Mammal species of the world: A taxonomic and geographic reference, 3rd ed. Johns Hopkins University Press. Baltimore, USA.

Hack, A.M., R. East, and D.I. Rubenstein. 2002. Status and action plan for the plains zebra. Pages 43–60 *in* P.D. Moehlman (Ed.), 2002. Equids: Zebras, asses and horses. Status survey and conservation action plan. International Union for Conservation of Nature, Gland, Switzerland.

Higuchi, A., B. Bowman, M. Freiberger, O.A. Ryder, and A.C. Wilson. 1984. DNA sequences from the quagga, an extinct member of the horse family. Nature 312:282–284.

Hill, R.A. 2009. Is isolation the major genetic concern for endangered equids? Animal Conservation 12:518–519.

Hoffmann, M., C. Hilton-Taylor, A. Angulo, M. Bohm, T.M. Brooks, *et al.* 2010. The impact of conservation on the status of the world's vertebrates. Science 330:1503–1509.

Hrabar, H., and G.I.H. Kerley. 2013. Conservation goals for the Cape mountain zebra *Equus zebra zebra*—Security in numbers? Oryx 47:403–409.

Hrabar, H., C. Birss, D. Peinke, S.R.B. King, P. Novellie, and S. Schultz. 2015. National red list assessment: *Equus zebra zebra*, Cape mountain zebra. South African National Biodiversity Institute and Endangered Wildlife Trust, Pretoria and Johannesburg, South Africa.

Ito, T.Y., B. Lhagvasuren, A. Tsunekawa, M. Shinoda, S. Takatsuki, B. Buuveibaatar, and B. Chimeddorj. 2013. Fragmentation of the habitat of wild ungulates by anthropogenic barriers in Mongolia. PLoS ONE 8:e56995.

IUCN. International Union for Conservation of Nature. 2013. Guidelines for using the IUCN red list categories and criteria. Version 10.1. Prepared by the Standards and Petitions Subcommittee. Gland, Switzerland. Available at www.iucnredlist.org/documents/RedListGuidelines.pdf.

IUCN. International Union for Conservation of Nature. 2015. IUCN red list categories and criteria. Version 5.2. Gland, Switzerland. Available at www.iucnredlist.org.

Joubert, E. 1973. Habitat preference, distribution, and status of the Hartmann zebra *Equus zebra hartmannae* in south west Africa. Madoqua 7:5–15.

Kaczensky, P., O. Ganbaatar, N. Altansukh, N. Enkhsaikhan, C. Stauffer, and C. Walzer. and 2011*a*. The danger of having all your eggs in one basket—Winter crash of the re-introduced Przewalski's horses in the Mongolian Gobi. PLoS ONE 6:e28057.

Kaczensky, P., R. Kuehn, B. Lhagvasuren, S. Pietsch, W. Yang, and C. Walzer. 2011*b*. Connectivity of the Asiatic wild ass population in the Mongolian Gobi. Biological Conservation 144:920–929.

Kaczensky, P., B. Lkhagvasuren, O. Pereladova, H. Mahmoud-Reza, and A. Bouskila. 2015. *Equus hemionus*. IUCN red list of threatened species. Version 2015.2. International Union for Conservation of Nature, Gland, Switzerland. Available at www.iucnredlist.org.

Kebede, F. 1999. Ecology and conservation of the African wild ass (*Equus africanus*) in the Danakil, Ethiopia. Thesis, University of Kent, Canterbury, UK

Kebede, F. 2013. Ecology and community-based conservation of Grevy's zebra (*Equus grevyi*) and African wild ass (*Equus africanus*) in the Afar region, Ethiopia. Thesis, University of Addis Ababa, Addis Ababa, Ethiopia.

Kebede, F., L. Berhanu, and P.D. Moehlman. 2007. Distribution and population status of the African wild ass (*Equus africanus*) in Ethiopia. Report to Saint Louis Zoo. Saint Louis, MO, USA.

Kebede, F., A. Bekele, P.D. Moehlman, and P.H. Evangelista. 2012. Endangered Grevy's zebra in the Alledeghi Wildlife Reserve, Ethiopia: Species distribution modeling for the determination of optimum habitat. Endangered Species Research 17:237–244.

Kebede, F., S. Rosenbom, L. Khalatbari, P.D. Moehlman, A. Beja-Pereira, and A. Bekele. 2014. Genetic diversity of the Ethiopian Grevy's zebra (*Equus grevyi*) populations that includes a unique population of the Alledeghi Plain. Mitochondrial DNA doi:10.3109/19401736.2014.898276.

Kenya Wildlife Service. 2012. Conservation and management strategy for Grevy's zebra (*Equus grevyi*) in Kenya (2012–2016). Nairobi, Kenya.

King, S.R.B., L. Boyd, W. Zimmermann, and B. Kendall. 2015. *Equus ferus*. IUCN red list of threatened species. Version 2015.2. International Union for Conservation of Nature, Gland, Switzerland. Available at www.iucnredlist.org.

Klein, R.G., and K. Cruz-Uribe. 1999. Craniometry of the genus *Equus* and the taxonomic affinities of the extinct South African quagga. South African Journal of Science 95:81–86.

Klingel, H. 1972. Somali wild ass: Status survey in the Danakel region, Ethiopia. WWF Project No. 496. Final Report to EWCO, Addis Ababa, Ethiopia.

Kuznetsov, V. 2014. Return of onager. International Journal Turkmenistan 1–2:84–92.

Leonard, J.A., N. Rohland, S. Glaberman, R.C. Fleischer, A. Caccone, and M. Hofreiter. 2005. A rapid loss of stripes: The evolutionary history of the extinct quagga. Biology Letters 1:291–295.

Liu, G., A.B.A. Shafer, W. Zimmermann, D. Hu, W. Wang, H. Chu, J. Cao, and C. Zhao. 2014. Evaluating the reintroduction project of Przewalski's horse in China using genetic and pedigree data. Biological Conservation 171:288–298.

Lkhagvasuren, B. 2007. Population assessment of khulan (*Equus hemionus*) in Mongolia. Exploration into the Biological Resources of Mongolia (Halle/Saale) 10:45–48.

Lloyd, P.H. 2002. Cape mountain zebra conservation success: An historical perspective. Pages 39–41 *in* B.L. Penzhorn (Ed.), Proceedings of a symposium on relocation of large African mammals. SAVA Wildlife Group, Onderstepoort, South Africa.

Low, B., S.R. Sundaresan, I.R. Fischhoff, and D.I. Rubenstein. 2009. Partnering with local communities to identify conservation priorities for endangered Grevy's zebra. Biological Conservation 142:1548–1555.

Meldebekov, A.M., M.K. Bajzhanov, A.B. Bekenov, and A.F. Kovshar. 2010. The red data book of the Republic of Kazakhstan, vol. 1: Animals, part 1: Vertebrates, 4th ed. Ministry of Education and Science / Ministry of Agriculture, Almaty, Kazakhstan.

Moehlman, P.D. (Ed.), 2002. Equids: Zebras, asses and horses. Status survey and conservation action plan. International Union for Conservation of Nature, Gland, Switzerland.

Moehlman, P.D. 2005. Endangered wild equids. Scientific American. 292:86–93.

Moehlman, P.D, V.A. Runyoro, and H. Hofer. 1997. Wildlife population trends in the Ngorongoro Crater, Tanzania. Pages 59–69 *in* M. Thomson (Ed.), Multiple land-use: The experience of the Ngorongoro Conservation Area, Tanzania. Protected Areas Series Publication. International Union for Conservation of Nature, Gland, Switzerland.

Moehlman, P.D., F. Kebede, and H. Yohannes. 1998. The African wild ass (*Equus africanus*): Conservation status in the Horn of Africa. Applied Animal Behaviour Science 60:115–124.

Moehlman, P.D., H. Hrabar, R. Smith, and R. Hill. 2008*a*. *Equus zebra*. IUCN Red list of threatened species. Version 2015.2. International Union for Conservation of Nature, Gland, Switzerland. Available at www.iucnredlist.org.

Moehlman, P.D., D.I. Rubenstein, and F. Kebede. 2008*b*. *Equus grevyi*. IUCN red list of threatened species. Version 2009.2. International Union for Conservation of Nature, Gland, Switzerland. Available at www.iucnredlist.org.

Moehlman, P.D., N. Shah, and C. Feh. 2008*c*. *Equus hemionus*. IUCN red list of threatened species. Version 2009.2. International Union for Conservation of Nature, Gland, Switzerland. Available at www.iucnredlist.org.

Moehlman, P.D., H. Yohannes, R. Teclai, and F. Kebede. 2008*d*. *Equus africanus*. IUCN red list of threatened species. Version 2009.2. International Union for Conservation of Nature, Gland, Switzerland. Available at www.iucnredlist.org.

Moehlman, P.D., F. Kebede, and H. Yohannes. 2013. *Equus africanus*, African wild ass. Pages 414–417 *in* J.S. Kingdon, and M. Hoffmanm (Eds.), The mammals of Africa, vol. 5: Carnivores, pangolins, rhinos and equids. Academic Press, San Diego, CA, USA.

Moodley, Y., and E.H. Harley. 2005. Population structuring in mountain zebras (*Equus zebra*): The molecular consequences of divergent demographic histories. Conservation Genetics 6:953–968.

Muoria, P.K., P. Muruthi, W.K. Kariuki, B.A. Hassan, D. Mijele, and N.O. Oguge. 2007. Anthrax outbreak among Grevy's zebra (*Equus grevyi*) in Samburu, Kenya. African Journal of Ecology 45:483–489.

Mwasi, S., and E. Mwangi. 2007. Proceedings of the National Grevy's Zebra Conservation Strategy Workshop, 11–14 April 2007. KWS Training Institute, Naivasha, Kenya.

Norton-Griffiths, M., H. Frederick, D.M. Slaymaker, and J. Payne. 2013. Preliminary estimates of wildlife and livestock populations in the Oyu Tolgoi Area of the south-eastern Gobi Desert, Mongolia, May–July 2013. Unpublished preliminary report to Oyu Tolgoi, Mongolia.

Novellie, P.A. 2008. *Equus zebra*. IUCN red list of threatened species. Version 2009.2. International Union for Conservation of Nature, Gland, Switzerland. Available at www.iucnredlist.org.

Novellie, P.A, P.H. Lloyd, and E. Joubert. 1992. Mountain zebras. Pages 6–9 *in* P. Duncan (Ed.), Zebras, asses and horses: An action plan for the conservation of wild equids. International Union for Conservation of Nature, Gland, Switzerland.

Novellie, P.A., M. Lindeque, P. Lindeque, P. Lloyd, and J. Koen. 2002. Status and action plan for the mountain zebra (*Equus zebra*). Pages 28–42 *in* P. D. Moehlman (Ed.), Equids: Zebras, asses and horses. A status survey and conservation action plan. International Union for Conservation of Nature, Gland, Switzerland.

Oakenfull, E.A., H.N. Lim, and O.A. Ryder. 2000. A survey of equid mitochondrial DNA: Implications for the evolution, genetic diversity and conservation of *Equus*. Conservation Genetics 1:341–355.

Oakenfull, E.A., H. Yohannes, F. Kebede, J. Swinburne, M. Binns, and P.D. Moehlman. 2002. Conservation genetics of African wild asses. Final Report. Zoological Societies of Chicago and San Diego, USA.

Ogutu, J.O., and N. Owen-Smith. 2003. ENSO, rainfall and temperature influences on extreme population declines among African savanna ungulates. Ecology Letters 6:412–419.

Ogutu, J.O., N. Owen-Smith, H.-P. Peipho, and M.Y. Said. 2011. Continuing wildlife population declines and range contraction in the Mara region of Kenya during 1977–2009. Journal of Zoology 285:99–109.

Paklina, N., and M.K. Pozdnyakova. 1989. Why the Przewalski horses of Mongolia died out. Przewalski Horse 24:30–34.

Penzhorn, B. 2013. *Equus zebra*. Pages 438–443 *in* J.S. Kingdon, and M. Hoffmanm (Eds.), The mammals of Africa, vol. 5: Carnivores, pangolins, rhinos and equids. Academic Press, San Diego, CA, USA.

Plakhov, K.N., S.V. Sokolov, V.F. Levanov, and A.Z. Akylbekova, 2012. Pages 151–153 *in* A.M. Meldebekov, A.B. Bekenov, Y.A. Grachev *et al.* (Eds.), News in Kulan reintroduction in Kazakhstan: Zoological and game management researches in Kazakhstan and adjacent countries (Almaty, 1–2 March 2012). [In Russian.] Almaty, Kazakhstan.

Ransom, J.I., P. Kaczensky, B.C. Lubow, O. Ganbaatar, and N. Altansukh. 2012. A collaborative approach for estimating terrestrial wildlife abundance. Biological Conservation 153:219–226.

Reading, R.P., H. Mix, B. Lhagvasuren, C. Feh, D.P. Kane, S. Dulamtseren, and S. Enkhbold. 2001. Status and distribution of khulan (*Equus hemionus*) in Mongolia. Journal of Zoology 254:381–389.

Robert, N., C. Walzer, S.R. Ruegg, P. Kaczensky, O. Ganbaatar, and C. Stauffer. 2005. Pathologic findings in reintroduced Przewalski's horses (*Equus caballus przewalskii*) in southwestern Mongolia. Journal of Zoo and Wildlife Medicine 36:273–285.

Rosenbom, S., V. Costa, S. Chen, L. Khalatbari, G.H. Yusefi, *et al.* 2015. Reassessing the evolutionary history of ass-like equids: Insights from patterns of genetic variation in contemporary extant populations. Molecular Phylogenetics and Evolution 85:88–96.

Rowen, M., and J.R. Ginsberg. 1992. Grevy's zebra (*Equus grevyi* Oustalet). Pages 10–12 *in* P. Duncan (Ed.), Zebras, asses, and horses: An action plan for the conservation of wild equids. International Union for Conservation of Nature, Gland, Switzerland.

Ryder, O.A. 1993. Przewalski's horse: Prospects for reintroduction into the wild. Conservation Biology 7:13–15.

Schaller, G.B., Z. Lu, H. Wang, and T. Su. 2005. Wildlife and nomads in the eastern Chang Tang Reserve, Tibet. Memorie della Società Italiana di Scienze Naturali e del Museo Civico di Storia Naturale di Milano 33:59–67.

Schipper, J., J.S. Chanson, F. Chiozza, N.A. Cox, and M. Hoffmann, *et al.* 2008. The status of the world's land and marine mammals: Diversity, threat, and knowledge. Science 322:225–230.

Shah, N. 2002. Status and action plan for the kiang (*Equus kiang*). Pages 72–81 *in* P. D. Moehlman (Ed.), Equids: Zebras, asses and horses. A status survey and conservation action plan. International Union for Conservation of Nature, Gland, Switzerland.

Shah, N., and Q. Qureshi. 2007. Social organisation and determinants of spatial distribution of Khur (*Equus hemionus khur*). Exploration into the Biological Resources of Mongolia (Halle / Salle) 10:189–200.

Shah, N., A. St-Louis, Z. Huibin, W. Bleisch, J. van Gruissen, and Q. Qureshi. 2008. *Equus kiang*. The IUCN red list of threatened species. Version 2014.3. International Union for Conservation of Nature, Gland, Switzerland.

Sokolov, V.E., G. Amarsanaa, M.W. Paklina, M.K. Posdnjakowa, E.I. Ratschkowskaja, and N. Chotoluu. 1992. Das Letzte Przewalskipferd areal und seine Geobotanische Characteristik. Pages 213–218 *in* S. Seifert (Ed.), Proceedings of the 5th international symposium on the preservation of the Przewalski horse. Zoologischer Garten Leipzig, Leipzig, Germany.

Stears, K, A. Shrader, J. Selier, S.R.B. King, and P.D. Moehlman. 2015. South Africa national red list assessment *Equus quagga* plains zebra. South African National Biodiversity Institute and Endangered Wildlife Trust, Pretoria and Johannesburg, South Africa.

St-Louis, A., and S.D. Côté. 2008. *Equus kiang* (Perissodactyla: Equidae). Mammalian Species 835:1–11.

Teclai, R. 2006. Conservation of the African wild ass (*Equus africanus*) on Messir Plateau (Asa-ila), Eritrea: The role of forage availability and diurnal activity pattern during the wet season and beginning of the dry season. Thesis, University of Kent, Kent, UK.

Tsering, D., J. Farrington, and K. Norbu. 2006. Human–wildlife conflict in the Chang Tang region of Tibet. World Wide Fund for Nature China–Tibet Program. Lhasa, Tibet, China.

Usukhjargal, D., and N. Bandi. 2013. Reproduction and mortality of re-introduced Przewalski's horse *Equus przewalskii* in Hustai National Park, Mongolia. Journal of Life Sciences 7:623–629.

van Dierendonck, M.C., and M.F. Wallis de Vries. 1996. Ungulate reintroductions: Experiences with the takhi or Przewalski horse (*Equus ferus przewalskii*) in Mongolia. Conservation Biology 10:728–740.

Williams, S.D. 1998. Grevy's zebra: Ecology in a heterogeneous environment. Thesis, University College London, London, UK.

Williams, S.D. 2002. Status and action plan for Grevy's Zebra (*Equus grevyi*). Pages 11–27 *in* P.D. Moehlman (Ed.), Equids: Zebras, asses and horses. A status survey and conservation action plan, International Union for Conservation of Nature, Gland, Switzerland.

Williams, S.D. 2013. *Equus Grevyi* Grevy's zebra. Pages 422–428 *in* J.S. Kingdon and M. Hoffmanm (Eds.), The mammals of Africa, vol. 5: Carnivores, pangolins, rhinos and equids. Academic Press, Waltham, MA, USA.

Wingard, J.R., and P. Zahler. 2006. Silent steppe: The illegal wildlife trade crisis in Mongolia. Mongolia Discussion Papers. World Bank, East Asia and Pacific Environment and Social Development Department, Washington, DC, USA.

Xia, C., J. Cao, H. Zhang, X. Gao, W. Yang, and D. Blank. 2014. Reintroduction of Przewalski's horse (*Equus ferus przewalskii*) in Xinjiang, China: The status and experience. Biological Conservation 177:142–147.

Yalden, D.W., M.J. Largen, and D. Kock. 1986. Catalogue of the mammals of Ethiopia. 6. Perissodactyla, Proboscidea, Hyracoidea, Lagomorpha, Tubulidentata, Sirenia and Cetacea. Italian Journal of Zoology Supplement 4:31–103.

Zhao, T., and G. Liang. 1992. On returning to its native place and conservation of the Przewalski horse. Pages 227–231 *in* S. Seifert (Ed.), Proceedings of the 5th international symposium on the preservation of the Przewalski Horse. Zoologischer Garten Leipzig, Leipzig, Germany.

13

Challenges and Opportunities for Conserving Equid Migrations

L. STEFAN EKERNAS
AND JOEL BERGER

The field of conservation biology is undergoing a paradigm shift from "numerical minimalism" to defining success as conserving ecological processes and functions (Soulé *et al.* 2003; Redford *et al.* 2013). Simply put, conserving ecological function means maintaining large populations over entire landscapes (Redford *et al.* 2011). Among the planet's most dramatic examples of ecologically functioning populations are mass migrations. Migrating animals integrate ecological communities at continental scales and provide key ecological services such as nutrient transport and cycling (Bauer and Hoye 2014). Fully conserving a species requires that migrations be maintained.

Our goals herein are three. First, we describe what is known of equid migrations. Because little is known about precise migrations per se, we offer a general overview of long-distance and wide-ranging movements of wild equids. Second, we investigate factors that predict extirpations by synthesizing information on global equid migrations and contrasting migrations that persist to those that have been lost. This approach facilitates an understanding of the adequacy and limitations of protected areas at a landscape level to reveal what humans must do to accomplish equid conservation. Finally, we contextually frame results from our analyses for other large terrestrial mammals and emphasize what scientists can bring to the conservation table. It is clear that the conservation of long-distance migration in terrestrial mammals is fraught with challenges, particularly for equids and other large species (Berger 2004). Protected areas are nearly always too small to encompass entire migration paths (Harris *et al.* 2009). If migrations are to persist, they need to be maintained in human-dominated landscapes. In these areas, threats are often profound: a burgeoning human population, poverty, habitat loss, fragmentation, and few incentives for people to conserve wild lands and wildlife (Bolger *et al.* 2008). These challenges are largely intractable, especially to individuals and organizations whose primary goal is to conserve equid populations. Conservationists need to be creative in finding solutions.

A natural starting point for identifying conservation opportunities is to recognize gaps in knowledge about human impacts and species needs. These vary greatly across continents, local regions, and species. Securing information about human impacts and species needs at varying scales is a crucial starting point, and this initial stage is where scientific inquiry makes its greatest contribution to any conservation effort.

Finding solutions hinges on recognizing factors that can be controlled. These factors vary from management agencies making land-use decisions, changing harvest rates (both legal and illegal), and configuring infrastructure to accommodate wildlife, for example, by limiting fencing, placing wildlife crossings on roads and railroads, and incorporating wildlife into urban planning. The common denominator is that these factors are local, site-specific, and can improve conservation outcomes even in the face of unyielding threats such as climate change and a rapidly growing human population.

Defining Migrations

Migrations are typically defined as predictable seasonal movements with geographically separate home ranges, which are distinguished from nomadic movements that are not predictable, not seasonal, and encompassed in a contiguous area (Milner-Gulland *et al.* 2011; Singh *et al.* 2012). Nonetheless, nomadic movements can exceed the spatial scale of migrations. For example, Asiatic wild ass (*Equus hemionus*) in Mongolia's Gobi Desert have home ranges as large as 70,000 km^2 (Kaczensky *et al.* 2011), which is more than twice the size of the entire Serengeti-Mara ecosystem. For the purpose of conserving ecological processes and functions, there is little distinction between long-distance nomadic and migratory movements: both transport nutrients between biological communities, influence trophic dynamics at large spatial scales, and increase fitness for moving animals that have increased potential for improving foraging opportunities, reducing predation, and improving reproductive output (Bauer and Hoye 2014). The important point is that these movements are large, encompassing linear distances of tens to hundreds of kilometers. For brevity, we refer to all such movements as "migration."

Wild equids are found in arid and semiarid regions of Asia and Africa, and migrations have been documented across this range (fig. 13.1). We investigate three factors that may predict extirpation of equid migration: reserve size, human density both inside and outside protected areas, and fencing. This list is by no means exhaustive, as other factors also influence the persistence of migrations; we chose these three primarily because of data availability and ability to directly compare these factors across vast spatial scales spanning multiple continents. Effects of reserve size and human density might be correlated, and we therefore also investigate whether reserve sizes tend to be smaller at relatively higher human densities.

We used Google Scholar, Web of Science, the gray literature, reference books (Estes 1991; Duncan 1992; Nowak 1999), and two review articles on migrations (Bolger *et al.* 2008; Harris *et al.* 2009) to identify existing and extirpated equid migrations. Most equid migrations lack empirical data on distances moved, geographic routes, and number of animals involved. Even movements of relatively well-studied equid populations are poorly known in many cases; for example, it is unclear whether ~200-km linear-distance movements of *E. h. khur* in India's Rann of Kutch represent dispersal, migratory, or nomadic movements (Shah and Qureshi 2007; Moehlman *et al.* 2008). In addition to incomplete biological information, our literature search was restricted to English-language sources, meaning we may have neglected information in regional reports from Asian countries in particular. Therefore our data set probably overlooked some migrations, and poorly studied equid populations exhibiting nonseasonal movements are especially likely to have been missed.

We categorized migrations as extirpated if animal movements have been severely curtailed, even if animals still move relatively long distances (e.g., as is the case in Namib-Naukluft National Park, Namibia, and Kruger National Park, South Africa). We do not have exact knowledge on movements from some migrations (table 13.1), which makes a precise quantitative definition difficult, but all migrations that we classify as extirpated have been identified as such in peer-reviewed literature.

To gauge the effect of fencing, we classified blockages along the perimeter of each protected area into three categories: unfenced (<10% of perimeter fenced), partially fenced (10–90% fenced), or fully fenced (>90% of perimeter fenced). Fencing inside and outside reserve boundaries can of course also impede animal movements, but data on fences do not exist for most rural areas in Asia and Africa. Perimeter fencing was the only factor for which we were able to obtain reliable information across protected areas. Our measure of fencing is therefore a substantial underestimate of actual fences that migrating equids encounter.

We estimated human densities from Global Rural-Urban Mapping Project v1 (Center for International Earth Science Information Network 2012) using human density in 2000. Polygons of protected areas (Protected Planet 2012) were overlaid on human density in ArcGIS 10.0 (Environmental Systems Resource Institute 2011) to provide a human density estimate for each protected area. Ideally, we would only include areas where migrating equids travel, but such detailed data are not available for most equid migrations. Instead,

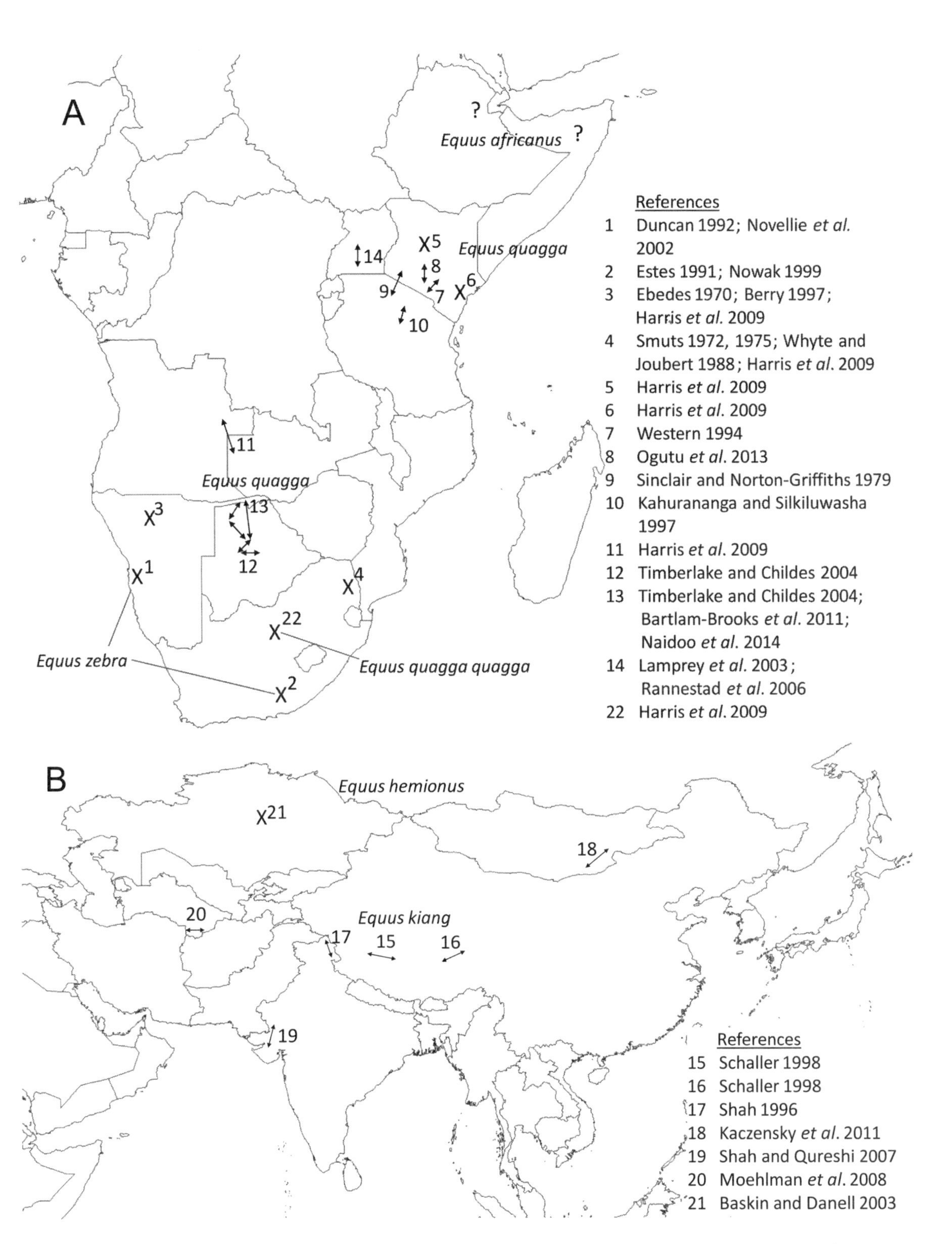

Fig. 13.1 Wild equid migrations that persist (arrows) and are extinct (X's) in Africa (A) and Asia (B). Numbers cross-reference table 13.1.

Table 13.1. **Wild equid migrations in relation to protected area size, fencing, and human density**

Species Name	Common Name		PA	Country	Migration Status	PA Size (km^2)	PA Fencing	Human Density	Current Migration Length (km)‡	Historical Migration Length (km)	Current Migration inside PA (%)
E. zebra	mountain zebra	1	Namib-Naukluft NP	Namibia	extinct*	49,768	partial	1.2	50?	200?	10
		2	Mountain Zebra NP	South Africa	extinct*	246	full	4.3	0	?	100
E. quagga	plains zebra	3	Etosha NP	Namibia	extinct	22,151	full	1.8	50	200–320	100
		4	Kruger NP	South Africa	extinct	19,485	full	36.2	30	60	100
		5	Laikipia District	Kenya	extinct*	9,473	partial	89.7	10	100–150	0§
		6	Tsavo NP	Kenya	extinct*	22,812	partial	15.0	?	50	100
		7	Amboseli NP	Kenya	persists	392	unfenced	35.8	50	50	5
		8	Nairobi NP	Kenya	persists	117	partial	427.5	30	30	5
		9	Serengeti NP	Tanzania	persists	14,763	unfenced	44.8	100	120	100
			Masai Mara GR	Kenya	persists	1,510	unfenced	62.0			
		10	Tarangire NP	Tanzania	persists	2,600	unfenced	31.8	60	60	50
		11	Liuwa Plains NP	Zambia	persists*	3,660	unfenced	5.7	250	250	20
			Cameia NP	Angola	persists*	1,445	unfenced	2.0			
		12	Makgadikgadi NP	Botswana	persists	7,300	unfenced	1.2	120	120	50
		13	Chobe NP	Botswana	persists	11,700	unfenced	2.2	250	250	30
			Nxai Pan NP	Botswana	persists	2,578	unfenced	1.6			
		14	Lake Mburo NP	Uganda	persists*	369	unfenced	126.8	20	20	50
E. kiang	kiang	15	Chang Tang NP	China	persists*	334,000	unfenced	0.4	>100	>100	100
		16	Qinghai	China	persists*	152,300	unfenced	2.8	>100	>100	100
		17	Changthang Sanctuary	India	persists*	4,000	unfenced	1.4	50	?	75
E. hemionus	Asiatic wild ass	18	Small Gobi SPA	Mongolia	persists	18,392	partial†	1.3	400	?	10
		19	Indian Wild Ass Sanctuary	India	persists	7,506	unfenced	46.8	200	200	50
		20	Badkhyz State NR	Turkmenistan	persists	877	partial†	11.2	100	100	30

Note: Protected areas in Botswana harbor multiple equid migrations; all other protected areas represent single migrations. References are listed in figure 13.1. Human density is calculated as protected area with 50-km buffer (not donut shaped). Human density is from GRUMPv1 (Center for International Earth Science Information Network 2012) overlaid with shapefiles of protected areas in ArcGIS (Environmental Systems Resource Institute 2011). Two extinct equid migrations were excluded because of insufficient historical information: 21, *E. hemionus* in Kazakhstan (Baskin and Danell 2003), and 22, *E. q. quagga* in South Africa (Harris et al. 2009). *E. grevyi* and *E. f. przewalskii* are not in the data set because of insufficient information on historical movements. *E. africanus* is not in the data set because of insufficient information on current movements. Abbreviations are as follows: GR, Game Reserve; NP, National Park; NR, Nature Reserve; PA, protected area; SPA, Strictly Protected Area.

☐ = extinct migration

*Migration has not been formally studied.

†Fencing is along international border.

‡Straight-line one-way trip.

§Current movements are 100% contained within Laikipia District, which does not have any formally designated protected area.

we used the protected area as a proxy for each migration, with buffers placed around the protected area to account for animals moving outside reserves. We estimated human density at three scales: inside the protected area only, inside the protected area plus a 10-km buffer, and inside the protected area plus a 50-km buffer (buffers created with Geospatial Modeling Environment; Beyer 2012). Analyses were done with human density at all three scales, but because results did not substantively differ when we changed scales, we only report results from human density with a 50-km buffer around the protected area. A 50-km buffer provides an ecologically relevant scale because migrating equids frequently travel tens of kilometers outside protected area boundaries.

We compiled a comprehensive list of documented wild equid migrations and in our analyses used protected area as the unit of measurement (table 13.1; fig. 13.1). To avoid replications, protected areas that harbor multiple migrations (e.g., two equid species, or two populations with different migrations) were included only once. Three extinct migrations never encompassed a protected area: plains zebra (*E. quagga*) in Kenya's Laikipia District, quagga (*E. q. quagga*) in South Africa, and Asiatic wild ass in Kazakhstan. Plains and Grevy's zebra (*E. grevyi*) persist in good numbers in the Laikipia District but no longer migrate. Because no formally designated protected area has been designated in Laikipia, and no data exist on routes used by migrating animals, we used data for the entire district in our analysis. Overhunting decimated quagga and Asiatic wild ass, and we did not include these migrations in our analyses because we lack historical information on human density, fencing, and exact locations of movements. We also removed from our analysis Asiatic wild ass in Great Gobi A and B Strictly Protected Areas, Mongolia. GPS collars indicate that animals move large distances within these protected areas (Kaczensky *et al.* 2011). Animals almost never left the protected areas, which are also fenced along the international border to China. These factors suggest that movements have been restricted and the migration should be classified as extinct; unfortunately, we have no data on historical movements to evaluate that claim. Because we are unable to confidently classify this migration as persisting or extinct, we removed it from our analysis.

Migration Persistence and Constraints

Reserve Size and Human Density

Human density is strongly associated with reserve size: protected areas are exponentially smaller in areas with higher human density (log-log regression, $\beta = -0.71$, $p < 0.001$, $r^2 = 0.32$; fig. 13.2). While our analysis omits consideration of the period of time that an area has been protected (hence we cannot consider potential causation), the negative exponential nature of this relationship indicates that small increases in human density are associated with much smaller protected areas.

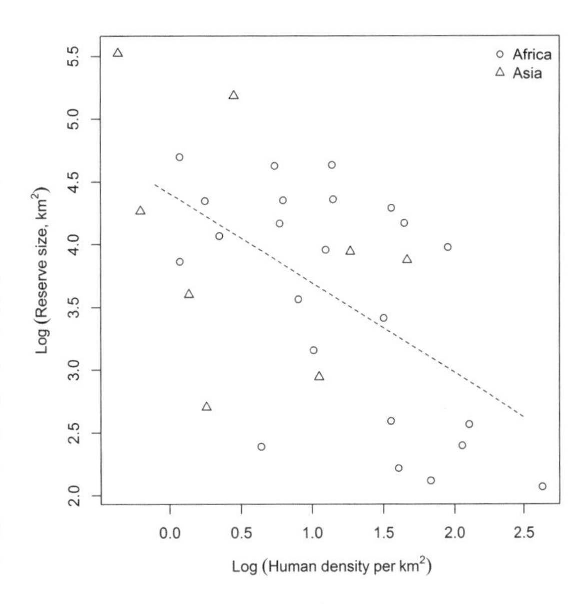

Fig. 13.2 Log-log plot of human density against protected area size ($\beta = -0.71$, $p < 0.001$, $r^2 = 0.32$).

Factors Predicting Migration Extirpation

Reserve Size. Protected area size is weakly if at all associated with the persistence of migrations (Mann Whitney U test, $n = 6$ extinct, $n = 14$ persist, $p = 0.21$). Equid migrations persist where protected areas are as small as 117 km^2 (Nairobi National Park; Ogutu *et al.* 2013) and have been reduced from protected areas as large as 49,700 km^2 (Namib-Naukluft National Park; Duncan 1992).

Human Density. Human density is not associated with migration persistence (Mann Whitney U test, $n = 6$ extinct, $n = 14$ persist, $p = 1.0$). Equid migrations persist at human densities exceeding 100 people per km^2 in two locations (Nairobi National Park in Kenya and Lake Mburo in Uganda), and conversely have been extirpated from areas with <2 people per km^2 (Etosha and Namib-Naukluft National Parks in Namibia).

Fencing. Fencing is strongly associated with migration extirpations, even though our measure of fencing severely underestimates actual fences that migrating equids encounter. All extirpations ($n = 6$) occurred in protected areas that are partially or fully fenced, and every fully fenced protected area in our data set has seen equid migrations extirpated or restricted. By contrast, 11 of 14 protected areas that maintain equid migrations are unfenced, with the remaining 3 (Nairobi National Park, Small Gobi Strictly Protected Area, and Badkhyz State Nature Reserve) being partially fenced (Fisher's exact test, unfenced vs. partial or full fencing, $n = 20$, $p = 0.002$).

Migration in the Modern Context

Across equid migrations, the most identifiable factor associated with the loss of migrations is fencing. Counter to our expectations, neither human density nor reserve size was a strong predictor of migration persistence. Migrations continue in small reserves, at high human densities, and sometimes in small reserves and high human density. We find great optimism in these results, because eliciting change in fences is much easier than altering human density or creating large reserves in areas with many people.

The value of fencing to conservation has been the subject of much recent debate (Creel *et al.* 2013; Packer *et al.* 2013; Woodroffe *et al.* 2014). Under the right circumstances, fencing can be an effective tool for reducing human–wildlife conflict and improving conservation outcomes (Packer *et al.* 2013; Woodroffe *et al.* 2014). But fencing fragments landscapes and has many unintended consequences, for example, providing a readily available supply of wire for snares and changing local communities' relationship with the land (Woodroffe *et al.* 2014). Even well-planned fences can inadvertently sever unknown migration routes; the recent discovery of Africa's longest known land migration (by plains zebra; Naidoo *et al.* 2014) emphasizes how little we still know about animal movements. Conversely, removing fences has the potential to rapidly restore equid migrations. The 2004 removal of a ~100-km-long veterinary fence, bifurcating a historical migration route in northwest Botswana since the fence was erected in 1968, led to plains zebra reestablishing the migration within four years (Bartlam-Brooks *et al.* 2011).

Several of our findings are of note. Of eight documented equid migration extinctions that have been documented, seven occurred in Africa and only one in Asia. Three countries account for all seven African extirpations: South Africa (3), Namibia (2), and Kenya (2). These three countries have two primary commonalities. First, all have fenced some of their protected areas, but especially in South Africa and Namibia. Namibia has retained some long-distance equid migrations inside fenced reserves (Etosha and Namib-Naukluft), but these migrations have been restricted from their historical baseline (Ebedes 1970; Duncan 1992), and we thus classified them as extirpated (table 13.1). Second, compared to other African and Asian countries, all three have relatively well-developed ecological knowledge and thus have documented both the existence and disappearance of equid migrations.

Lack of ecological knowledge from Asia almost certainly deflates the number of extirpated migrations there; Przewalski's horse (*E. ferus przewalskii*), for example, do not appear in our data set because they became extinct in the wild before their movements were documented. Similarly, dramatic range reductions of Asiatic wild ass in parts of Mongolia, China, India, Iran, and Russia as a result of overhunting have severed a capacity to migrate (see chap. 12), yet these do not appear in our data set because no information exists on prior movements. Overhunting has probably played a stronger role in shaping equid distributions than any of the factors we analyze.

Details from individual case studies offer insights about migrations. Namib-Naukluft National Park (49,700 km^2) as well as Kruger and Etosha National Parks (each ~20,000 km^2) reflect the difficulties of conserving long-distance movements inside even very large protected areas. Encompassing migrations inside protected areas may require sizes almost an order of magnitude larger, such as Changthang in western China, a protected area roughly the size of Italy. Barring these rare circumstances, conserving connectivity and equid habitat outside protected areas is critical.

Connectivity is easier to maintain when human density is low. Zebra migrating between eastern Angola and Liuwa Plains National Park, Zambia, have certainly benefited from low human density, and the migration persists despite a lack of strong conservation efforts. But, as Namibia vividly demonstrates, low human density is no guarantee for conserving migrations. Nor is high human density guaranteed to doom migrations. Equid migrations might be easier to conserve at low human density, but our data demonstrate that human density is not destiny.

Nairobi National Park is illustrative. As the smallest protected area in our data set, it also has the highest human density, is fenced on three sides, and still plains zebra migrate. Zebra spend the dry season in the 117-km^2 national park, but in the rainy season, they migrate through the southern unfenced border to the Athi-Kaputiei Plains, which covers 2,200 km^2 and has little formal protection (Ogutu *et al.* 2013). It is unclear how much longer that migration will persist. Pastoralist communities, urbanization, and crop agriculture are all rapidly expanding in the Athi-Kaputiei Plains, reducing tolerance for wildlife. Coupled with an upsurge in fencing, the prospects for continued zebra migration routes are low (Ogutu *et al.* 2013). Nairobi National Park is demonstrative of the broader trends we report; neither reserve size nor human density per se determines outcomes for conserving migrations. Instead, what drives conservation outcomes is human tolerance, especially when that (lack of) tolerance manifests in physical infrastructure. Fences that prohibit wildlife movements are an outcome of local communities' antipathy or unconcern for wildlife. On the flip side, wildlife-friendly fencing that allows passage by wild ungulates as well as under- and overpasses for wildlife across roads are physical manifestations of communities' desire to maintain wildlife on the landscape, a prospect with increasing support for long-distance migration in at least some rural areas (Berger *et al.* 2006; Berger and Cain 2014), though far from the haunts of wild equids.

With that understanding, how can we move forward with conserving equid migrations? As stated above, a key to defining successful strategies is to identify factors that can be controlled. Fencing and other infrastructure, as our analysis shows, are often crucial. Harvest can also have a big effect, particularly when it is illegal or unregulated. A survey of 25 imperiled and extinct large-animal migrations attributed either infrastructure development or overharvest as the primary threat in 96% of the cases (Bolger *et al.* 2008). Underlying these proximate drivers is low tolerance, or at best disregard, for wildlife.

Conservation programs are most likely to be effective when they simultaneously target proximate threats, formal legislation and rules governing land use, and local attitudes that ultimately drive conservation outcomes (Berger and Cain 2014). To implement such a scheme, both top-down and bottom-up approaches are required, which if done well can be mutually reinforcing. Community events to bolster local support for wildlife can help influence public officials, for example; conversely, state and federal agencies, municipal governments, and public officials are crucial partners in disseminating information to the public (Berger and Cain 2014).

Conclusion

What role does science play in this framework? In and of itself, science plays only a small role. It is critical in the initial stages: locating migration routes, identifying and quantifying threats, and estimating sustainable harvest levels. This type of information is crucial for moving conservation forward, but the information itself does not achieve conservation. Poaching does not stop, and fences do not come down. Publication in a scientific journal is not conservation. To be valuable in conservation, scientific knowledge should be leveraged in all aspects of conservation efforts. Collecting, analyzing, and disseminating data provide scientists with credibility that can open doors to policymakers and industry that are unavailable to pure advocates (Berger and Cain 2014). Scientific knowledge can also play an important role in building public support. Baba Dioum's (1969) dictum "We will conserve only what we love, we will love only what we understand, and we will understand only what we are taught" means that scientific knowledge is a necessary component of conservation efforts, but to actually achieve local support, scientific knowledge needs to be communicated. Vucetich and Nelson (2013) argue that the central purpose of conservation science should be to instill people with a sense of wonder about the natural world, as appreciation of nature is the cornerstone for all conservation.

Our point is simple: science in isolation does little to advance conservation. When leveraged, it can. While the details of how this might best be achieved will vary from case to case, the key is to identify which factors on the ground can be changed and which cannot. Equid migrations face a deluge of threats that individual conservationists, or even major conservation organizations, have no ability to alter. But local on-the-ground realities can be altered to mitigate these threats. Once we acknowledge these constraints, realistic actions and opportunities emerge.

ACKNOWLEDGMENTS

The authors thank J. Ransom, P. Kaczensky, and an anonymous reviewer for valuable insights and comments that improved this chapter.

REFERENCES

Bartlam-Brooks, H.L.A., M.C. Bonyongo, and S. Harris. 2011. Will reconnecting ecosystems allow long-distance mammal migrations to resume? A case study of a zebra *Equus burchelli* migration in Botswana. Oryx 45:210–216.

Baskin, L., and K. Danell. 2003. Ecology of ungulates: A handbook of species in eastern Europe and northern and central Asia. Springer Berlin, Heidelberg, Germany.

Bauer, S., and B.J. Hoye. 2014. Migratory animals couple biodiversity and ecosystem functioning worldwide. Science 344:doi:10.1126/science.1242552.

Berger, J. 2004. The longest mile: How to sustain long distance migration in mammals. Conservation Biology 18:320–332.

Berger, J., and S. Cain. 2014. Moving beyond science to protect a mammalian migration corridor. Conservation Biology 28:1142–1150.

Berger, J., S.L. Cain, and K. Berger. 2006. Connecting the dots: An invariant migration corridor links the Holocene to the present. Biology Letters 2:528–531.

Berry, H.H. 1997. Aspects of wildebeest *Connochaetes taurinus* ecology in the Etosha National Park—A synthesis for future management. Madoqua 20:137–148.

Beyer, H.L. 2012. Geospatial modelling environment. Available at www.spatialecology.com/gme/.

Bolger, D.T., W.D. Newmark, T.A. Morrison, and D.F. Doak. 2008. The need for integrative approaches to understand and conserve migratory ungulates. Ecology Letters 11:63–77.

Center for International Earth Science Information Network. 2012. Global rural–urban mapping project, version 1 (GRUMPv1): Population density grid. Columbia University, International Food Policy Research Institute, World Bank, Centro Internacional de Agricultura Tropical, and NASA Socioeconomic Data and Applications Center, Palisades, NY, USA. Available at http://sedac.ciesin.columbia.edu/data/collection/grump-v1.

Creel, S., M.S. Becker, S.M. Durant, J. M'soka, W. Matandiko, *et al.* 2013. Conserving large populations of lions—The argument for fences has holes. Ecology Letters 16:1413–e3.

Dioum, B. 1969. Address to the Tenth General Assembly of the International Union for the Conservation of Nature and Natural Resources, 24 November to 1 December, New Delhi, India.

Duncan, P. (Ed.). 1992. Zebras, asses and horses: An action plan for the conservation of wild equids. International Union for Conservation of Nature, Gland, Switzerland.

Ebedes H. 1970. The nomadic plains zebra *Equus burchelli antiquorum* H Smith 1941 of the Etosha Salina. Nature Conservation, Windhoek, Namibia.

Environmental Systems Resource Institute. 2011. ArcGIS. Redlands, CA, USA.

Estes, R. 1991. The behavior guide to African mammals: Including hoofed mammals, carnivores, and primates. University of California Press, Berkeley, USA.

Harris, G., S. Thirgood, J.G.C. Hopcraft, J.P. Cromsigt, and J. Berger. 2009. Global decline in aggregated migrations of large terrestrial mammals. Endangered Species Research 7:55–76.

Kaczensky, P., R. Kuehn, B. Lhagvasuren, S. Pietsch, W. Yang, and C. Walzer. 2011. Connectivity of the Asiatic wild ass population in the Mongolian Gobi. Biological Conservation 144:920–929.

Kahurananga, J., and F. Silkiluwasha. 1997. The migration of zebra and wildebeest between Tarangire National Park and Simanjiro Plains, northern Tanzania, in 1972 and recent trends. African Journal of Ecology 35:179–185.

Lamprey, R., E. Buhanga, and J. Omoding. 2003. A study of wildlife distributions, wildlife management systems, and options for wildlife-based livelihoods in Uganda. International Food Policy Research Institute, USAID, Kampala, Uganda.

Milner-Gulland, E.J., J.M. Fryxell, and A.R.E Sinclair. 2011. *Animal migration: A synthesis*. Oxford University Press, Oxford, UK.

Moehlman, P.D., N. Shah, and C. Feh. 2008. *Equus hemionus*. IUCN red list of threatened species. Version 2015.2. International Union for Conservation of Nature, Gland, Switzerland. Available at www.iucnredlist.org.

Naidoo, R., M.J. Chase, P. Beytell, P. Du Preez, K. Landen, G. Stuart-Hill, and R. Taylor. 2014. A newly discovered wildlife migration in Namibia and Botswana is the longest in Africa. Oryx doi:10.1017/S0030605314000222.

Novellie, P., M. Lindeque, P. Lindeque, P. Lloyd, and J. Koen. 2002. Status and action plan for the mountain zebra (*Equus zebra*). Pages 28–42 *in* P.D. Moehlman (Ed), Equids: Zebras, asses and horses. Status survey and conservation action plan. International Union for Conservation of Nature, Gland, Switzerland.

Nowak, R.M. (Ed.). 1999. Walker's mammals of the world. Johns Hopkins University Press, Baltimore, USA.

Ogutu, J.O., N. Owen-Smith, H.P. Piepho, M.Y. Said, S.C. Kifugo, R.S. Reid, H. Gichohi, P. Kahumbu, and S. Andanje. 2013. Changing wildlife populations in Nairobi National Park and adjoining Athi-Kaputiei plains: Collapse of the migratory wildebeest. Open Conservation Biology Journal 7:11–26.

Packer, C., A. Loveridge, S. Canney, T. Caro, S.T. Garnett, *et al.* 2013. Conserving large carnivores: Dollars and fence. Ecology Letters 16:635–641.

Protected Planet. 2012. World Database on Protected Areas. World Conservation Union, International Union for Conservation of Nature, World Conservation Monitoring Centre, United Nations Environment Programme, Gland, Switzerland. Available at www.protectedplanet.net.

Rannestad, O.T., T. Danielsen, S.R. Moe, and S. Stokke. 2006. Adjacent pastoral areas support higher densities of wild ungulates during the wet season than the Lake Mburo National Park in Uganda. Journal of Tropical Ecology 22:675–683.

Redford, K.H., G. Amato, J. Baillie, P. Beldomenico, E.L. Bennett, *et al.* 2011. What does it mean to successfully conserve a (vertebrate) species? BioScience 61:39–48.

Redford, K.H., J. Berger, and S. Zack. 2013. Abundance as a conservation value. Oryx 47:157–158.

Schaller, G.B. 1998. Wildlife of the Tibetan Steppe. University of Chicago Press, Chicago, USA.

Shah, N. 1996. Status and distribution of Western kiang (*Equus kiang kiang*) in Changthang Plateau, Ladakh, India. Thesis, University of Baroda and Gujarat Nature Conservation Society, Baroda, Gujarat, India.

Shah, N., and Q. Qureshi. 2007. Social organization and determinants of spatial distribution of Khur (*Equus hemionus khur*). Exploration into the Biological Resources of Mongolia 10:189–200.

Sinclair, A.R.E., and M. Norton-Griffiths (Eds.). 1979. Serengeti: Dynamics of an ecosystem. University of Chicago Press, Chicago, USA.

Singh, N.J., L. Börger, H. Dettki, N. Bunnefeld, and G. Ericsson. 2012. From migration to nomadism: Movement variability in a northern ungulate across its latitudinal range. Ecological Applications 22:2007–2020.

Smuts, G.L. 1972. Seasonal movements, migration and age determination of Burchell's zebra (*Equus burchelli antiquorum*,

H. Smith, 1841) in the Kruger National Park. Thesis, University of Pretoria, Pretoria, South Africa.
Smuts, G.L. 1975. Home range sizes for Burchell's zebra *Equus burchelli antiquorum* from the Kruger National Park. Koedoe 18:139–146.
Soulé, M.E., J.A. Estes, J. Berger, and C.M. Del Rio. 2003. Ecological effectiveness: Conservation goals for interactive species. Conservation Biology 17:1238–1250.
Timberlake, J.R., and S.L. Childes (2004). Biodiversity of the Four Corners area, vol. 2: Technical reviews (chapters 5–15). Occasional Publications in Biodiversity No. 15. Biodiversity Foundation for Africa, Bulawayo / Zambezi Society, Harare, Zimbabwe.
Vucetich, J.A., and M.P. Nelson. 2013. The infirm ethical foundations of conservation. Pages 9–25 *in* M. Beckoff (Ed.), Ignoring nature no more: The case for compassionate conservation. University of Chicago Press, Chicago, USA.
Western, D. 1994. Ecosystem conservation and rural development: The case of Amboseli. Pages 15–52 *in* D. Western and R.M. Wright (Eds.), Natural connections: Perspectives on community based conservation. Island Press, Washington, DC, USA.
Whyte, I.J., and S.C.J. Joubert. 1988. Blue wildebeest population trends in the Kruger National Park and the effects of fencing. South African Journal of Wildlife Research 18:78–87.
Woodroffe, R., S. Hedges, and S.M. Durant. 2014. To fence or not to fence. Science 344:46–48.

14

Reintroduction of Wild Equids

PETRA KACZENSKY, HALSZKA HRABAR, VICTOR LUKAREVSKIY, WALTRAUT ZIMMERMANN, DORJ USUKHJARGAL, OYUNSAIKHAN GANBAATAR, AND AMOS BOUSKILA

Most wild equids have lost major parts of their distribution range in historic times, and the Przewalski's horse (*Equus ferus przewalskii*) even went extinct in the wild, with few animals surviving in captivity (Volf 1986; Boyd and Houpt 1994). Interest in equid restoration has also been high, however, and before the last Przewalski's horses disappeared in the wild in the 1960s, the first Asiatic wild asses (*E. hemionus*) had already been reintroduced to Barsa-Kelmes Island in Kazakhstan in the 1950s, and additional projects were to follow (Pavlov 1996; Feh *et al.* 2002). Reintroductions of Przewalski's horses, using captive-bred stock, began in 1992 in Mongolia (van Dierendonck and Wallis de Vries 1996; Bandi and Dorjraa 2012), followed by initiatives in northern China and southeastern Kazakhstan (Wakefield *et al.* 2002; Zimmermann 2005). Reintroductions of African wild equids have been restricted to Cape mountain zebra (*E. zebra zebra*) in South Africa, one of two subspecies of the mountain zebra (Novellie *et al.* 2002).

Many of the earlier equid reintroduction projects were initiated when little formal knowledge about the ecology or spatial requirements of the respective species was available, and when experience with and scientific literature on animal reintroductions were scarce. But we have come a long way since these first pioneer initiatives. In 1988, the Reintroduction Specialist Group of the International Union for Conservation of Nature Species Survival Commission (IUCN/SSC) was founded, and decades of experience and a large body of scientific literature have become available and integrated into various recommendations (e.g., Sarrazin and Barbault 1996; Fischer and Lindenmayer 2000; Amstrong and Seddon 2007; Seddon *et al.* 2007), including the recently updated IUCN/SSC "Guidelines on Reintroductions and Other Conservation Translocations" (IUCN/SSC 2013). Consequently, we want to stress that any ongoing or planned reintroductions, be it of equids or other animals, need to be planned, executed, and evaluated according to these guidelines (box 14.1).

In this chapter, we also use the IUCN/SSC (2013) definitions for conservation translocations—"the human-mediated movement of living organisms from one area, with release in another"—but focus entirely on population restorations—"conservation translocation to within the indigenous range." There are many more initiatives aiming to establish wild equid species living under seminatural conditions outside their range (conservation introduction) for educational, scientific, or management purposes (Zimmermann 2005). Within the context of Asiatic wild ass and Przewalski's horse conservation translocations, we focus entirely on reintroductions—"the intentional movement and release of an organism inside its indigenous range from which it has disappeared." In the context of Cape mountain zebra, we also include reinforcements—"the intentional movement and release of an organism into an existing population of conspecifics"—in this particular context meaning the human-assisted movement of animals among small, isolated (fenced) populations managed as one metapopulation, with the aim to reinforce population size or enhance or maintain genetic variability (Hrabar and Kerley 2013). We subjectively labeled a reintroduction as "success" if the current population size was at least ~100 individuals and suggested a stable or increasing trend.

By compiling the existing knowledge and experiences from equid reintroductions going back as far as 60 years, spanning three species with different ecological niches and covering two continents as well as a multitude of different countries with a broad diversity of sociopolitical realities, we hope to contribute further to the existing body of knowledge and ultimately help improve ongoing or future wild equid reintroduction initiatives.

BOX 14.1 IUCN REINTRODUCTION GUIDELINES

The requirements for conservation translocations according to IUCN/SSC's (2013) "Guidelines for Reintroductions and Other Conservation Translocations" are as follows.

- Any translocation needs rigorous justification. Feasibility assessment should include a balance of the conservation benefits against the costs and risks of both the translocation and alternative conservation actions.
- Any proposed translocation should have a comprehensive risk assessment with a level of effort appropriate to the situation. Where risk is high or uncertainty remains about risks and their impacts, a translocation should not proceed.
- Social, economic, and political factors must be integral to translocation feasibility and design.
- Design and implementation of conservation translocations should follow the standard stages of project design and management, including gathering baseline information and analysis of threats, and iterative rounds of monitoring and management adjustment once the translocation is underway.
- Finally, translocations should be fully documented, and their outcomes made publicly and suitably available to inform future conservation planning.

Asiatic Wild Ass Reintroductions

Kazakhstan

The Asiatic wild ass (*E. h. kulan*) became extinct in Kazakhstan at the end of the 1930s as a result of overhunting and competition with livestock (Heptner *et al.* 1988; Kaczensky *et al.*, 2015). However, reintroductions began in 1953, when an initial eight wild asses from Turkmenistan were brought to the Barsa-Kelmes Island in the former Aral Lake (Bannikov 1981). Eleven more animals followed until 1964 (Pavlov 1996). The population increased and provided the stock for further reintroduction initiatives to Altyn Emel National Park, Aktau-Buzachy Sanctuary on Buzachy Peninsula, and Andassay Sanctuary (table 14.1; fig. 14.1).

The newly established wild ass population in Altyn Emel National Park grew rapidly, but the species status in the Andassay Sanctuary remained unclear (Meldebekov *et al.* 2010). Consequently, further releases were initiated since 2006 from Altyn Emel National Park to the Andassay Sanctuary within the framework of the governmental wild ungulate conservation program (Levanov *et al.* 2013). Postrelease monitoring has been poor, the fate of the reintroduced animals remains largely unknown, and there is concern that poaching levels may be high (S. Sokolov, personal communication). In order to prevent wild ass from quickly dispersing with the risk of losing contact with each other, the first animals to be reintroduced were kept in adaptation enclosures prior to release (soft release) for four to five months at Altyn Emel National Park and the Andassay Sanctuary. Wild asses were released immediately (hard release) at the Aktau-Buzachy Sanctuary (Pavlov 1996). The second wave of reintroductions to the Andassay Sanctuary initiated in 2006 initially held wild asses in adaptation enclosures. Despite feeding, however, body condition of captive animals deteriorated and injuries occurred, and thus from 2010 on, wild asses were immediately released upon arrival (S. Sokolov, personal communication).

Out of four reintroduction initiatives in Kazakhstan, two have been successful: (1) the Barsa-Kelmes

***Table 14.1.* Overview on wild ass (*E. hemionus*) reintroductions**

	Number on Map	Date	Area	Number Transported or Released	Origin	Estimated Current Population	Status of Reintroduction	Source
Turkmenistan								
	1	1979–1989	Meana Chaacha (eastern Kopetdag)	48	Badkhys, Turkmenistan	100	success	V.S. Lukarevskiy, personal communication 2014, O. Pereladova, personal communication 2014
	2	1981	Kuruhhaudan/Kalinin	18	Badkhys, Turkmenistan	10–15	future unclear	Pavlov 1996, V.S. Lukarevskiy, personal communication 2014
	3	**1982**	**Kurtusu**	**7-9**	**Badkhys, Turkmenistan**	**extinct**	**failure**	**Pavlov 1996, V.S. Lukarevskiy, personal communication 2014**
	4	**1982**	**Mirzadag/Germab**	**12-13**	**Badkhys, Turkmenistan**	**extinct**	**failure**	**Pavlov 1996, V.S. Lukarevskiy, personal communication 2014**
	5	1988–1989	western Kopetdag	47	Badkhys, Turkmenistan	13	future unclear	V.S. Lukarevskiy, personal communication 2014
	6	1983–1987	Kaplankyr Reserve at Sarykamish Lake/ Ustyurt Plateau	~100	Badkhys, Turkmenistan	350–400	success (spreading into Uzbekistan)	Pavlov 1996, Kuznetsov 2014, N. Marmazinkaja, unpublished data 2012–2013
Kazakhstan								
	7	1953–1964	Barsa-Kelmes Island	19	Badkhys, Turkmenistan	347	success	Pavlov 1996, Meldebekov *et al.* 2010
	8	1982–1984	Altyn Emel National Park	32–38	Barsa-Kelmes Island, Kazakhstan	2,500–3,000	success	Pavlov 1996, Plakhov *et al.* 2012
	9	**1991**	**Aktau-Buzachy Sanctuary on Buzachy Peninsula**	**31-35**	**Barsa-Kelmes Island, Kazakhstan**	**most likely extinct**	**failure**	**Pavlov 1996, Levanov et al. 2013**
	10	1986–1990	Andassay Sanctuary	120	Barsa-Kelmes Island, Kazakhstan	<50	unclear	Pavlov 1996, Cheremnov and Sokolov 2012
		2006–2011	Andassay Sanctuary	95	Altyn Emel NP	30–40	ongoing	Levanov *et al.* 2013

	11	*planned*	*Altyn Dala*	*NA*			*possibility discussed*	*Kaczensky 2011*
	12	*planned*	*Arganaty Mountains*	*NA*			*possibility discussed*	*Cheremnov and Sokolov 2012*
Uzbekistan								
	13	1978–1979	Dzheiran Ecocentre	7	Barsa-Kelmes Island, Kazakhstan	98	success (51-km^2 fenced area)	Dzheiran Ecocentre, unpublished data 2013, O. Pereladova, personal communication 2014
Iran								
	14	**1973**	**Khosh-Yeilaq Protected Area**	**11**	**Touran**	**extinct**	**failure**	**Ziaie 2008**
	15	2008	Kalmand Protected Area	11	Gourab breeding center	12	more releases planned	Akbar *et al*. 2012, Hemami and Momeni 2013, M.-R. Hemami, personal communication 2014
Israel								
	16	1982–1993	Negev	38	European and Iranian zoos	250	success	Davidson *et al*. 2013, Gueta *et al*. 2014
Saudi Arabia								
	17	**2000**	**Taif**	**5**	**European zoos**	**extinct**	**failure**	**Denzau and Denzau 1999, Pohle 2010**
Ukraine								
	18	1982	Birjutschii Peninsula	11–14	Askania Nova breeding center	91	success (fenced off toward mainland; area ~100 km^2)	Pavlov 1996, O. Yaremchenko, personal communication 2014

Note: Boldface entries indicate failed reintroduction initiatives; italic entries indicate present discussion of the possibility of reintroduction. Success was defined as current population of ~100 individuals and trend stable or increasing. NA, not applicable.

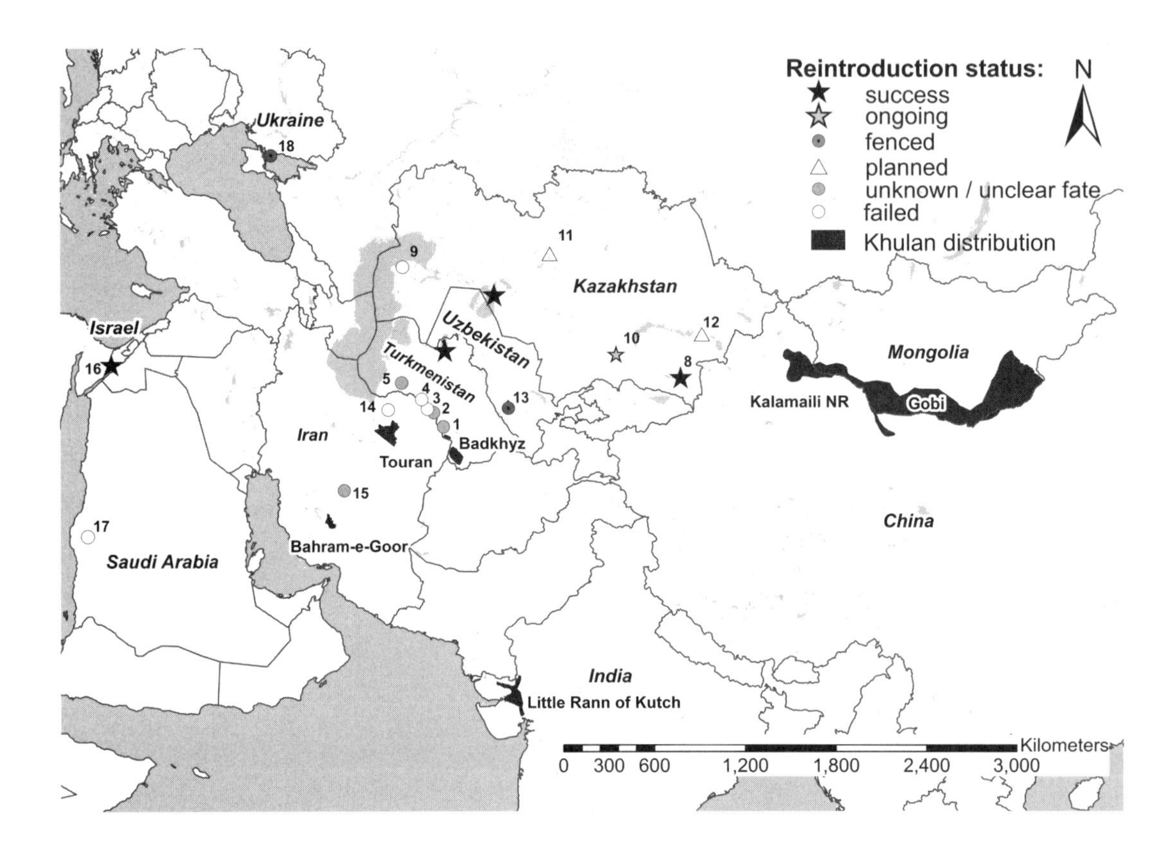

Fig. 14.1 Global wild ass (*E. hemionus*) distribution and location of reintroduction sites. Numbers refer to sites in table 14.1. NR, Nature Reserve.

population, which started out as a heavily managed island population (Bannikov 1981), became free-ranging with the drying up of the Aral Lake and (2) the Altyn Emel National Park population, which is currently the largest reintroduced equid population anywhere (Plakhov *et al.* 2012). One reintroduction eventually failed, and the outcome of the forth remains unclear (table 14.1; fig. 14.1). Although reintroduction programs managed to reestablish the Asiatic wild ass in Kazakhstan, the species still occupies only ~0.25% of its historic range in the country.

The potential for further reintroductions is currently being discussed, with particular focus on the central Kazakh Steppe. This area has become largely depopulated after the breakdown of the Soviet Union, is located within the range of the Betpak-Dala saiga (Saiga *tatarica*) population in the Altyn Dala region, and has a large network of existing and planned protected areas. The area has the potential for the reestablishment of the entire former large-mammal assemblage in one of the largest remaining largely intact steppe ecosystems globally (>500,000 km²; Kaczensky 2011; Frankfurt Zoological Society 2012). Another area discussed is the Arganaty Mountain range in southeastern Kazakhstan (Cheremnov and Sokolov 2012).

Turkmenistan

In Turkmenistan, the Asiatic wild ass survived only in the southernmost parts by the 1930s and had become confined to the Badkhyz area by the 1940s (Lukarevskiy and Gorelov 2007). From 1979 to 1989, a total of ~200 wild asses were transported from Badkhyz to six different reintroduction areas within Turkmenistan (table 14.1; fig. 14.1). (Note that different numbers may be found in different reports/publications as animals were not necessarily released at the same site within a wider region. We chose to combine sites by regions which we believe constitute different [sub]populations.) Soft-release techniques were initially used in most areas, and wild asses were kept in adaptation enclosures for up to nine months. Some mortalities occurred both during transport (e.g., 7.2% mortality during transport for the reintroduction to Meana Chaacha) and in the adaptation enclosure (including aborts in pregnant females; Pavlov 1996). Wild ass mares also gave birth in the adaptation enclosures, however, thus often increasing the actual number of released animals (Pavlov 1996).

Although population development in several of the reintroduction regions was initially positive, only the reintroduction to the Kaplankyr Reserve at Sarykamish Lake has been a true success, with animals now spreading even into adjacent Uzbekistan (northern bank of Sarykamish Lake; Kuznetsov 2014; N. Marmazinkaja, unpublished data). Several other sites are likely still home to small numbers of wild ass, but in three areas wild ass went extinct, and for one area the state of the population is unknown. The autochthonous population in Badkhyz has also seen a lot of fluctuations, from as low as ~200 in 1942 up to a peak population of ~5,000 in 1993–1996 (Denzau and Denzau 1999; Lukarevskiy 1999). In 1996, poaching pressure increased dramatically and numbers dropped to 2,400 by 1998 and to ~500 by the beginning of the 2000s. Conservation measures started in 2000, and the population grew back to ~850–900 wild asses in 2005, but it is believed to have dropped again to ~600 animals in 2010 and potentially as low as 200–420 by 2013 (N. Khudaykuliyev, personal communication; V. Kuznetsov, personal communication). Consequently, population numbers of reintroduced wild ass may now outnumber the autochthonous source population. Despite reintroduction projects, wild ass numbers in Turkmenistan are likely <900 individuals, occupy only a fraction of their former range, and are still (or again) threatened by illegal hunting.

Israel

Israel was home to the Syrian wild ass (*E. h. hemippus*), the smallest subspecies of the Asiatic wild ass that inhabited the Middle East region (Geigl and Grange 2012). The Syrian wild ass became extinct both in the wild and in captivity by the 1930s (Blank 2007). In the 1960s, captive breeding was initiated in Hai-Bar Yotvata Reserve, in southern Israel, with captive stock (11 animals) originating from Iran (*E. h. onager*) and Turkmenistan (*E. h. kulan*). From 1982 to 1987, 28 wild asses in four groups were released in the Makhtesh Ramon Reserve (Saltz and Rubenstein 1995), and in 1992–1993, an additional 10 wild asses in two groups were released at the Paran Wadi, both areas located in the Negev Desert (Blank 2007). Initial population growth was slow (Saltz and Rubenstein 1995), but eventually the population increased both in size and distribution area. In 1997, the population was estimated at >100 individuals over a total area of 2,500 km² (Saltz *et al.* 2006).

At present, there are three main population nuclei, one each around the original release sites and one in the Negev Highlands (natural expansion), with a total estimated population of ~250 individuals roaming over a core area of ~900 km² (Davidson *et al.* 2013; Gueta *et al.* 2014). The reintroduced population lives in an extremely arid environment prone to droughts, and in summer there are only three permanent water sources (all artificial) within the current distribution area of the wild ass. In winter, wild ass can also make use of ephemeral pools. The population is fully protected, and presently the major known cause of mortality is vehicle collisions; from July 2009 to July 2013, 24 individuals were killed by traffic (Israel Nature and Parks Authority, unpublished data 2013). Although population size has doubled since 1999, an increase in the probability of extreme events under the predicted climate change scenario may still make this population vulnerable to extinction as modeled by Saltz *et al.* (2006). Another simulation model, based on changes in allele frequencies, suggests that a strong polygynous mating system is leading to increased genetic drift (Renan *et al.* 2015).

Iran

Wild ass (*E. h. onager*) disappeared from western Iran in the 1930s, but were still widespread in the central and eastern arid and semiarid plains until the 1950s (Denzau and Denzau 1999). By the 1980s, only four populations were left. No wild asses have been reported from the Kavir National Park since 1986, however, and none in recent years from the once-transboundary Sarakhs population along the border to Turkmenistan (Iranian Department of Environment (DoE), unpublished data). Currently, Iran contains only two autochthonous wild ass populations, one in Touran Biosphere Reserve in northeastern Iran and one in Bahram-e-Goor protected area in southwestern Iran. Both populations are small, with the highly threatened Touran population being estimated at 145 individuals and showing a decreasing trend, and the Bahram-e-Goor population numbering 632 but recently showing an increasing trend (Hemami *et al.* 2012; Hemami and Momeni 2013; DoE, unpublished data 2014).

Interest in wild ass restoration started in the 1970s with the well-documented (featured in the 1974 *Wild Kingdom* episode "Chase of the Onager") reintroduction of 11 wild asses (captured in Touran) to the Khosh-Yeilaq Wildlife Refuge. Initially a success, protection measures eventually deteriorated, and no wild asses have been seen in this area since 2006 (Ziaie 2008). Given the precarious conservation status of wild asses in Iran, the

Gourab breeding center near Yazd was established in 1997 (Hamadanian 2005). Until 2008, the captive population reached 40 individuals, and subsequently animals from Gourab were used to establish another 7 breeding centers. As of 2013, there were 33 wild asses in these captive facilities (B. Shahriari, personal communication). Unfortunately, the entire breeding stock in Iran is highly inbred, as it is based on only four wild-caught founders from Touran, and there is a limited number of breeding stallions. Consequently, breeding success has been poor in recent years, and the DoE keeps trying to capture additional animals, particularly stallions, from the wild populations (B. Shahriari, personal communication).

There has been one recent, although unintended, release of captive-bred wild ass into the wild. In 2010, 11 Asiatic wild asses (4 stallions and 7 mares) escaped from the Gourab breeding center into the 2,290-km² Kalmand-Bahadoran protected area (Akbar *et al.* 2012; Hemami and Momeni 2013). By the end of 2013, six foals were born, four animals were captured and transferred to other areas, and three animals died. By 2014, the population numbered 12, and there are plans to release additional wild asses. At present, however, captive breeding and reintroduction initiatives do little to help secure the future of the wild ass in Iran, but rather threaten to divert the attention from the urgent conservation needs of the autochthonous populations. More focus is needed on understanding and addressing the threats to wild ass conservation in Iran before attempting to reintroduce the species (Tatin *et al.* 2003; Hemami and Momeni 2013).

Other Regions

Wild ass in Uzbekistan had completely disappeared by the 1930s. In 1977/1978, five wild asses (*E. h. kulan*) from the reintroduced population on Barsa-Kelmes Island were brought to the newly established 51-km² Dzheiran (also Jeyran) Ecocenter (also known as the Bukhara Breeding Centre; Bahoul *et al.* 2001). In 2013, the population numbered about 100 wild asses. In 2012, five wild asses were released in the much larger (204 km²) unfenced part of the Ecocenter, with the aim to restore the species in the southwestern Kyzylkum Desert. Although unfenced, the area is an isolated landscape with the Amu-Bukhara Canal, a railroad, and the Kuyumazar Reservoir forming the boundaries (N. Soldatova, unpublished data; O. Pereladova, personal communication).

Wild asses are believed to have lived in Ukraine until the twelfth century (Denzau and Denzau 1999) and have been reintroduced to one location in Ukraine. In 1982, 14 wild asses (*E. h. kulan*) from the Askania Nova Breeding Centre (in Ukraine) were released in the ~100-km² part of the Azovo-Sivashskiy National Park on the Birjutschii Peninsula (the protected area is fenced against the mainland). This population is presently estimated at 92 individuals (O. Yaremchenko, personal communication).

In Saudi Arabia, five animals from European zoos (*E. h. onager*) were brought to an acclimatization enclosure between 1989 and 1995. But the animals were never released, and the last captive individual died in 2004 (Pohle 2010).

Lessons learned from wild ass reintroductions:

- Success of wild ass reintroductions has been mixed. Out of 18 reintroduction initiatives, 5 successfully reestablished substantial (≥100) free-ranging populations (2 in Turkmenistan, 2 in Kazakhstan, and 1 in Israel), 2 reestablished the species in large fenced areas within the historical distribution range in Uzbekistan and Ukraine, 4 are very small and their future is unsure, and 5 have failed.
- The 4 successful reintroduction initiatives that released wild asses into unrestrained areas (not counting Barsa-Kelmes, as it was initially an island) all used >30 founders.
- In Kazakhstan, Turkmenistan, and Iran, reintroduced (as well as autochthonous) wild ass populations are still subject to illegal hunting—one of the key factors that led to their extinction in the first place. The reintroductions that failed likely succumbed to the combined effect of small population size and illegal hunting.
- Many approaches were used to capture (e.g., by lasso out of a pursuing jeep, traps at water points, drives into a corral), transport (by air and land, in individual boxes and in groups in containers), and release (both hard and soft releases) wild asses. Mortalities occurred but were not systematically documented, making a retrospective quantitative assessment impossible.

Przewalski's Horse Reintroductions

The Przewalski's horse, or *takhi* in Mongolian, became extinct in the wild during the 1960s. Extinction of the Przewalski's horse is attributed to the combined effects of overhunting, pasture competition with livestock, and extreme weather events (Heptner *et al.* 1988; Wakefield *et al.* 2002). The last confirmed sighting of a Przewalski's horse in the wild occurred in the

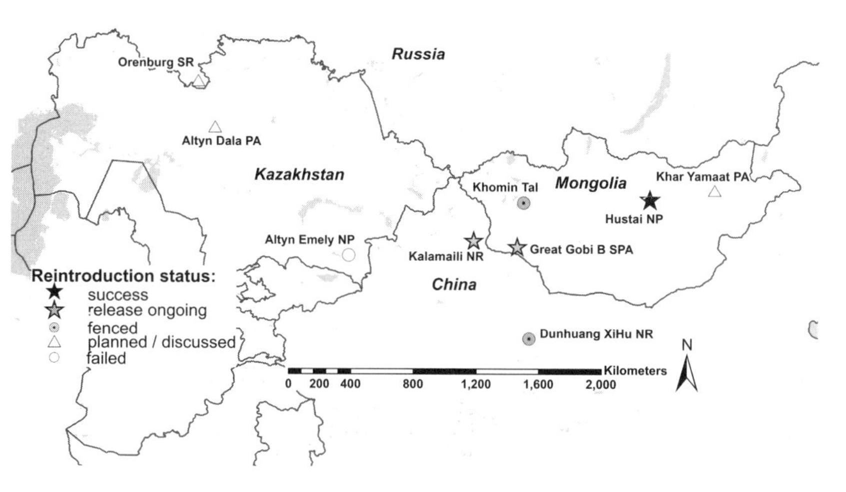

Fig. 14.2 Przewalski's horse (*E. f. przewalskii*) reintroduction sites. NP, National Park; NR, Nature Reserve; PA, protected area; SPA, Strictly Protected Area; SR, steppe reserve (International Union for Conservation of Nature category I).

Dzungarian Gobi in southwestern Mongolia in 1969 (Sokolov and Orlov 1986), but wild-born Przewalski's horse foals were transported to captive facilities in Europe at the end of the nineteenth and beginning of the twentieth centuries. Carefully managed captive breeding resulted in >1,000 individuals by the mid-1980s and thus provided a large enough reservoir for potential reintroductions (Boyd and Houpt 1994; Volf 1996; see also chap. 11) (fig. 14.2).

In reaction to the extinction of the Przewalski's horse in southwestern Mongolia and northwestern China, the Food and Agricultural Organization (FAO) of the United Nations and the United Nations Environment Programme (UNEP) organized an expert consultation. The group passed a general resolution to reintroduce Przewalski's horses to their former range in Mongolia and China as the main method of recovery (FAO 1986). Subsequently, a global management plan for Przewalski's horse reintroductions was proposed (Seal 1992).

Mongolia

Recommendations sponsored by the Mongolian Government and UNEP for the reintroduction of the Przewalski's horse suggested establishing multiple sites with a primary free-ranging, self-sustaining population in the Dzungarian basin of the greater Gobi Desert (Mongolian Takhi Strategy and Plan Work Group 1993). In 1990, two reintroduction projects were simultaneously initiated, one in Takhin Tal in the Great Gobi B Strictly Protected Area (SPA) in the Dzungarian Gobi and one in the Hustai Nuruu in the central Mongolian mountain steppe. In 1992, the first captive-born animals arrived from Europe and were released into the adaptation enclosures in both areas (Bandi and Dorjraa 2012).

Takhin Tal: Great Gobi B Strictly Protected Area

In the early phase, the project was plagued by shortcomings in infrastructure and training, which resulted in rather high mortality rates of introduced Przewalski's horses (van Dierendonck and Wallis de Vries 1996). In the Gobi Desert, population growth could only be achieved by introducing additional captive animals (Slotta-Bachmayr *et al.* 2004). Initial project focus was only on the Przewalski's horse and the immediate release area surrounding the Takhin Tal research station (Kaczensky *et al.* 2008*a*). The spatial requirements of a self-sustaining population and the interactions with wildlife and local people were greatly underestimated (Kaczensky *et al.* 2007, 2008*b*, 2011*a*).

Project management was reorganized in 1999/2000 with the International Takhi Group becoming the new executing organization, dedicated to running the program based on best scientific practices and in accordance with the IUCN/SSC (2013) "Guidelines for Reintroductions and Other Conservation Translocations." With the introduction of routine veterinary procedures and postmortem pathologic examination in 1999, piroplasmosis, a tick-borne blood disease endemic in central Asia, was discovered as a mortality factor of newly imported horses (Robert *et al.* 2005). To reduce the mortality risk, all newly arrived animals need to be treated with a subtherapeutic dose of Imidocarb (Carbesia, Schering-Plough, France) prior to being exposed to ticks. Furthermore, retrospective assessment suggested that the relative safety of the adaptation enclosure (e.g., allowing for daily health checks, supplementary feeding, and veterinary treatment) was counterbalanced by an elevated injury risk and reduced reproduction. Consequently, time in the adaptation enclosure was reduced to one year, which allowed

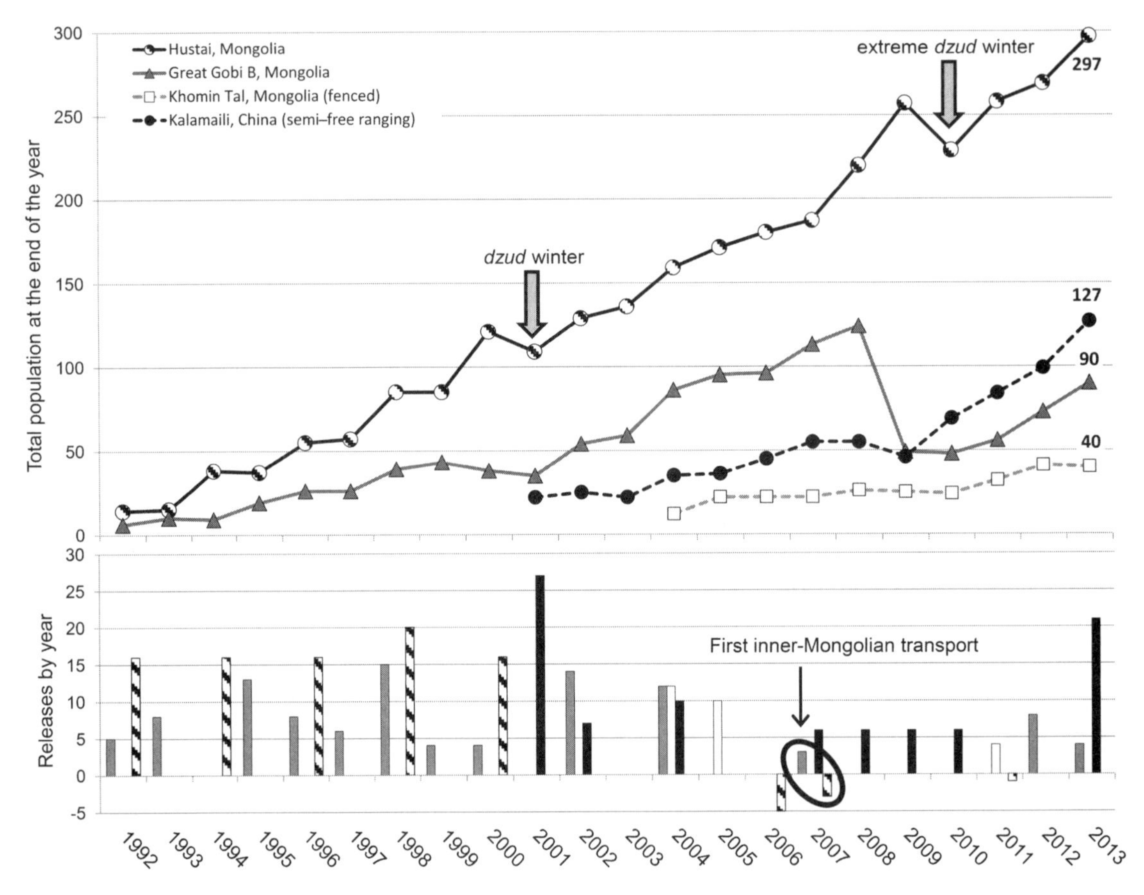

Fig. 14.3 Successful and ongoing Przewalski's reintroductions in Mongolia and China. Lines show population development, and bars show the number of horses released (or removed) per year. Numbers for Hustai, Khomin Tal, and Kalamaili show population size at the end of the calendar year, but for Great Gobi B at the end of the horse year (from 1 May until the end of April the next year).

piroplasmosis treatment. This seems to have been sufficient to allow Przewalski's horses to adapt to local climate conditions and inhibit large-scale postrelease movements (O. Ganbaatar, unpublished data; C. Walzer, unpublished data).

The new management was not immediately met by success, as in 2000/2001 the area was hit by severe winter conditions (called *dzud* in Mongolia), and the small Przewalski's horse population suffered a net loss of 21%, with almost no foals produced the following spring (Slotta-Bachmayr *et al.* 2004). Since 2002/2003, the Przewalski's horse population finally started to show positive population growth independent of released animals. By December 2009, the Great Gobi B SPA population had reached 138 animals and was coming close to a minimal viable population (MVP) assuming a scenario of "low-severity catastrophes" (Slotta-Bachmayr *et al.* 2004). In the winter of 2009/2010, however, one of the worst *dzud* winters in modern history hit Mongolia and northern China. While millions of livestock died throughout Mongolia, the Przewalski's horse population in the Great Gobi B SPA crashed to a mere 49 individuals, providing a textbook example of the risk that small populations and spatially confined species in unpredictable environments face (Kaczensky *et al.* 2011*b*). To speed up recovery, transporting captive-bred Przewalski's horses resumed in 2012 (fig. 14.3). By the end of 2013, natural reproduction and the transport of an additional eight captive-bred individuals has resulted in a population of 90 Przewalski's horses (Ganbaatar *et al.* 2014).

Demographic data and habitat use patterns suggest that the semideserts and desert steppes of the Gobi constitute an edge rather than prime Przewalski's horse habitat (van Dierendonck and Wallis de Vries 1996; Kaczensky *et al.* 2008*b*). Yet the remoteness of the area and the low pasture productivity also results in rather low human and livestock presence. In addition, the Great Gobi B SPA covers a large area of ~9,000 km², and a possible extension to ~23,000 km² is presently being discussed. Main threats to the reintroduced population in the Great Gobi B SPA are (1) the small population size and the limited spatial extent of the current population, and (2) conflicts over sustainable resource use (illegal collection of firewood, grazing, illegal mining, poaching). But because of the remoteness of the area and the poor infrastructure, the problems are rather minor when compared to other areas of Mongolia, and the likelihood of hybridization with domestic horses, although possible, is not a foremost concern (Smith 2010).

Hustai National Park

Population development in the more productive mountain steppe habitat at Hustai was much more rapid, with a mean annual population growth of 10% after the transports had stopped (fig. 14.3) (Usukhjargal and Bandi 2013). Home ranges are about 10 times smaller than in the desert steppe of the Gobi, ranging from 7 to 60 km² (Nandintsetseg 2008). Effects of the *dzud* winters (2000/2001 and 2009/2010) could also be seen but were much less severe than in the Great Gobi B population (fig. 14.3). By the end of 2013, the population in Hustai had reached 297 Przewalski's horses. It is largely because of the success of this population that the Przewalski's horse was down-listed to Endangered in the 2008 IUCN Red List of Threatened Species.

The protection status of Hustai National Park is closely linked to the Przewalski's reintroduction project. Contrary to Great Gobi B SPA, which had already been set aside in the 1970s for wildlife protection (particularly wild equids; Zhirnov and Ilyinsky 1986), the 510-km²-large Hustai Nuruu region was only declared a protected area in 1993, one year after the reintroduction of the first Przewalski's horses. From 1993 to 2003, Hustai National Park was initially managed by the Mongolian Association for the Conservation of Nature and the Environment, but since 2003 it has been run by the Hustai National Park Trust (HNPT), the only protected area in Mongolia where the Mongolian Government has delegated the management to a nongovernmental organization.

Hustai National Park is only ~100 km from the Mongolian capital Ulaanbaatar and has become a popular tourist destination, being highly successful in communicating and promoting the case of the Przewalski's horse in Mongolia. Furthermore, it allows a multitude of national and international students to study various aspects of mountain steppe ecology (Bandi and Dorjraa 2012). But the closeness to the capital and the higher productivity of the region also makes the region attractive for livestock grazing and agriculture, resulting in Hustai becoming somewhat of an island with little potential for wildlife to expand their ranges into the surrounding steppe. The threat of human encroachment into the protected area also challenges landscape-scale conservation. Intensive ranger patrols are necessary to combat poaching and illegal grazing of livestock. Increasing intraspecific aggression, including infanticide and the dispersal of bachelor stallions outside of the protected area, suggests that the Przewalski's horse population may soon reach carrying capacity.

With little potential for expansion, management strategies will need to be adapted. The potential of using Przewalski's horses from Hustai National Park to supplement or initiate new reintroduction projects in Mongolia is presently discussed, particularly by the World Wildlife Fund for Nature (WWF) Mongolia for the Khar Yamaat protected area in eastern Mongolia (D. Nandintsetsed, personal communication) (fig. 14.2). The feasibility of such an approach was tested in 2007 when three stallions from Hustai National Park were successfully transported to the Great Gobi B SPA, the first inner-Mongolian Przewalski's horse translocation. Main threats to the reintroduced population in Hustai National Park are (1) conflicts over sustainable resource use (grazing, poaching); (2) hybridization with domestic horses, particularly at the fringes of the protected area; and (3) change in the water regime due to climate change.

Khomin Tal

In 2004, the association TAKH in cooperation with WWF in Khomin Tal started a third reintroduction in Mongolia in the Depression of the Great Lakes (fig. 14.2). The project has leased a 140-km² area in the buffer zone of the Khar Us Nuur National Park, which was fenced to separate Przewalski's horses and livestock (Joly *et al.* 2012). In 2004 and 2005, a total of 22 Przewalski's horses were transported from Le Villaret, France, to Khomin Tal. Initial population growth was poor owing to a lack of reproduction caused largely by delayed reversibility of porcine zona pellucida (PZP) treatment prior to transport (Feh 2012). In 2011, another transport of four mares was initiated, but population size remained rather stagnant and numbered 40 individuals by the end of 2013 (D. Nandintsetseg, personal communication). Although the ultimate goal of the project is to have a truly free-ranging population, it is presently unclear if this goal can be achieved, as grazing pressure has been increasing in the area and there are unresolved conflicts over land use. Furthermore, the Przewalski's horse population is still too small to make predictions for the future.

China

China established three breeding centers for Przewalski's horses: the Jimsar Wild Horse Breeding Centre (WHBC) in Xingiang Province in 1985, the Gansu Breeding and Research Centre for Rare and Threatened Wild Animal Species in Gansu Province in 1989, and the Anxi Breeding Centre also in Gansu Province in 1992 (Wakefield *et al.* 2002; Zimmermann 2005). Initial breeding stock in the three centers came primarily from captive facilities in Europe, with some additional horses from the United States and Beijing Zoo (Zimmermann 2005). A first attempt to release horses to a large fenced enclosure was made in 1995 at the Gansu Breeding and Research Centre, but it failed (Zimmermann 2005). A new attempt was launched in 2010 at the 200-km² fenced Dunhuang Xihu National Nature Reserve (also Shuanghu), however; 7 horses were transported in 2010, and 21 more in 2012 (Chaohua and Zhu 2012; Q. Cao, personal communication). In 2012, the first foal was born by a mare from the 2010 transport (Wang *et al.* 2012).

Releases of Przewalski's horses into the wild have so far only happened from the WHBC to the Kalamaili Nature Reserve (Chen *et al.* 2008; Xia *et al.* 2014) (fig. 14.2). First releases were undertaken in 2001/2002 at Jiliekuduke at the northern edge of Kalamaili Nature Reserve. But because 4 out of the 27 Przewalski's horses in the first released group died as a result of harsh winter conditions, the rest were brought back to the 200-ha adaptation enclosure for the rest of the winter. Subsequently, more Przewalski's horses were released at Jiliekuduke, but in order to avoid winter losses and interaction with domestic horses (which return to the reserve with local herders in winter), they were always herded back into the acclimatization enclosures in winter, and thus they were only semi–free ranging (Zimmermann 2005; Xia *et al.* 2014).

In 2007, the upgrading of National Road 216, running through Kalamaili Nature Reserve to connect the cities of Altai and Urumqi, was completed. Traffic volume and speed on the new road greatly increased, and five Przewalski's horses were killed in vehicle collisions between August and November. Speed limits were introduced and warning signs were erected near the Jiliekuduke release site, while at the same time the WHBC and Kalamaili Nature Reserve looked for alternative release sites. In 2007, six horses were released in Karmust, 80 km south of Jiliekuduke. Within the first few weeks this group moved 100 km to the west, however, settling in the Sangequan area (west of Kalamaili Nature Reserve). In 2008, another harem was released in Karmust but disappeared without a trace, and thus the area was abandoned as a further release site (Q. Cao, personal communication).

In the summer of 2008, the Sangequan region experienced a severe drought, and the Przewalski's horses moved south into the Gurbantünggüt Desert in search of water along a paved road. One week later, they were spotted drinking water at a ditch near an oil-drilling platform 50 km away from Sangequan. Two mares died a few days later due to malnutrition; the rest of the herd was eventually driven back to Sangequan and has remained there since. In 2013, WHBC, Altai Forestry Bureau, and Sangequan station built a 5-km aqueduct, linking Sanquanquan to a nearby canal and thus providing permanent water for the Przewalski's horses in this area (Q. Cao, personal communication).

In April 2009, most Przewalski's horses from Jiliekuduke were translocated (in crates and by truck) to a new release site at Qiaomuxibai, 120 km to the south and 40 km away from National Road 216 in the center of Kalamaili Nature Reserve. The horses were still managed in the same way, however, corralled in a 50-ha pen during winter and fed with alfalfa. International collaboration intensified around this time, particularly the involvement of the Smithsonian National Zoological Park, which resulted in many new projects and approaches. In 2013, Chinese authorities announced that they would (1) stop the winter roundup for all newly released Przewalski's horses and gradually abandon the winter roundup for the Qiaomuxibai population, (2) expand the Sangequan population by releasing more harems and improving the monitoring capacities, and (3) release horses at another new release site in the Yekesituobie area >100 km north of Kalamaili Nature Reserve (C. Jie, personal communication via Q. Cao, personal communication).

Following this announcement, two harems with nine horses total were released in Sangequan, and a harem of six horses was released in Yekesituobie. These newly released groups have remained free ranging in winter. Monitoring was strengthened in these areas by capacity building and community involvement initialized by the Smithsonian's National Zoological Park. By January 2014, all released individuals had survived their first winter without assistance. Population growth of Przewalski's horses in Kalamaili Nature Reserve was initially low because of poor reproduction and high mortality, but it has improved in recent years (Chen *et al.* 2008) (fig. 14.3). By the end of 2013, the Przewalski's horse population in Kalamaili, free ranging and semi–free ranging, numbered 127 individuals (Xia *et al.* 2014) (fig. 14.3).

Main threats to the reintroduced population in Kalamaili include (1) the low protection status of the Kalamaili Nature Reserve, which has already resulted in the reduction of the 18,000-km² reserve by 4,000 km² for coal-related industries in 2012; (2) quick expansion of human developments, such as mining and constructions in the surrounding area; and (3) limited water sources and competition over pasture with local nomads and their livestock (Cao *et al.* 2012; Xia *et al.* 2014; Q. Cao, personal communication).

Kazakhstan

In 2003, eight captive-bred Przewalski's horses were transported from Europe to the 5,200-km² Altyn Emel National Park in southeastern Kazakhstan for reintroduction (Neumann-Denzau and Chirikova 2004; Zimmermann 2005). Initial mortalities were high and no reproduction occurred, so that by the end of 2006, only three Przewalski's horses were left. In 2007, two foals were born and a second transport brought another six horses, increasing the population to eleven animals. By June 2011, only five free-ranging Przewalski's horses and two stallions in captivity (they had come close to a village and were captured to avoid conflicts) remained in Altyn Emel National Park (Kaczensky 2011). In 2012, two stallions were transferred to Almaty Zoo and two were transferred to a new enclosure within the protected area. Only one Przewalski's horse group remains free ranging (but are fed hay on a regular basis), consisting of four individuals in 2012 and six in 2013 (S. Zuther, personal communication), and thus the reintroduction has clearly failed.

The habitat in Altyn Emel National Park is rather unproductive, supporting *Artemisia* and other desert-steppe plants, whereas grasses like *Stipa* are rare. In this kind of habitat, home ranges can be expected to be in the same magnitude as in the Mongolian Gobi, that is, averaging 500 km² (Kaczensky *et al.* 2008*b*). The oasis complex of the initial release site certainly provides a "mesic island" in the dry desert steppe for Przewalski's horses (see Kaczensky *et al.* 2008*b*), but the overall suitable area is small (also see Neumann-Denzau and Chirikova 2004). Realistically, the area available for Przewalski's horses in the western core area of the National Park is only about 15 × 20 km. Thus, despite the park covering a total area of 5,200 km², it provides only limited space suitable for Przewalski's horses (Kaczensky 2011).

There are presently preparations for a new reintroduction project in the Altyn Dala region of central Kazakhstan (see also Asiatic Wild Ass Reintroductions above; Frankfurt Zoological Society 2012). Two acclimation enclosures have been built in the Altyn Dala conservation area, and a first transport bringing Przewalski's horses from captive facilities in Europe is planned for 2016.

Russia

Another project in the planning phase is a reintroduction initiative to the 200-km² Orenburg Steppe Reserve (fig. 14.2). The project is integrated into Russia's Ministry of Natural Resources state strategy of threatened ecosystems conservation and rare species reconstruction and is supported by the provincial (oblast) administration. Scientific guidance of the project is provided by the Institute of Steppe of the Ural Branch of the Russian Academy of Science (RAS), the Severtsov Institute of Ecological and Evolution Problems of RAS, and the Zoological Museum of the Lomonosov Moscow State University. Several acclimatization enclosures (40 ha each) have been built, and the project received the first horses in summer 2015 from association TAKH in Le Villaret, France (N. Spasskaya, personal communication). The Orenburg Steppe Reserve and the surrounding area are sparsely populated, and the goal is to develop a free-ranging population in the region.

Other Regions

The Dzheiran Ecocenter in Uzbekistan also keeps a small number of Przewalski's horses together with Asiatic wild asses (Bahoul *et al.* 2001); sixteen are present in one fenced territory and eight in another territory (see fig. 14.2 for location; N. Soldatova, unpublished data; O. Pereladova personal communication).

Lessons learned from Przewalski's horse reintroductions:

- Reintroduction has reestablished this once "extinct in the wild" species back to the wild. Out of six reintroduction initiatives, however, only one is a clear success, two are ongoing with positive population development, one is stagnating in a very large enclosure, and two failed (of which one is reattempting a release).
- The species is still confined to a tiny fraction of its former distribution range (~3,000 km² in Mongolia and ~1,000 km² in China).
- Pasture competition and interaction with livestock (potentially resulting in hybridization with

domestic horses and disease transmission) have not been fully eliminated in any of the reintroduction sites, and will likely intensify with increasing success.

- Three of the Przewalski's horse reintroductions have happened in marginal desert-steppe habitat. One has failed (Altyn Emel, Kazakhstan), one has suffered major setbacks (Great Gobi B SPA, Mongolia), and one only just started to truly release Przewalski's horses (Kalamaili, China). Thus the potential of these marginal habitats remains to be seen, particularly in light of climate change scenarios predicting a higher frequency of extreme events.
- The successful project in Hustai and the two ongoing projects that show a positive population development (Great Gobi B and Kalamaili) all have transported >80 individuals over a period of several years and initially kept newly arrived Przewalski's horses in adaptation enclosures.
- In newly reintroduced Przewalski's horses, mortality rates are higher and reproduction rates are lower than in established groups or second-generation horses.
- Przewalski's horses imported from Europe are naïve to the tick-transmitted disease equine piroplasmosis, requiring treatment prior to being exposed to ticks.

Zebra Reintroductions

Reintroductions have played a variable role in the conservation and management of the three zebra species. No reintroductions of Grevy's zebra (*E. grevyi*) have occurred (B. Low, personal communication), but a few efforts were directed toward reinforcing small populations (e.g., Meru National Park's population of two individuals was supplemented with thirteen individuals in 2002; Franceschini *et al.* 2008). In contrast, plains zebra (*E. quagga*) reintroductions have been copious, as the species occurs extensively on game ranches throughout southern Africa (particularly in South Africa, Namibia, and Zimbabwe), many of which have relied upon reintroductions to return the species (e.g., after becoming locally extinct as a result of farming; Hack *et al.* 2002). Reintroductions have undoubtedly had the largest impact on Cape mountain zebra conservation (Novellie *et al.* 2002; Hrabar and Kerley 2013).

Cape Mountain Zebra

Mountain zebra once ranged from the southern parts of South Africa through Namibia into the extreme southwest of Angola. Two subspecies are recognized—Hartmann's mountain zebra (*E. z. hartmannae*) and Cape mountain zebra—and have been confirmed by recent genetic work (Moodley and Harley 2005). Endemic to South Africa, Cape mountain zebra were once widespread in the mountains of the Western and Eastern Cape Provinces (Millar 1970*a*,*b*) (fig. 14.4). Excessive hunting and habitat loss to agriculture left their numbers in a critical state by the 1930s, confining the subspecies to just five localities. Two of these subpopulations subsequently went extinct, Outeniqua in the early 1970s and Baviaanskloof in the late 1980s, leaving only three natural populations that survived to the present day: the Mountain Zebra National Park, the Kammanassie Nature Reserve, and the Gamka Nature Reserve populations (Millar 1970*a*; Lloyd 1984) (fig. 14.4).

During the 1960s and 1970s, the Mountain Zebra National Park population increased steadily to a point where, in 1979, the first reintroduction of Cape mountain zebra became possible. Twenty-three individuals were reintroduced to the newly established Karoo National Park. The Mountain Zebra National Park population continued to increase during the 1980s and 1990s, enabling a further 25 reintroductions to other protected areas and game ranches within the subspecies' historic range. Despite further regular removals, the Mountain Zebra National Park population is currently in excess of 750 individuals (South African National Parks, unpublished data). The most recent census in 2009 counted 2,790 animals in 52 subpopulations (fig. 14.4), thus actually exceeding the target size of 2,500 Cape mountain zebra set by the IUCN Equid Specialist Group (Novellie *et al.* 2002).

By 2009, there were twice as many of mountain zebra populations on private lands (hereinafter referred to as private populations, $n = 35$) compared to populations on formally protected government-owned lands (hereinafter referred to as formally protected populations, $n = 17$), yet the majority of all populations (68%) still occurred on formally protected land. Analysis of 36 questionnaires with sufficient information revealed that performance of 27 populations (16 private and 11 formally protected) was reported as good, 6 (2 formally protected and 4 private) were stable, while 3 (2 formally protected and 1 private) had decreased in size. Of the last, one population was subjected to drought conditions soon after release, and one suffered from poaching and emigration (Harbar and Kerley 2013). Postreintroduction population performance of Cape mountain zebra populations was higher for populations with ≥14 founders, a number that had been established as the recommended minimum number by

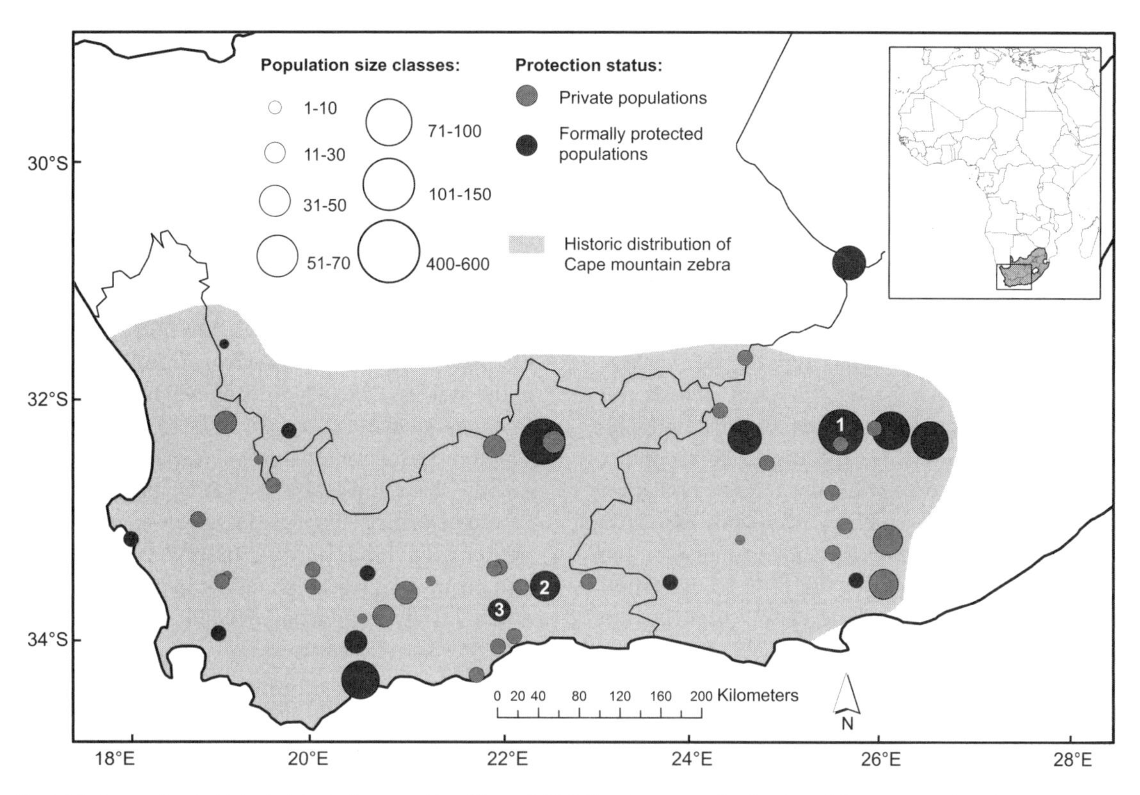

Fig. 14.4 Approximate historic (shaded region; Novellie *et al.* 2002) and current distribution of Cape mountain zebra subpopulations identified in 2009 in South Africa. Numbers mark the location of the last autochthonous populations in Mountain Zebra National Park (1), Kammanassie Nature Reserve (2), and Gamka Nature Reserve (3). Adapted from Hrabar and Kerley 2013

the IUCN Equid Specialist Group (Novellie *et al.* 1996; Hrabar and Kerley 2013).

Records from 54 separate Cape mountain zebra translocation events (between 1981 and 2009), involving 472 individuals, showed a total mortality rate of 10.38%. Twenty-eight percent of deaths occurred during capture or transport (e.g., individuals being fatally kicked while confined in the truck). Mortality in Cape mountain zebra also appears to be dependent upon release method, as the mortality rate from soft releases was 14% ($n = 92$ individuals released from 9 reinforcements) compared to a 7% mortality rate after hard releases (deaths within a couple of months after release, $n = 246$ individuals released from 32 translocations). The (more commonly used) hard-release method therefore appears to be more successful in terms of mortalities, although the apparent lower mortality rate might be an underestimation owing to the difficulty in detecting postrelease deaths. A more subtle cost is the negative impact on reproductive performance in newly released Cape mountain zebra: the rate of annual increase over the first three to five years after reintroduction was significantly lower than during the subsequent three to five years (0.4% vs. 9.3%, respectively; Novellie *et al.* 1996).

The recent burgeoning investment in private nature reserves in South Africa (Sims-Castley *et al.* 2006) is partly responsible for the significant increase in habitat available to Cape mountain zebra, as is the expansion of the Mountain Zebra National Park (from 65 to 410 km^2) and Karoo National Park (from 284 to 881 km^2). Consequently, the movement of Cape mountain zebra from existing subpopulations into these new areas has ensured continued population growth (a growth rate of 10% between 2002 and 2009; Hrabar and Kerley 2013) by reducing density-dependent effects (as observed in the De Hoop population; Smith *et al.* 2007). Considering that all properties in South Africa (private and formally protected) are completely isolated from one another because of fencing, reintroductions have played a pivotal role in enabling the dispersion and growth of the subspecies.

Yet the permanent isolation of subpopulations does mean that successful future conservation efforts, potentially aiming for 12,000 individuals (Hrabar and Kerley 2013), will continue to rely on reintroductions as current populations become saturated and reinforcements within the metapopulation matrix are needed to maintain genetic variability. In 2009, 11 (21%) of 52 subpopulations had <14 individuals, meaning they faced a higher risk of inbreeding depression and increased loss of genetic diversity from genetic drift (Frankham 1996). Private initiatives are somewhat limited by the expense of purchasing sufficient individuals for initial reintroduction or subsequent reinforcements, also because Cape mountain zebra have little profit potential for trophy hunting because they are listed in CITES Appendix I, which bans the export of body parts for commercial purposes.

Two-thirds of the entire genotype is presently represented in just two populations (in Kammanassie and Gamka Nature Reserves), as all reintroduced populations (except for the De Hoop National Park population) originate from only one natural relic population (Mountain Zebra National Park) and already have little genetic variability (see chap. 7). Efforts to increase the other two relic populations in situ prior to utilization (translocations) have been unsuccessful to date but are urgently required, as low genetic variability seems to also correlate with susceptibility to sarcoid tumors (Sasidharan *et al.* 2011).

Lessons learned from Cape mountain zebra reintroductions:

- Reintroductions and reinforcements brought the Cape mountain zebra back from the brink of extinction. However, all 52 Cape mountain zebra subpopulations are fenced and managed as one metapopulation via reintroduction to new sites and reinforcements among sites.
- Genetic variability in Cape mountain zebra is low and remains unevenly distributed. Many populations on private land are small and thus remain highly susceptible to demographic and genetic stochasticity.
- Postreintroduction population performance of Cape mountain zebra populations was higher for populations with ≥14 founders.
- Reintroductions and reinforcements came at a cost for the animals. The average translocation related mortality was 10.38%, and reproductive performance in newly released Cape mountain zebra was lower in the first 3–5 years after translocation.

Conclusion

Reintroductions have played a pivotal role for the conservation of three of the seven recent equid species:

- Reintroduction of wild Asiatic wild ass has successfully reestablished the species in Kazakhstan and Israel and somewhat increased the species range in Turkmenistan (recently even spreading into adjacent Uzbekistan). In sum, the reintroduced populations number ~3,700 individuals and thus make up ~7% of the global Asiatic wild ass population.
- Reintroduction of captive-bred Przewalski's horse has successfully reestablished this species in Mongolia and China, and the free-ranging population currently numbers >500 in three locations. The growing free-ranging Przewalski's horse population has resulted in the delisting from Extinct in the Wild to Critically Endangered in 2008, and Endangered in 2011.
- The reintroduction of the Przewalski's horse greatly promoted awareness for wildlife conservation of desert-steppe and steppe ecosystems in Mongolia. It has become an umbrella species, also promoting research and conservation of much less liked and iconic wildlife, including the Asiatic wild ass. Przewalski's horse reintroduction has led to the creation of Hustai National Park and helps encourage enlargement of the Great Gobi B SPA.
- Reintroduction and reinforcement of Cape mountain zebra in South Africa has resulted in 52 subpopulations with a total of 2,790 animals (by 2009), thus exceeding the target size of 2,500 initially recommended by the IUCN Equid Specialist Group.

Despite these achievements, reintroduction projects are costly, logistically challenging, and require long-term commitment (IUCN/SSC 2013). Reintroductions are potentially risky to the lives, well-being, and reproductive potential of individuals moved (Letty *et al.* 2007; Harrington *et al.* 2013), while reinforcements may also put the receiving population at risk (Champagnon *et al.* 2012). Setbacks have to be expected, particularly at the beginning, when populations are small and when the original causes of extinction have not been fully eliminated. Reintroductions with wild-caught stock may threaten the conservation status of autochthonous populations, whereas captive breeding programs risk change to the gene pool (Williams and Hoffman 2009) or morphological (O'Regan and Kitchner 2005) and behavioral traits (McDougall *et al.* 2006) in a species. A strong focus on reintroduction can even threaten to divert much needed attention, funds, and efforts from conserving the last autochthonous populations. Consequently, reintroduction should be seen as a last resort, and priority needs to be given to conserving wild equids and their habitats in situ.

ACKNOWLEDGMENTS

This chapter would not have been possible without the help and input of many people sharing their data and experience. The authors thank (in alphabetical order): Shirli Bar-Davis (Ben-Gurion University of the Negev), Qing Cao (Princeton University), Gertrud and Helmut Denzau (Panker, Germany), Saeideh Esmaeili (Isfahan University of Technology), Boaz Freifeld (Israel Nature and Park Authority), Mahmoud-Reza Hemami (Isfahan University of Technology), Tatiana Kuzmina (Schmalhausen Institute of Zoology of the National Academy of Sciences of Ukraine), Belinda Low (Grevy's Zebra Trust), Natalia Marmazinskaja (Samarkand, Uzbekistan), Anna Merska (Local Park and Zoological Garden Foundation in Krakow), Hossein Mohammadi (DoE), Dejid Nandintsetseg (National University of Mongolia), Olga Pereladova (World Wildlife Fund Russia), Sharon Renan (Ben-Gurion University of the Negev), Bahareh Shahriari (DoE), Kateryna Slivinska (Schmalhauzen Institute of Zoology, National Academy of Sciences of the Ukraine), Sergey V. Sokolov (Okhotprojekt Ltd.), Natalia Soldatova (Ecocenter Djeiran), Natalia Spasskaya (Moscow State University), Chris Walzer (University of Veterinary Medicine, Vienna), Canjun Xia (Xinjiang Institute of Ecology and Geography, Chinese Academy of Sciences), Olga Yaremchenko (Ukrainian Society for the Protection of Birds), and Steffen Zuther (Association for the Conservation of Biodiversity of Kazakhstan and Frankfurt Zoological Society). Thank you all.

REFERENCES

Akbar, H., A. Habibipoor, and M. Abedini. 2012. Viability analysis of the re-introduced onager (*Equus hemionsus onager*) population in Iran. Middle East Journal of Scientific Research 11:242–245.

Armstrong, D.P., and P.J. Seddon. 2007. Directions in reintroduction biology. Trends in Ecology and Evolution 23:19–25.

Bahloul, K., O.B. Pereladova, N. Soldatova, G. Fisenko, E. Sidorenko, and A.J. Sempere. 2001. Social organisation and dispersion of introduced kulans (*Equus hemionus kulan*) and Przewalski horses (*Equus przewalskii*) in the Bukhara Reserve, Uzbekistan. Journal of Arid Environments 47:309–323.

Bandi, N., and O. Dorjraa (Eds.). 2012. Takhi: Back to the wild. International Takhi Group, Hustai National Park, Great Gobi B Strictly Protected Area, and Association pour le cheval de Przewalski-TAKH, Ulaanbaatar, Mongolia.

Bannikov, A.G. 1981. The Asian wild ass. [In Russian.] Lesnaya Promyshlennost, Moscow, Russia. [English translation by M. Proutkina, Zoological Society of San Diego, San Diego, CA, USA.]

Blank, D.A. 2007. Asiatic wild ass in Israel. Exploration into the Biological Resources of Mongolia 10:261–266.

Boyd, L., and K.A. Houpt (Eds.). 1994. Przewalski's horse. State University of New York Press, Albany, NY, USA.

Cao, Q., M. Songer, Y.J. Zhang, D.F. Hu, D.I. Rubenstein, and P. Leimgruber. 2012. Resource use and limitations for released Przewalski's horses at Kalamaili Nature Reserve, Xinjiang, China. Presented at the International Wild Equid Conference (IWEC), 18–22 September 2012, Vienna, Austria.

Champagnon, J., J. Elmberg, M. Guillemain, M. Gauthier-Clerc, and J.-D. Lebreton. 2012. Conspecifics can be aliens too: A review of effects of restocking practices in vertebrates. Journal for Nature Conservation 20:231–241.

Chaohua, X., and J. Zhu. 2012. Endangered horses released into the wild. China Daily September 7. Available at http://europe.chinadaily.com.cn/china/2012-09/07/content_15741025.htm.

Chen, J., Q. Weng, J. Chao, D. Hu, and K. Taya. 2008. Reproduction and development of the released Przewalski's horses (*Equus przewalskii*) in Xinjiang, China. Journal of Equine Science 19:1–7.

Cheremnov, D., and S.V. Sokolov. 2012. Reintroduction of the Koulan in the territory of Arganaty Mountains. Presented at the International Wild Equid Conference (IWEC), 18–22 September 2012, Vienna, Austria.

Davidson, A., Y. Carmel, and S. Bar-David. 2013. Characterizing wild ass pathways using a non-invasive approach: Applying least-cost path modelling to guide field surveys and a model selection analysis. Landscape Ecology 28:1465–1478.

Denzau, G., and H. Denzau. 1999. Wildesel. [In German.] Jan Thorbecke Verlag, Stuttgart, Germany.

FAO. Food and Agricultural Organization of the United Nations. 1986. The Przewalski horse and restoration to its natural habitat in Mongolia. Animal Production and Health Paper 61. Rome, Italy.

Feh, C. 2012. Delayed reversibility of PZP (porcine zona pellucida) in free-ranging Przewalski's horse mares. Presented at the International Wild Equid Conference (IWEC), 18–22 September 2012, Vienna, Austria.

Feh, C., N. Shah, M. Rowen, R.P. Reading, and S.P. Goyal. 2002. Status and action plan for the Asiatic wild ass (*Equus hemionus*). Pages 62–71 *in* P.D. Moehlman (Ed.), Equids: Zebras, asses and horses. Status survey and conservation action plan. International Union for Conservation of Nature, Gland, Switzerland.

Fischer, J., and D.B. Lindenmayer. 2000. An assessment of the published results of animal relocations. Biological Conservation 96:1–11.

Franceschini, M.D., D.I. Rubenstein, B. Low, and L.M. Romero. 2008. Fecal glucocorticoid metabolite analysis as an indicator of stress during translocation and acclimation in an endangered large mammal, the Grevy's zebra. Animal Conservation 11:263–269.

Frankfurt Zoological Society. 2012. Annual report 2012 and prospects 2013. Kazakhstan: Altyn Dala—Serengeti of the North. Report 30-33. Frankfurt, Germany. Available at https://fzs.org.

Frankham, R. 1996. Relationship of genetic variation to population size in wildlife. Conservation Biology 10:1500–1508.

Ganbaatar, O., N. Altansukh, and P. Kaczensky. 2014. Monitoring of Przewalski's horses and other plains ungulates in the

Great Gobi B Strictly Protected Area in SW Mongolia in 2013. Unpublished field report. Available at www.vetmeduni.ac.at/fileadmin/v/fiwi/Projekte/Gobi_Research_Project/Report_Takhin_tal_01_2014_final_corrected.pdf.

Geigl, E.-M., and T. Grange. 2012. Eurasian wild asses in time and space: Morphological versus genetic diversity. Annals of Anatomy 194:88–102.

Gueta, T., A.R. Templeton, and S. Bar-David. 2014. Development of genetic structure in a heterogeneous landscape over a short time frame: The reintroduced Asiatic wild ass. Conservation Genetics 15:1231–1242.

Hack, M.A., R. East, and D.I. Rubenstein. 2002. Status and action plan for the plains zebra (*Equus burchellii*). Pages 43–60 *in* P.D. Moehlman (Ed.), Equids: Zebras, asses and horses. Status survey and conservation action plan. International Union for Conservation of Nature, Gland, Switzerland.

Hamadanian, A. 2005. Onagers (*Equus hemionus onager*) in Iran, wild and captive. Zoologischer Garten 75:126–128.

Harrington, L.A., A. Moehrenschlager, M. Gelling, R.P.D. Atkinson, J. Hughes, and D.W. MacDonald. 2013. Conflicting and complementary ethics of animal welfare considerations in reintroductions. Conservation Biology 27:486–500.

Hemami, M.-R., and M. Momeni. 2013. Estimating abundance of the endangered onager Equus hemionus onager in Qatruiyeh National Park, Iran. Oryx 47:266–272.

Hemami, M.-R., S. Esmaeili, M. Momeni, and M. Bagheri. 2012. The status of Persian wild ass: Threats and the conservation needs. Presented at the International Wild Equid Conference (IWEC), 18–22 September 2012, Vienna, Austria.

Heptner, V.G., A.A. Nasimovich, and A.G. Bannikov. 1988. Mammals of the Soviet Union, vol. 1: Artiodactyla and Perissodactyla. Smithsonian Institution Libraries and the National Science Foundation, Washington, DC, USA. [English translation of the original book published in 1961 by Vysshaya Shkola, Moscow, Russia].

Hrabar, H., and G. Kerley. 2013. Conservation goals for the Cape mountain zebra: Security in numbers? Oryx 47:403–409.

IUCN/SSC. International Union for Conservation of Nature Species Survival Commission. 2013. Guidelines for reintroductions and other conservation translocations. Version 1.0. Gland, Switzerland.

Joly, F., S. Saïdi, T. Begz, and C. Feh. 2012. Key resource areas of an arid grazing system of the Mongolian Gobi. Mongolian Journal of Biological Sciences 10:13–24.

Kaczensky, P. 2011. First assessment of the suitability of the Altyn Dala and Altyn Emel region of Kazakhstan for Przewalski's horse re-introduction. Technical report. Research Institute of Wildlife Ecology, University of Veterinary Medicine, Vienna, Austria.

Kaczensky, P., N. Enkhsaihan, O. Ganbaatar, R. Samjaa, and C. Walzer. 2007. Identification of herder–wildlife conflicts in the Gobi B Strictly Protected Area in SW Mongolia. Exploration into the Biological Resources of Mongolia 10:99–116.

Kaczensky, P., O. Ganbaatar, H. von Wehrden, N. Enksaikhan, D. Lkhagvasuren, and C. Walzer. 2008*a*. Przewalski horse re-introduction in the Great Gobi B SPA—From species to ecosystem conservation. Pages 125–130 *in* B. Boldgiv (Ed.), Proceedings of the international conference "Fundamental and Applied Issues of Ecology and Evolutionary Biology," April 25, 2008, in Ulaanbaatar. Ecology Department, Faculty of Biology, University of Mongolia, Ulaanbaatar, Mongolia.

Kaczensky, P., O. Ganbaatar, H. von Wehrden, and C. Walzer. 2008*b*. Resource selection by sympatric wild equids in the Mongolian Gobi. Journal of Applied Ecology 45:1662–1769.

Kaczensky, P., C. Walzer, O. Ganbaatar, N. Enkhsaikhan, N. Altansukh, and C. Stauffer. 2011*a*. Re-introduction of the "extinct in the wild" Przewalski's horse to the Mongolian Gobi. Pages 214–219 *in* P.S. Soorae (Ed.), Global re-introduction perspectives 2011: More case studies from around the globe. International Union for Conservation of Nature Species Survival Commission, Re-introduction Specialist Group, Gland, Switzerland, and Environment Agency-Abu Dhabi, Abu Dhabi, UAE.

Kaczensky, P., O. Ganbaatar, N. Altansukh, N. Enkhsaikhan, C. Stauffer, and C. Walzer. 2011*b*. The danger of having all your eggs in one basket—Winter crash of the re-introduced Przewalski's horses in the Mongolian Gobi. PLoS ONE 6:e28057.

Kaczensky, P., B. Lkhagvasuren, O. Pereladova, M.-R. Hemami, and A. Bouskila. 2015. *Equus hemionus*. IUCN red list of threatened species. Version 2015.2. International Union for Conservation of Nature, Gland, Switzerland.

King, S.R.B., L. Boyd, W. Zimmermann, and B.E. Kendall. 2015. *Equus ferus*. IUCN red list of threatened species. Version 3.1. International Union for Conservation of Nature, Gland, Switzerland. Available at www.iucnredlist.org.

Kuznetsov, V. 2014. Return of onager. International Journal Turkmenistan 1–2:84–92.

Letty, J., S. Marchandeau, and J. Aubineau. 2007. Problems encountered by individuals in animal translocations: Lessons from field studies. Ecoscience 14:420–431.

Levanov, V.F., S.V. Sokolov, and P. Kaczensky. 2013. Corral mass capture device for Asiatic wild asses. Wildlife Biology 19:325–334.

Lloyd, P.H. 1984. The Cape mountain zebra 1984. African Wildlife 38:144–149.

Lukarevskiy, V.S. 1999. Large mammals of southern Turkmenistan and problems of their conservation. Pages 216–231 *in* Rare species of mammals of Russia and bordering territories. [In Russian.] Theriological Society, Moscow, Russia.

Lukarevskiy, V.S., and Y.K. Gorelov. 2007. Khulan (*Equus hemionus* Pallas 1775) in Turkmenistan. Exploration into the Biological Resources of Mongolia 10:231–240.

McDougall, P.T., D. Réale, D. Sol, and S.M. Reader. 2006. Wildlife conservation and animal temperament: Causes and consequences of evolutionary change for captive, reintroduced, and wild populations. Animal Conservation 9:39–48.

Meldebekov, A.M., M.K. Bajzhanov, A.B. Bekenov, and A.F. Kovshar. 2010. The red data book of the Republic of Kazakhstan, vol. 1: Animals. 1. Vertebrates, 4th ed. Ministry of Education and Science / Ministry of Agriculture, Almaty, Kazakhstan.

Millar, J.C.G. 1970*a*. Census of Cape mountain zebra. African Wildlife (1) 24:17–25.

Millar, J.C.G. 1970*b*. Census of Cape mountain zebra. African Wildlife (2) 24:105–114.

Mongolian Takhi Strategy and Plan Work Group. 1993. Recommendations for Mongolia's takhi strategy and plan.

Mongolian Government, Ministry of Nature and Environment, Ulaanbaatar, Mongolia.
Moodley, Y., and E.H. Harley. 2005. Population structuring in mountain zebras (*Equus zebra*): The molecular consequences of divergent demographic histories. Conservation Genetics 6:953–968.
Nandintsetseg, D. 2008. The home range and habitat use of takhi (*Equus przewalskii* Poljakov, 1881) reintroduced to Hustai National Park. Unpublished annual research report. Hustai National Park Trust, Ulaanbaatar, Mongolia.
Neumann-Denzau, G.P., and M. Chirikova. 2004. First steps to establish a wild population of Przewalski's horses in the Altyn-Emel National Park, Kazakhstan, along with historical excursus. Der Zoologische Garten 74:365–370.
Novellie, P., P.S. Millar, and P.H. Lloyd. 1996. The use of VORTEX simulation models in a long term programme of reintroduction of an endangered large mammal, the Cape mountain zebra (*Equus zebra zebra*). Acta Oecologica 17:657–671.
Novellie, P., M. Lindeque, P. Lindeque, P.H. Lloyd, and J. Koen. 2002. Status and action plan for the mountain zebra (*Equus zebra*). Pages 28–42 *in* P.D. Moehlman (Ed.), Equids: Zebras, asses and horses. Status survey and conservation action plan. International Union for Conservation of Nature, Gland, Switzerland.
O'Regan, H.J., and A. Kitchner. 2005. The effects of captivity on the morphology of captive, domesticated and feral mammals. Mammal Review 35:215–230.
Pavlov, M.P. 1996. Translocations of kulans in the former Soviet Union. Reintroduction News 12:5–16.
Plakhov, K.N., S.V. Sokolov, V.F. Levanov, and A.Z. Akylbekova. 2012. News in kulan reintroduction in Kazakhstan. Pages 151–153 *in* A.M. Meldebekov, A.B. Bekenov, Y.A. Grachev *et al.* (Eds.), Zoological and game management researches in Kazakhstan and adjacent countries (Almaty, 1–2 March 2012). [In Russian.] Almaty, Kazakhstan.
Pohle, C. 2010. Asiatic wild ass: International studbook. Tierpark, Berlin-Friedrichsfelde, Germany.
Renan, S., G. Greenbaum, N. Shahar, A.R. Templetond, A. Bouskila, and S. Bar-David. 2015. Stochastic modelling of shifts in allele frequencies reveals a strongly polygynous mating system in the re-introduced Asiatic wild ass. Molecular Ecology 24:1433–1446.
Robert, N., C. Walzer, S.R. Ruegg, P. Kaczensky, O. Ganbaatar, and C. Stauffer. 2005. Pathological investigations of reintroduced Przewalski's horse (*Equus caballus przewalskii*) in Mongolia. Journal of Zoo and Wildlife Medicine 36:273–285.
Saltz, D., and D.I. Rubenstein. 1995. Population dynamics of a reintroduced Asiatic wild ass (*Equus hemionus*) herd. Ecological Application 5:327–335.
Saltz, D., D.I. Rubenstein, and G.C. White. 2006. The impact of increased environmental stochasticity due to climate change on the dynamics of Asiatic wild ass. Conservation Biology 20:1402–1409.
Sarrazin, R., and R. Barbault. 1996. Reintroduction: Challenges and lessons for basic ecology. Trends in Ecology and Evolution 11:474–478.
Sasidharan, S.P., A. Ludwig, C. Harper, Y. Moodley, H.J. Bertschinger, and A.J. Guthrie. 2011. Comparative genetics of sarcoid tumour-affected and non-affected mountain zebra (*Equus zebra*) populations. South African Journal of Wildlife Research 41:36–49.
Seal, U.S. 1992. The draft global Przewalski horse conservation plan: A summary and comments on goals of captive propagation for conservation. Pages 107–110 *in* S. Seifert (Eds.), Proceedings of the 5th international symposium on the preservation of the Przewalski horse. Zoologischer Garten Leipzig, Leipzig, Germany.
Seddon, P.J., D.P. Armstrong, and R.F. Maloney. 2007. Developing the science of reintroduction biology. Conservation Biology 21:303–312.
Sims-Castley, R., G.I.H. Kerley, B.G.S. Geach, and J. Langholz. 2006. The socio-economic significance of eco-tourism-based private game reserves in South Africa's Eastern Cape Province. Parks 15:6–18.
Slotta-Bachmayr, L., R. Boegel, P. Kaczensky, C. Stauffer, and C. Walzer. 2004. Use of population viability analysis to identify management priorities and success in reintroducing Przewalski's horses to southwestern Mongolia. Journal of Wildlife Management 68:790–798.
Smith, R.K., A. Marais, P. Chadwick, P.H. Lloyd, and R.A. Hill. 2007. Monitoring and management of the endangered Cape mountain zebra *Equus zebra zebra* in the Western Cape, South Africa. African Journal of Ecology 46:207–213.
Smith, S. 2010. Genetic analysis of a potential hybrid from a domestic mare and Przewalski stallion. Unpublished report. Research Institute of Wildlife Ecology, University of Veterinary Medicine, Vienna, Austria.
Sokolov, V.E., and V.N. Orlov. 1986. Introduction of Przewalski horses into the wild. Pages 77–88 *in* The Przewalski horse and restoration to its natural habitat in Mongolia. Animal Production and Health Paper 61. Food and Agricultural Organization of the United Nations, Rome, Italy.
Tatin, L., B.F. Darreh-Shoori, C. Tourenq, and B. Azmayesh. 2003. The last populations of the critically endangered onager *Equus hemionus onager* in Iran: Urgent requirements for protection and study. Oryx 37:488–491.
Usukhjargal, D., and N. Bandi. 2013. Reproduction and mortality of re-introduced Przewalski's Horse *Equus przewalskii* in Hustai National Park, Mongolia. Journal of Life Sciences 7:623–629.
van Dierendonck, M.C., and M.F. Wallis de Vries. 1996. Ungulate reintroductions: Experiences with the takhi or Przewalski horse (*Equus ferus przewalskii*) in Mongolia. Conservation Biology 10:728–740.
Volf, J. 1996. Das Urwildpferd. Die Neue Brehm-Bücherei 249. Westarp Wissenschaften, Magdeburg, Germany.
Wakefield, S., J. Knowles, W. Zimmermann, and M. van Dierendonck. 2002. Status and action plan for the Przewalski's horse (*Equus ferus przewalskii*). Pages 82–92 *in* P.D. Moehlman (Ed.), Equids: Zebras, asses and horses. Status survey and conservation action plan. International Union for Conservation of Nature, Gland, Switzerland.
Wang, H., Z.Q. He, H.J. Wang, and X.Y. Niu. 2012. Study on survival status of reintroduced *Equus przewalskii* in Dunhuang West Late National Nature Reserve. [In Chinese.] Journal of Gansu Forestry Science and Technology 37:44–46.

Wild Kingdom. 1974. Chase of the onager. YouTube video, 23:22. Posted January 17, 2013. Available at www.youtube.com/watch?v=B46R4rZZPZc.

Williams, S.E., and E.A. Hoffman. 2009. Minimizing genetic adaptation in captive breeding programs: A review. Biological Conservation 142:2388–2400.

Xia, C., J. Cao, H. Zhang, X. Gao, W. Yang, and D. Blank. 2014. Reintroduction of Przewalski's horse (*Equus ferus przewalskii*) in Xinjiang, China: The status and experience. Biological Conservation 177:142–147.

Zhirnov, L.V., and V.O. Ilyinsky. 1986. The Great Gobi National Park—A refuge for rare animals in the central Asian deserts. Centre for International Projects, Moscow, Russia.

Ziaie, H., 2008. A field guide to the mammals of Iran, 2nd ed. [In Persian.] Wildlife Center, Tehran, Iran.

Zimmermann, W. 2005. Przewalskipferde auf dem Weg zur Wiedereinbürgerung—Verschiedene Projekte im Vergleich. Sonderdruck aus Zeitschrift des Kölner Zoo 4:183-209. [Translated into English as Przewalski's horses on the track to reintroduction—Various projects compared. Available at www.lpv-augsburg.de/files/downloads/Przewalskis_horse_Reserves_EN.pdf.]

Epilogue

JASON I. RANSOM AND
PETRA KACZENSKY

Modern equids are survivors. The species in this small taxonomic group are highly adapted herbivores that successfully compete with other, much more numerous, ungulates on the world's grasslands. Equids evolved to be resilient herd animals, migrating between resources with the seasons. They are long-lived, and populations are able to persist through droughts and harsh winters if their numbers are sufficiently large and interconnected. This resiliency allows them to thrive on some of the most marginal grazing habitats so long as they have regular access to water and room to roam. Modern equids are limited, however, in their ability to thrive in a world increasingly dominated by humans.

In historic times, we lost the wild relative of the domestic horse, and we drove the Przewalski's horse to extinction in the wild. We are rapidly doing the same to the African wild ass in northern Africa. The quagga (*Equus quagga quagga*) and the Syrian wild ass (*E. hemionus hemippus*) have gone from the earth, with the last individuals dying in zoos in 1883 and 1927, respectively (fig. E.1). Many of the remaining wild equid populations are small or highly fragmented, and their survival is far from guaranteed. In some areas, old threats—competition with humans and livestock for water and pasture, and illegal hunting—are intensifying or reemerging. In other areas, new threats are emerging, such as busy transportation corridors, mining and associated large-scale infrastructure development, large-scale fencing to protect pastures or secure international borders, and new diseases and parasites. These threats are affecting habitat quality, connectivity, and equid survival. Climate change adds a new challenge to equids, potentially reducing their future suitable habitat (Lou *et al.* 2015).

When given a chance, equids have shown a remarkable ability to recover and reclaim former ranges. The Przewalski's horse was eventually brought back to the wild after extensive captive breeding efforts, Cape mountain zebra are recovering following a human-facilitated metapopulation approach, and plains zebras in Botswana have resumed migration with the removal of a veterinary cordon fence (Bartlam-Brooks *et al.* 2011). New protected areas where equids persist have also been established, like the vast Chang Tang reserve on the Tibetan Plateau (Schaller 1998). Concurrently, new developmental schemes like biodiversity offsetting and no-net-loss policies are being developed, and more flexible protection schemes are being discussed (Bull *et al.* 2013; Evans *et al.* 2015).

Fig. E.1 The quagga (*E. quagga quagga*), a subspecies of plains zebra, photographed in 1870 by Frederick York at the London Zoo. This is one of only five known photographs of the now-extinct quagga. The last individual died in 1883 at the Amsterdam Zoo. The last Syrian wild ass (*E. hemionus hemippus*) (not pictured), a subspecies of Asiatic wild ass, died in 1927 at the Vienna Zoo.

There is no doubt that wild equids have been one of the most fundamental and important elements of the world's grasslands for millennia and that their vast numbers and seasonal movements have been vital to the diversity of the associated plant, herbivore, and predator communities (as still observed with plains zebra in the Serengeti). But in an era of increasing human needs, how does protecting equids relate to marginalized pastoral societies that want more grass and water for their domestic herds and fewer predators to worry about (Kideghesho 2010)? Good science can help reveal important relationships, such as how cattle mixed with zebra may actually produce healthier range and benefit herders and zebra (Odadi *et al.* 2011). Science alone, however, will not save the world's perishing wildlife.

Effective conservation requires collaborative efforts with a holistic lens. It must include the people who share resources with the species and ecosystems we aim to protect. Their livelihoods, health, beliefs, and visions must form part of the conservation solution rather than being a secondary consideration. Ignoring their needs means losing important allies in the battle to maintain traditional or natural landscapes and their associated wildlife. The conservation solutions must therefore better the people as well as the wildlife, for there are not enough exclusive resources for either (Sanderson *et al.* 2002).

Increases in such efforts are beginning to produce lessons learned; scientists providing conservation tools to stakeholders is not the most effective approach, but rather full participatory engagement in action is what makes the difference (Reed 2008). Organizations like Grevy's Zebra Trust are working with local people to develop community action plans and are recruiting and training scouts that collect valuable citizen science data (Low *et al.* 2009) (fig. E.2). Combining modern science with traditional ecological knowledge in a collaborative way leads to mutual understanding and new models for joint data gathering, and will allow for surprising leaps forward in resource management and conservation (Berkes 2008; Spoon and Arnold 2012).

Such collaboration must grow, for in the end it will not be scientists who save wild equids. It will be the pastoral herders that embrace collaboration to enhance their livelihoods while making room for the wildlife that enriches their world. It will be the schoolchildren whose computers span classrooms across the globe to engage with worlds so very different than their own. It will be the politicians who recognize the importance of a wild equid to their constituents, and then hold it up as an icon of conservation. It will be *people* who help wild equids survive in the modern world.

Fig. E.2 Conservation staff discuss results of the Grevy's zebra scout program with members of the Ngaroni community, Kenya. Such meetings and workshops give communities an opportunity to learn about zebra conservation, develop a community action plan, and define their own core group to lead local efforts. Photo by Daniel I. Rubenstein

REFERENCES

Bartlam-Brooks, H.L.A., M.C. Bonyongo, and S. Harris. 2011. Will reconnecting ecosystems allow long-distance mammal migrations to resume? A case study of a zebra *Equus burchelli* migration in Botswana. Oryx 45:210–216.

Berkes, F. 2008. Sacred ecology: Traditional ecological knowledge and resource management. Taylor and Francis, Philadelphia, PA, USA.

Bull, J.W., K.B. Suttle, N.J. Singh, and E.J. Milner-Gulland. 2013. Conservation when nothing stands still: Moving targets and biodiversity offsets. Frontiers in Ecology and the Environment 11:203–210.

Evans, D.M., R. Altwegg, T.W.J. Garner, M.E. Gompper, I.J. Gordon, J.A. Johnson, and N. Pettorelli. 2015. Biodiversity offsetting: What are the challenges, opportunities and research priorities for animal conservation? Animal Conservation 18:1–3.

Kideghesho, J.R. 2010. "Serengeti shall not die": Transforming an ambition into a reality. Tropical Conservation Science 3:228–248.

Low, B., S.R. Sundaresan, I.R. Fischhoff, and D.I. Rubenstein. 2009. Partnering with local communities to identify conservation priorities for endangered Grevy's zebra. Biological Conservation 142:1548–1555.

Luo, Z., Z. Jiang, and S. Tang. 2015. Impacts of climate change on distributions and diversity of ungulates on the Tibetan Plateau. Ecological Applications 25:24–38.

Odadi, W.O., M.K. Karachi, S.A. Adulrazak, and T.P. Young. 2011. African wild ungulates compete with or facilitate cattle depending on season. Science 333:1753–1755.

Reed, M. 2008. Stakeholder participation for environmental management: A literature review. Biological Conservation 141:2417–2431.

Sanderson, E.W., M. Jaiteh, M.A. Levy, K.H. Redford, A.V. Wannebo, and G. Woolmer. 2002. The human footprint and the last of the wild. BioScience 52:891–904.

Schaller, G.B. 1998. Wildlife of the Tibetan Steppe. University of Chicago Press, Chicago, USA.

Spoon, J., and R. Arnold. 2012. Collaborative research and co-learning: Integrating Nuwuvi (Southern Paiute) ecological knowledge and spirituality to revitalize a fragmented land. Journal for the Study of Religion, Nature, and Culture 6:477–500.

Contributors

Cheryl Asa, PhD
St. Louis Zoo
USA

Shirli Bar-David, PhD
The Jacob Blaustein Institutes for Desert Research
Ben-Gurion University of the Negev
Israel

Albano Beja-Pereira, PhD
Research Center in Biodiversity and Genetic Resources
University of Porto
Portugal

Joel Berger, PhD
Colorado State University
Wildlife Conservation Society
USA

David Berman, PhD
Australian Wild Horse Management Services
Australia

Amos Bouskila, PhD
The Jacob Blaustein Institutes for Desert Research
Ben-Gurion University of the Negev
Israel

Lee Boyd, PhD
Washburn University
USA

Qing Cao, MS, PhD candidate
Princeton University
USA

Justine Chiu, BS
Princeton University
USA

E. Gus Cothran, PhD
Texas A&M University
USA

L. Stefan Ekernas, PhD
University of Montana
USA

Oyunsaikhan Ganbaatar, MS, PhD candidate
Great Gobi B Strictly Protected Area
Mongolia

Eva-Maria Geigl, PhD
Institut Jacques Monod
Centre National de la Recherche Scientifique–University Paris Diderot
France

Elena Giulotto, PhD
Laboratory of Molecular and Cellular Biology
University of Pavia
Italy

Katherine A. Houpt, PhD
Cornell University
USA

Halszka Hrabar, PhD
Centre for African Conservation Ecology
Nelson Mandela Metropolitan University
South Africa

Petra Kaczensky, PhD
Research Institute for Wildlife Ecology
University of Veterinary Medicine Vienna
Austria

Albert J. Kane, DVM, MPVM, PhD
Animal and Plant Health Inspection Service
US Department of Agriculture
USA

Fanuel Kebede, PhD
Ethiopian Wildlife Conservation Authority
Ethiopia

Sarah R.B. King, PhD
Natural Resources Ecology Laboratory
Colorado State University
USA

Laura Lagos, PhD
Instituto de Investigación y Análisis Alimentarios
Universidad de Santiago de Compostela
Spain

Nicolas Lescureux, PhD
Centre d'Ecologie Fonctionnelle and Evolutive
Centre National de la Recherche Scientifique
France

John D.C. Linnell, PhD
Norwegian Institute for Nature Research
Norway

Victor Lukarevskiy, PhD
Russian Academy of Science
Russia

Patricia D. Moehlman, PhD
IUCN / SCC Equid Specialist Group Chair
Tanzania

Dejid Nandintsetseg, MS, PhD candidate
Senckenberg Biodiversity and Climate Research Centre
Goethe University
Germany

Megan K. Nordquist, MS
Brigham Young University
USA

Haniyeh Nowzari, PhD
Islamic Azad University
Iran

Cassandra M.V. Nuñez, PhD
Iowa State University
USA

Sandra L. Olsen, PhD
Division of Archaeology, Biodiversity Institute
University of Kansas
USA

Tsendsuren Oyunsuren, PhD
Institute of Biology
Mongolian Academy of Sciences
Mongolia

Jan Pluháček, PhD
Institute of Animal Science
Czech Republic

David M. Powell, PhD
Wildlife Conservation Society
USA

Mélanie Pruvost, PhD
Institut Jacques Monod
Centre National de la Recherche Scientifique–University Paris Diderot
France

Jason I. Ransom, PhD
Natural Resources Ecology Laboratory
Colorado State University
USA

Daniel I. Rubenstein, PhD
Princeton University
USA

Kathryn A. Schoenecker, PhD
US Geological Survey
Natural Resources Ecology Lab
Colorado State University
USA

Mandi Wilder Schook, PhD
Cleveland Metroparks Zoo
USA

Alberto Scorolli, PhD
Universidad Nacional del Sur
Argentina

Natalia N. Spasskaya, PhD
Zoological Museum of Moscow
Lomonosow State University
Russia

Dorj Usukhjargal, MS, PhD candidate
Hustai National Park
Mongolia

Waltraut Zimmermann, PhD
Przewalski's Horse International Studbook Keeper
Equid Taxon Advisory Group
European Association of Zoos and Aquariums
Germany

Index